Finite Size Scaling and Numerical Simulation of Statistical Systems

Finite Size Scaling and Numerical Simulation of Statistical Systems

Editor

V Privman

Department of Physics
Clarkson University

World Scientific

Singapore • New Jersey • London • Hong Kong

Published by

World Scientific Publishing Co. Pte. Ltd.
P O Box 128, Farrer Road, Singapore 9128
USA office: 687 Hartwell Street, Teaneck, NJ 07666
UK office: 73 Lynton Mead, Totteridge, London N20 8DH

FINITE SIZE SCALING AND NUMERICAL SIMULATION OF STATISTICAL SYSTEMS

ISBN 981-02-0108-7

Printed in Singapore by Utopia Press.

PREFACE

This book presents a collection of review articles providing both an introduction and a survey of recent advances in the field of *Finite Size Scaling* in phase transitions and related disciplines. Both theoretical foundations and numerical methods are covered. This includes scaling theory, the renormalization group approach, Monte Carlo and transfer matrix numerical applications, and recent uses of finite size scaling in Lattice Gauge Theory and for random systems.

Finite size scaling theory attempts to describe how *long-scale, collective phenomena* associated with the onset of large fluctuations near critical points, at first order phase transitions, in polymer systems, etc., manifest themselves in small samples (capillaries, pores), and in *numerical computer simulations* which are always done on limited-size lattice or continuum models. The latter application of finite size scaling theory has grown in importance in the last decade. Indeed, the advances in large-scale computing, based largely on the Monte Carlo method, have allowed accurate evaluation of bulk, surface, and interfacial properties of statistical mechanical models, as well as applications in particle physics, polymers, and random systems. All these studies employ finite size scaling ideas and also provide stimuli for theoretical advances, suggesting new emphases, topics, and testing the existing theoretical predictions.

Reviews in this book offer convenient reference sources, introductions, and guides to current research, with an emphasis on various numerical methods and their relations to finite size effects. Each review has a substantial introductory component, and the book as a whole should be accessible to readers with no special prior knowledge of finite size scaling theory or details of its applications in conjunction with numerical methods. However, a general background in phase transitions or a related field is assumed.

The first three reviews (Chapters I-III), by Privman, Jasnow, and Rudnick, are theoretical. They provide an introduction to the modern theory of finite-size effects. Chapter I summarizes *scaling theory* and related approaches, for critical points, for interfacial properties, and also for first-order transitions. In Chapter II, field-theoretical techniques are surveyed, culminating in the formulation of the renormalization group method for finite systems. Then, Chapter III presents results for spherical models which serve to illustrate and test the general theoretical predictions.

The next three chapters (Chapters IV-VI), by Binder, Landau, and Mon, review applications of the finite size scaling theory in *Monte Carlo numerical studies* of critical phenomena. All three reviews describe quantities, geometries, and scaling results in the formulation appropriate for Monte Carlo data analyses. Specific results for selected models and details of their numerical derivation followed by finite-size analysis are described, illustrating the general versatility and growing importance of the Monte Carlo method. The emphasis in Chapter IV is on general definitions, anisotropic systems, and other recent results. Chapter V presents diverse examples of applications of Monte Carlo methods to critical points and first-order transitions. The focus of Chapter VI is on universal finite-size amplitudes and associated geometry related properties such as surface and corner free energies, etc.

Recent theoretical developments related to conformal invariance and other new exact results obtained for *two-dimensional models* have generated further interest in *numerical techniques based on the transfer matrix method.* In two dimensions, the transfer matrix approach, termed phenomenological renormalization, is among the most powerful methods of estimating critical-point quantities. Foundations of the phenomenological renormalization method, selected results, and their finite size scaling analyses are reviewed in Chapters VII and VIII by Nightingale and Henkel. Chapter VII is centered on more conventional isotropic model results, as well as on a general overview of the field. Chapter VIII is devoted to the quantum Hamiltonian variant of the method. Both chapters discuss connections with conformal invariance and other exact results in two dimensions, and also recent applications of the phenomenological renormalization method to three-dimensional systems.

The last three chapters of the book (Chapters IX-XI), by Bhanot, Young, and Schulman, are devoted to finite size effects in systems for which the theoretical framework, and the appropriate nomenclature, are outside the more "traditional" uses of finite size scaling ideas in phase transitions. Thus, Chapter IX describes applications of finite size scaling, including a survey of recent numerical results, in Lattice Gauge Theories of *particle physics.* Chapter X is devoted to spin glasses. Generally, for *random systems*, new aspects of the finite size behavior enter, related to long equilibration times, and to averaging over randomness. Finally, Chapter XI introduces finite size effects associated with metastable phases.

The editor wishes to express his thanks to C.R. Doering and M.E. Fisher for their interest in this project and valuable suggestions, and to the contributing authors for a job well done. He hopes that in addition to being a comprehensive summary/introduction, this book will convey the excitement and dynamics of a rapidly growing and developing field of science.

Vladimir Privman

October 1989

Potsdam, New York

CONTENTS*

*Each review has a separate, detailed table of contents.

I. FINITE-SIZE SCALING THEORY

Vladimir Privman

TABLE of CONTENTS

(continued)

(continued)

FINITE-SIZE SCALING THEORY

Vladimir Privman

Department of Physics, Clarkson University
Potsdam, New York 13676, USA

1. INTRODUCTION

1.1. *Opening Remarks*

In this chapter we review the finite-size scaling theory of continuous phase transitions (critical points), as well as certain results on finite-size effects at first-order transitions, and also for systems with fluctuating interfaces. Studies of finite-size effects are important in interpreting experimental data and numerical results of Monte Carlo (MC) and transfer matrix calculations, and in connection with other theoretical developments, notably, conformal invariance and surface critical phenomena.

Our exposition will be biased towards numerical data analyses where a prototype finite-size system is d-dimensional hypercubic, L^d, or nearly hypercubic-shaped, or cylindrical, $L^{d-1} \times \infty$, (for transfer matrix studies). Much of our discussion will refer to the theoretically most studied case of *periodic boundary conditions* which are a natural choice in many MC and transfer matrix applications. Results for other boundary conditions will be described, when available. While most of the considerations will be quite general, we will use the nomenclature of the ferromagnetic n-vector models on regular d-dimensional lattices. Note that transfer matrix calculations are usually performed in $d = 2$. Only recently, some $d = 3$ studies have been reported. On the other hand, MC calculations can be carried out in dimensionalities as high as $d = 5$ or 6.

1.2. *Outline of the Review*

As previously mentioned, this chapter is devoted to scaling theories of finite-size effects. Other chapters of this book cover calculational methods, *e.g.*, the Renormalization Group (RG) approach, and numerical applications.

Since the introduction of the finite-size scaling ideas by Fisher and co-workers in early seventies, several comprehensive expository and review articles have appeared covering both the original scaling ideas and later advancements, see, *e.g.*, [1-7]. Some of the important contributions have been reprinted in [8]. In the following subsection of the introduction (Sect. 1.3) a simple finite-size scaling ansatz [1] will be explained.

The modern form of the critical-point finite-size scaling [9], reviewed in Sect. 2, incorporates *hyperscaling* ideas and leads to the identification of universal finite-size critical-point amplitudes. Universal amplitudes and, more generally, universal quantities derivable from finite-size scaling functions have been estimated by numerical

MC, exact $2d$ conformal invariance, and numerical transfer matrix calculations. These studies are selectively surveyed in Sect. 3: a more comprehensive review can be found, *e.g.*, in [10].

Sections 4 and 5 are devoted, respectively, to interfacial finite-size properties, and to first-order transitions. Size effects on interfacial fluctuations (Sect. 4) both near T_c and below T_c, are quite diverse and have been studied relatively recently. Finite-size rounding of first-order transitions (Sect. 5) involves an interesting phenomenon of the emergence of characteristic finite-size length scales. Finally, brief summary remarks are given in Sect. 6.

1.3. *Basic Scaling Postulate*

Near the critical point at $t = 0$, $H = 0$, where

$$t \equiv (T - T_c)/T_c \quad , \tag{1.1}$$

and H is the ordering field, various thermodynamic quantities diverge. The bulk ($L = \infty$) zero field critical behavior of, *e.g.*, the specific heat, $C(t, H; L)$, is given by

$$C_s(t, 0; \infty) \approx (A_\pm/\alpha)|t|^{-\alpha}, \tag{1.2}$$

where s denotes "singular part", and the exponent α is included in the amplitudes for historical reasons. The standard scaling expression in nonzero field (see, *e.g.*, a review [11]) is

$$C_s(t,H;\infty) \approx |t|^{-\alpha} \mathcal{C}_\pm \left(H|t|^{-\Delta}\right), \tag{1.3}$$

where $\mathcal{C}_\pm$ are certain scaling functions. As usual, in the above expressions the $+-$ refer to $t > 0$ and $t < 0$, respectively; the sign $\approx$ indicates that corrections to scaling have been omitted. The exponents α and $\Delta = \beta + \gamma$ are universal, as is the ratio A_+/A_-. The full scaling functions $\mathcal{C}_\pm$ can be made universal by introducing proper metric factors for t and H; see Sect. 2.1-2.2 and further below.

For finite-size systems it has been recognized [1,12,13] that the system size L "scales" with the correlation length $\xi(t,H;\infty)$ of the bulk system. (Strictly speaking, this is only true in $d = 2,3$, see Sect. 2.1.) Indeed, if $L >> \xi(t,H;\infty)$, no significant finite-size effects should be observed. On the other hand, for $L \leq \xi(t,H;\infty)$, the system size will cut-off long-distance correlations so that an appreciable finite-size rounding of critical-point singularities is to be expected. Since the bulk correlation length scales similarly to (1.3),

$$\xi(t,H;\infty) \approx |t|^{-\nu} \Xi_\pm \left(H|t|^{-\Delta}\right), \tag{1.4}$$

the finite-size scaling combination is naturally $L/|t|^{-\nu}$. Thus one assumes

$$C_s(t,H;L) \approx |t|^{-\alpha} \tilde{\mathcal{C}}_\pm \left(H|t|^{-\Delta}; L|t|^{\nu}\right), \tag{1.5}$$

$$\xi(t,H;L) \approx |t|^{-\nu} \tilde{\Xi}_\pm \left(H|t|^{-\Delta}; L|t|^{\nu}\right), \tag{1.6}$$

with similar expressions for other quantities [1]. It is interesting to note that no new critical exponents were introduced in (1.5)-(1.6).

Relations (1.5)-(1.6) already yield many useful predictions, for example,

$$C_s(0,0;L) \propto L^{\alpha/\nu} \quad \text{and} \quad \xi(0,0;L) \propto L\,. \tag{1.7}$$

However, it turns out that additional rearrangement is quite useful. It involves four steps. Firstly, we use the so-called L-scaled instead of the t-scaled relations, *i.e.*, we redefine the scaling functions to have L enter in nonanalytic powers, while t and H enter linearly in combinations $tL^{1/\nu}$ and $HL^{\Delta/\nu}$. Secondly, we note that for a finite-size system there is actually no singularity at $t, H = 0$. Therefore, the scaling function will be smooth, analytic at the origin $tL^{1/\nu}, HL^{\Delta/\nu} = 0$. No distinct $\pm$ functions are needed. The third step is to allow for nonuniversal metric factors for t and H, by using scaling combinations $atL^{1/\nu}$ and $bHL^{\Delta/\nu}$. Then the scaling functions will be universal. It turns out that no metric factor is needed for L, see [9]. This issue will be further explored in Sect. 2.1-2.2. The final point is to use *free-energy density*, f, measured per k_BT, instead of the specific heat. Note that in terms of the singular parts, we have a simple relation

$$C_s = -k_B \frac{\partial^2 f_s}{\partial t^2} \quad . \tag{1.8}$$

Various thermodynamic quantities follow from the free energy by differentiation.

After carrying out all the above redefinitions, and making use of the hyperscaling relation

$$2 - \alpha = d\nu \tag{1.9}$$

in the free-energy expression, we end up with

$$f_s(t, H; L) \approx L^{-d} Y\left(atL^{1/\nu}, bHL^{\Delta/\nu}\right), \tag{1.10}$$

$$\xi(t, H; L) \approx LX\left(atL^{1/\nu}, bHL^{\Delta/\nu}\right), \tag{1.11}$$

where the scaling functions X and Y are universal [9]: see also Sect. 2.1.

We did not actually detail the transformation from (1.5)-(1.6) to (1.10)-(1.11). The reason is that it is much easier to proceed from (1.10)-(1.11) back to (1.5)-(1.6). Thus, for instance, for ξ we put

$$\tilde{\Xi}_{\pm}(z_1; z_2) = z_2 X\left(\pm a z_2^{1/\nu}, b z_1 z_2^{\Delta/\nu}\right), \tag{1.12}$$

with a similar somewhat more complicated relation for $\tilde{C}_{\pm}$. Note that the precise choice of the metric factors $a > 0$ and $b > 0$ is not specified at this stage. They will be discussed in Sect. 2.4.

Relations (1.10)-(1.11) are convenient for deriving the behavior *at the critical point.* Thus, we have

$$f_s(0, 0; L) \approx Y(0, 0) L^{-d}, \tag{1.13}$$

$$\xi(0,0;L) \approx X(0,0)L, \tag{1.14}$$

involving two *universal amplitudes*, and also

$$C_s(0,0;L) \approx -k_B a^2 \left[\frac{\partial^2 Y(\tau,\omega)}{\partial \tau^2}\right]_{\tau,\omega=0} L^{\alpha/\nu}, \tag{1.15}$$

etc.

The behavior in the bulk limit of $L \to \infty$ with $(t,H) \not\equiv (0,0)$ is less apparent in (1.10)-(1.11). For example, for the free energy we obtain the bulk t-scaled scaling relation by requiring that

$$Y(\tau,\omega) \approx (\pm\tau)^{d\nu} Q_\pm \left(\omega(\pm\tau)^{-\Delta}\right), \quad \text{for} \quad \tau \to \pm\infty, \tag{1.16}$$

where the functions $Q_\pm$ are universal. This limiting behavior of Y implies

$$f_s\left(t,H;\infty\right) \approx a^{2-\alpha}|t|^{2-\alpha} Q_\pm \left(ba^{-\Delta}H|t|^{-\Delta}\right), \tag{1.17}$$

which is the usual bulk scaling law, with metric factors a and b making the scaling functions universal, and with $2-\alpha = d\nu$ "built in". Relation (1.3) is implicit in (1.17), via (1.8). The expression for $\mathcal{C}_\pm$ is, however, cumbersome and is not given here.

If the finite-size system is sufficiently high-dimensional to have its own critical point, as happens, *e.g.*, for $L \times \infty^2$ Ising "slabs" in $3d$, then this will be manifested by a singularity of $Y(\tau,\omega)$ and $X(\tau,\omega)$

at $\tau = \tau_0 \neq 0$, $\omega = 0$, see [1-2] for further discussion and references. For L^d and $L^{d-1} \times \infty$ systems one can define an *apparent shifted critical point, e.g.*, by a location of the specific heat maximum. Thus, we solve for τ_0 in

$$\left[\frac{\partial^3 Y(\tau,\omega)}{\partial \tau^3}\right]_{\tau=\tau_0,\omega=0} = 0. \tag{1.18}$$

The value τ_0 is universal. The apparent critical temperature shift is then given by

$$t_0 = a^{-1}\tau_0 L^{-1/\nu}. \tag{1.19}$$

2. FINITE-SIZE SCALING AT CRITICAL POINTS

2.1. Scaling Ansatz for $d < 4$

In this subsection we consider systems below the upper critical dimension d_c [11], where $d_c = 4$ for n-vector models. We also assume no logarithmic singularities, that is, we assume $\alpha \neq 0, -1, -2, \ldots$. For systems with *periodic boundary conditions* it has been established by heuristic scaling [9,14-15], as well as by explicit RG calculations [5,16-18], that the system size L enters with no corresponding metric factor, *i.e.*, as a macroscopic length scale leading naturally to relations (1.10)-(1.11). Thus the metric factors a and b contain all the nonuniversal system-dependent aspects of the leading critical behavior. More generally, the boundary conditions and system shape are also "macroscopic" geometric features in the sense that the scaling functions Y and X depend on the geometry.

For the free energy it is important to also consider the nonsingular "background" which can be chosen to have no field dependence,

$$f(t, H; L) = f_s(t, H; L) + f_{ns}(t; L). \tag{2.1}$$

Recent studies [5,9,14-18] indicate that for periodic boundary conditions one can assume $f_{ns}(t; L) = f_{ns}(t; \infty)$. For other boundary conditions, the background may lead to new interesting finite-size features (see below).

As already noted in deriving (1.17) from (1.10), the hyperscaling relation (1.9) is implicit in (1.10). In fact, relations (1.10)-(1.11) are rather strong forms of the scaling-with-hyperscaling postulate incorporating various properties which are known to hold for $d < 4$ within the RG formalism: they include the equality of critical exponents for $t < 0$ and $t > 0$; the use of the same metric factors for f_s and ξ; hyperscaling exponent relations; and the so-called two-scale-factor universality or hyperuniversality relations among critical point amplitudes [10]. The last property is of particular interest and warrants further discussion.

Consider the zero-field relation (1.17),

$$f_s(t, 0; \infty) \approx a^{d\nu} q_{\pm} |t|^{2-\alpha}, \tag{2.2}$$

where $q_{\pm} = Q_{\pm}(0)$ are universal. By considering the equivalent of (1.16) for the correlation length scaling function, we can also derive the relation

$$\xi(t,0;\infty) \approx a^{-\nu} k_{\pm} |t|^{-\nu}, \tag{2.3}$$

where $k_{\pm}$ are universal. We thus find that for small t, the $t > 0$ and $t < 0$ combinations

$$f_s(t \to 0^{\pm},0;\infty)\xi^d(t \to 0^{\pm},0;\infty) = q_{\pm} k_{\pm}^d \tag{2.4}$$

have universal constant limiting values. This is just the standard statement of hyperscaling [19-20] based on the hypothesis that for strongly fluctuating (non-mean-field) critical behavior there is a single divergent "macroscopic" length scale, ξ. The singular free energy in volume ξ^d is then a universal multiple of $k_B T_c$. The finite-size equivalent of this hypothesis [9] are relations (1.13)-(1.14), and more generally, the observation that the ratio of the "macroscopic" lengths L and ξ entails no new "microscopic" metric factor.

2.2. *RG and Corrections to Scaling*

The RG interpretation of the L-scaled relations (1.10)-(1.11) is particularly transparent. Indeed, the exponents of L in the scaling combinations, are just the *relevant* eigenexponents

$$\lambda_t = \frac{1}{\nu} > 0 \quad \text{and} \quad \lambda_H = \frac{\Delta}{\nu} > 0, \tag{2.5}$$

corresponding to the scaling fields [21] of the RG transformation,

$$g_t = al^{\lambda_t}t + o(t, H), \tag{2.6}$$

$$g_H = bl^{\lambda_H}H + o(t, H). \tag{2.7}$$

We introduced an arbitrary *fixed* (system independent) microscopic length scale l to have L enter in a dimensionless ratio when (1.10)-(1.11) are rewritten in terms of g_t and g_H,

$$f_s(t, H; L) \approx L^{-d} Y\left[g_t(L/l)^{\lambda_t}, g_H(L/l)^{\lambda_H}\right], \tag{2.8}$$

$$\xi(t, H; L) \approx LX\left[g_t(L/l)^{\lambda_t}, g_H(L/l)^{\lambda_H}\right]. \tag{2.9}$$

Note that we did not care to define H in dimensionless units. One can always chose b to have g_H dimensionless.

The RG relations (2.8)-(2.9) are actually more basic than the phenomenological scaling forms (1.10)-(1.11), and they can be extended to include arguments for additional relevant ($\lambda > 0$) scaling fields in the cases of *multicritical phenomena*, or associated with *surface couplings*.

Corrections to scaling result from the nonlinearity of the scaling fields $g(t, H)$ [*i.e.*, the $o(t, H)$ corrections in (2.6)-(2.7)], as well as from including additional arguments in (2.8)-(2.9) for the *irrelevant* ($\lambda < 0$) scaling fields. For finite-size systems only the leading irrelevant-variable correction to scaling contribution has been considered in detail in the literature and estimated numerically [22]. Let

θ denote the leading bulk correction to scaling exponent associated with the irrelevant RG field $g_u(t, H)$, taking the finite value ul^{λ_u} at $t = 0, H = 0$, with eigenexponent

$$\lambda_u = -\frac{\theta}{\nu} < 0. \tag{2.10}$$

Then the scaling relation (1.11) for ξ for instance, is replaced by

$$\xi \approx L\bar{X}\left(atL^{1/\nu}, bHL^{\Delta/\nu}; uL^{-\theta/\nu}\right), \tag{2.11}$$

which in turn can be expanded in the third argument to isolate the leading correction to scaling contribution due to the irrelevant variable g_u,

$$\xi \approx LX\left(atL^{1/\nu}, bHL^{\Delta/\nu}\right) + uL^{1-\frac{\theta}{\nu}}X_u\left(atL^{1/\nu}, bHL^{\Delta/\nu}\right). \tag{2.12}$$

The scaling functions $\bar{X}$ and X_u are universal. However, the correction term has the "front" nonuniversal coefficient u. A similar construction is possible for the free energy. For a systematic study of the bulk "nonlinear" and "irrelevant" corrections to scaling, and their interplay, consult [10,23].

2.3. Finite-Size Scaling in the Limit $\alpha \to 0$

Consider first the bulk zero-field critical-point free energy,

$$f_s(t; L = \infty) \approx F_\pm |t|^{2-\alpha}, \tag{2.13}$$

where we omitted the H-dependence ($H = 0$) for simplicity. We have

$$F_{\pm} = a^{2-\alpha} q_{\pm}, \tag{2.14}$$

see (2.2). Also, we expand $f_{ns}(t;\infty)$ about $t = 0$,

$$f_{ns}(t;\infty) = F_0 + F_1 t + F_2 t^2 + o(t^2). \tag{2.15}$$

As $\alpha \to 0$, poles develop in $F_{\pm}$ and F_2, yielding a logarithmic specific heat singularity: see, *e.g.*, [24] where similar effects for $\alpha \to -1, -2, -3, \ldots$, are also described. Thus, we have

$$q_{\pm} = -\frac{q_0}{\alpha} + \bar{q}_{\pm} + O(\alpha), \tag{2.16}$$

with universal $q_0 > 0$ and $\bar{q}_{\pm}$. Also, we have

$$F_2 = a^{2-\alpha}\frac{q_0}{\alpha} + \bar{F}_2 + O(\alpha), \tag{2.17}$$

so that the terms in α^{-1} cancel out. Collecting terms and assuming that F_0 and F_1 remain finite as $\alpha \to 0$, we get the leading terms in the free energy in the form

$$f(t;\infty) \approx F_0 + F_1 t + \left[\bar{F}_2 + a^2 \bar{q}_{\pm}\right] t^2 + a^2 q_0 t^2 \ln|t|. \tag{2.18}$$

Thus, the leading singular behavior of the free energy consists of the $t^2 \ln|t|$ and discontinuous t^2 contributions.

Extension of the above mechanism to finite-size systems with *periodic boundary conditions* (and $H = 0$) was proposed in [25]. Since the nonsingular background (2.15), with (2.17), remains the same for $L < \infty$, the α^{-1} divergence in (2.17) must be compensated for by a contribution from the scaling part,

$$f_s \approx L^{-d} Y\left(atL^{1/\nu}, 0\right). \tag{2.19}$$

As $\alpha \to 0$, we have

$$Y(\tau, 0) = -\frac{q_0}{\alpha}|\tau|^{2-\alpha} + \bar{W}_\pm(\tau) + O(\alpha), \tag{2.20}$$

which, after some algebra, yields the $\alpha = 0$ result

$$f(t; L) \approx f(t; \infty) + L^{-d}\bar{W}_\pm\left(atL^{d/2}\right), \tag{2.21}$$

where $f(t;\infty)$ is given by (2.18), and we used (1.9) with $\alpha = 0$. Because the leading term in (2.20) is nonanalytic in τ, however, we have to use separate scaling functions $\bar{W}_\pm$ for $\tau > 0$ and $\tau < 0$.

According to [25], a smooth finite-size scaling function can be defined at the expense of separating out logarithmic size dependence as follows,

$$f(t; L) \approx F_0 + F_1 t + \left[-a^2 q_0 \ln\left(aL^{d/2}\right) + \bar{F}_2\right] t^2 + L^{-d} W\left(atL^{d/2}\right). \tag{2.22}$$

The connection is given by

$$\bar{W}_{\pm}(\tau) = W(\tau) - \bar{q}_{\pm}\tau^2 - q_0\tau^2 \ln(\pm\tau). \tag{2.23}$$

2.4. *Selection of Metric Factors*

In this subsection we again consider zero-field critical behavior and discuss in detail the procedure for fixing the metric factor a entering various scaling relations introduced above. The discussion can in principle be extended to any other *relevant* scaling variable, *e.g.*, H, and the appropriate metric factor relating it to the RG scaling field [see (2.6)-(2.7)].

Consideration of relations (2.13)-(2.14), (2.19), as well as the $\alpha \simeq 0$ relations (2.17)-(2.18), indicates that $a^{2-\alpha}$ measures an overall nonuniversal free-energy scale or "strength" of, *e.g.*, the bulk $|t|^{2-\alpha}$ singularity, while all other parameters, including the bulk exponent α and amplitude ratio F_+/F_-, are universal and depend only on the universality class. Finite-size universal quantities, like $Y(\tau, 0)$, depend also on geometry (shape, and boundary conditions). The selection of a is actually quite arbitrary. It is convenient to define it in terms of the bulk parameters $F_{\pm}$ and α in order to have no geometry dependence. A particular choice of a implies certain restrictions on the universal scaling-function-related quantities which can be imposed without violating any of the general universality properties discussed in the preceding subsections. For example, if we take $a^{2-\alpha} = |F_+|$, then $q_+ = \text{sign}(F_+)$. We could also take $a^{2-\alpha} = |F_-|$, implying $q_- = \text{sign}(F_-)$, or a symmetric choice $a^{2-\alpha} = |F_+ + F_-|$, implying $q_+ + q_- = \text{sign}(F_+ + F_-)$, *etc.*

However, in some models $F_+ = 0$ while $F_- \neq 0$, or vice-versa. Similarly, $F_\pm$ diverge as $\alpha \to 0$. In fact, all three definitions mentioned in the preceding paragraph are inappropriate in certain cases. The guidance for a proper definition of a symmetric free-energy scale comes from a rather unexpected source: study of the complex-t plane zeros of the partition function,

$$Z(t, H = 0; L) = \mathrm{e}^{-Vf(t,0;L)}, \tag{2.24}$$

and the connection between the location of these zeros and the finite-size scaling relation (2.19). The following "natural" definition,

$$a = \left[F_+^2 + F_-^2 - 2F_+F_- \cos(\pi\alpha)\right]^{\frac{1}{2(2-\alpha)}}, \tag{2.25}$$

emerges in these studies [26] which are rather complicated and are not reviewed here. When a is defined by (2.25), it can be shown that it remains finite (nondivergent and nonzero) in all the "pathological" cases like $\alpha \to 0$, *etc.* Note that (2.25) implies the following relation among the universal scaling function coefficients,

$$q_+^2 + q_-^2 - 2q_+q_- \cos(\pi\alpha) = 1. \tag{2.26}$$

2.5. Nonperiodic Boundary Conditions

In this subsection we consider systems with *free boundary conditions* and review recent finite-size scaling results [27-29]. Extensions to other boundary conditions (*e.g.*, fixed, *etc.*) are quite straightforward. For simplicity, we put $H = 0$ in this subsection, and we restrict our discussion to the case of flat surfaces. With these restrictions, one can fully illustrate the new finite-size effects involved. Thus, we consider systems with flat surfaces, straight edges, and corners, in $3d$, and straight-edge polygons in $2d$. Extension to curved boundaries is also possible [27-28].

Consider first finite-size behavior away from T_c when the bulk correlation length,

$$\xi^{(b)} \equiv \xi(t; L = \infty), \tag{2.27}$$

is small compared to a characteristic system size L. For systems with no soft modes (spin waves), which we assume here for simplicity, it has been well established [1-2] that the free energy (and other thermodynamic quantities) can be expanded as

$$f(t; L) - f^{(b)}(t) = \frac{1}{L} f^{(s)}(t) + \frac{1}{L^2} f^{(e)}(t) + \frac{1}{L^3} f^{(c)}(t) \tag{2.28}$$
$$+ O\left(\mathrm{e}^{-L/\xi^{(b)}}\right), \qquad (d = 3),$$

$$f(t; L) - f^{(b)}(t) = \frac{1}{L} f^{(s)}(t) + \frac{1}{L^2} f^{(c)}(t) + O\left(\mathrm{e}^{-L/\xi^{(b)}}\right), \qquad (d = 2), \tag{2.29}$$

where

$$f^{(b)} \equiv f(t; \infty) \tag{2.30}$$

is the bulk free energy. The contributions proportional to $f^{(s)}, f^{(e)}$ and $f^{(c)}$ are attributed to surfaces, edges (in $3d$) and corners.

The bulk free energy $f^{(b)}(t)$ is the same for all boundary conditions. For each type of surface, one can define the surface free energy density, $f^{(\mathrm{surface})}$, (measured per k_BT and per unit surface area). Similarly, one can define the edge free energy for each type of edge, $f^{(\mathrm{edge})}$, specified by the opening angle and types of adjacent surfaces (and measured per k_BT and per unit length). Finally, for each corner, one can specify the corner free energy, $f^{(\mathrm{corner})}$, depending on the corner type (number of edges, opening angles, types of surfaces, and measured per k_BT). Unlike $f^{(s)}, f^{(e)}, f^{(c)}$, the quantities $f^{(\mathrm{surface})}, f^{(\mathrm{edge})}, f^{(\mathrm{corner})}$ depend only on the *local* geometry. These free energies are related up to sample shape dependent proportionality factors. For $3d$ slabs, $L \times \infty^2$, with free boundary conditions, we have $f^{(s)} = 2f^{(\mathrm{surface})}$ with $f^{(e)} = 0, f^{(c)} = 0$. Similar relations apply for $2d$ strips. For $3d$ cylinders with a square cross-section, $L^2 \times \infty$, we have $f^{(s)} = 4f^{(\mathrm{surface})}, f^{(e)} = 4f^{(\mathrm{edge})}, f^{(c)} = 0$. For $3d$ cubes, L^3, we have $f^{(s)} = 6f^{(\mathrm{surface})}, f^{(e)} = 12f^{(\mathrm{edge})}, f^{(c)} = 8f^{(\mathrm{corner})}$, while for $2d$ squares, L^2, we have $f^{(s)} = 4f^{(\mathrm{surface})}, f^{(c)} = 4f^{(\mathrm{corner})}$, *etc.* We will return to some of these relations in the next subsection.

Concerning the behavior near T_c, there is accumulating numerical and analytical evidence, reviewed in part in Sect. 3, that the *nonsingular part* $f_{ns}(t;L)$ has an expansion similar to (2.28)-(2.29). For example, in $2d$,

$$f_{ns}(t;L) = f_{ns}^{(b)}(t) + \frac{1}{L} f_{ns}^{(s)}(t) + \frac{1}{L^2} f_{ns}^{(c)}(t) + o\left(L^{-2}\right) . \qquad (2.31)$$

For the leading scaling of the *singular part* of the free energy in general $d < 4$, let us tentatively accept the form (2.19), although later we will find that modifications are necessary in some instances. The $o(L^{-2})$ corrections in (2.31) and similar terms, $o(L^{-3})$, in $3d$, are expected to be $\sim L^{-(d+1)}$, *etc.*, ($d = 2, 3$). However, the leading "geometry-associated" terms in the expansion of $f_{ns}(t; L)$ closely parallel the series (2.28)-(2.29). For example, for slabs or strips, as mentioned above, we anticipate that the edge and corner terms are not present, *i.e.*, only $f_{ns}^{(s)}$ is nonzero. To the extent that they are available, the field-theoretical ϵ-expansion results for nonperiodic finite-size systems [30-32] substantiate this phenomenological expectation which we already encountered in its simplest form when assuming $f_{ns}(t; L) \cong f_{ns}(t; \infty)$ for periodic boundary conditions. Presently, the field-theoretical studies are, however, limited due partly to difficulties in treating nonuniform order parameter profiles below T_c.

For systems with corners, the expansions of the form (2.31) have all the "geometry" terms present, which are proportional to inverse *integral* powers of $1/L$. On the other hand, the scaling term (2.19) is proportional to $1/L^d$ with, generally, varying d. When d passes through integer values 2 or 3, we anticipate "resonant" divergences in Y and $f_{ns}^{(c)}$ yielding logarithmic terms [27], reminiscent of the mechanism of logarithmic specific heat as $\alpha \to 0$ (Sect. 2.3). Thus, in the limit $d \to D = 2$ or 3, we take

$$Y(\tau, \omega) = \frac{y}{D - d} + \tilde{Y}(\tau, \omega) + O(d - D),$$

$$f_{ns}^{(c)}(t) = \frac{yl^{D-d}}{d-D} + \tilde{f}_{ns}^{(c)}(t) + O(d-D), \tag{2.32}$$

where for completeness we restored the magnetic-field dependence (second argument, ω, in Y). As before, l is a microscopic length which can be completely arbitrary here. (In Sect. 2.52 we required l to be system-independent. Here one can take, *e.g.*, $l = a^{-\nu}$.) After some algebra, one gets

$$f(t,H;L) \approx L^{-3}\tilde{Y}\left(atL^{1/\nu}, bHL^{\Delta/\nu}\right) + yL^{-3}\ln(L/l)$$
$$+ f_{ns}^{(b)}(t) + \frac{1}{L}f_{ns}^{(s)}(t) + \frac{1}{L^2}f_{ns}^{(e)}(t) + \frac{1}{L^3}\tilde{f}_{ns}^{(c)}(t), \quad (d=3), \tag{2.33}$$

$$f(t,H;L) \approx L^{-2}\tilde{Y}\left(atL^{1/\nu}, bHL^{\Delta/\nu}\right) + yL^{-2}\ln(L/l)$$
$$+ f_{ns}^{(b)}(t) + \frac{1}{L}f_{ns}^{(s)}(t) + \frac{1}{L^2}f_{ns}^{(c)}(t), \qquad (d=2), \tag{2.34}$$

where both the scaling function $\tilde{Y}(\tau,\omega)$ and the coefficient y are universal. Both depend on the universality class, *i.e.*, dimensionality, symmetry, *etc.*, but also, as is usual for finite-size systems, on the precise geometry and boundary conditions.

Thus far the scaling predictions (2.33)-(2.34) have been tested [28-29] only *at* T_c, as reviewed in Sect. 3.3.8 below. The universal and nonuniversal contributions L^{-D} cannot be separated at T_c, but one can study the $yL^{-D}\ln L$ term.

In fact, relations (2.33)-(2.34) apply generally to systems with curved boundaries [27-28] or in curved-space geometries. The amplitude y is then a global property [28]. However, if the only source of the $yL^{-D}\ln L$ contribution are *corners*, with otherwise flat geometry and boundaries, then one can attribute $y^{(\mathrm{corner})}$ to each corner [28] (see also Sect. 3), so that

$$y = \sum_{\mathrm{corners}} y^{(\mathrm{corner})} . \tag{2.35}$$

Universal terms proportional to $L^{-d}\ln L$ also arise in Gaussian models [28] which were investigated only for integer d (see Sect. 4.2). However, the mechanism of their emergence described above is not applicable in this case, in the least because the finite-size scaling form (1.10) does not apply.

2.6. Surface and Corner Free Energies

Consider again systems with nonperiodic boundary conditions, flat surfaces, and $H = 0$, so that relations (2.28)-(2.29) apply away from T_c, *i.e.*, when $L \to \infty$ for fixed $t \not\equiv 0$. These expansions define the functions $f^{(s)}, f^{(e)}, f^{(c)}$, which in turn are related to $f^{(\mathrm{surface})}$, $f^{(\mathrm{edge})}$, $f^{(\mathrm{corner})}$, with geometry-dependent rules relating the two sets, as explained in Sect. 2.5. After taking the $L \to \infty$ limit for fixed $t \not\equiv 0$, one can still consider the critical-point behavior of all the above free energies as $|t| \to 0$. For brevity we take $t \to 0^+$ here.

Singular contributions to the free energies can also be evaluated by considering the asymptotic form of the scaling function $\tilde{Y}(\tau, 0)$

in (2.33)-(2.34) for large positive τ, *i.e.*, in the regime $L >> (at)^{-\nu}$, or equivalently, $L >> \xi^{(b)}$. A standard argument [27] suggests

$$\tilde{Y}(\tau, 0) \approx y_3\tau^{3\nu} + y_2\tau^{2\nu} + y_1\tau^{\nu} - \nu y \ln \tau + y_0 + \ldots, \qquad (3d), \ (2.36)$$

$$\tilde{Y}(\tau, 0) \approx y_2\tau^{2\nu} + y_1\tau^{\nu} - \nu y \ln \tau + y_0 + \ldots, \qquad (2d), \qquad (2.37)$$

where the coefficients are universal and geometry dependent, except for $y_d \equiv q_+$. One thus obtains the standard results [1] [see also (2.2)]

$$f_s^{(b)}(t) \approx q_+(at)^{d\nu} \qquad \text{and} \qquad f_s^{(s)}(t) \approx y_{d-1}(at)^{(d-1)\nu}, \qquad (2.38)$$

as well as less familiar relations [27]

$$f_s^{(e)}(t) \approx y_1(at)^{\nu} \quad \text{in} \quad 3d, \qquad (2.39)$$

$$f_s^{(c)}(t) \approx -\nu y \ln|t|, \qquad d = 2 \text{ or } 3, \qquad (2.40)$$

where for completeness we also included the appropriate $t < 0$ expression in the leading order "corner" result (2.40).

Consideration of the connection between the set of coefficients $f^{(s)}$, $f^{(e)}$, $f^{(c)}$, and the surface, edge, corner free energies further leads to the conclusions that as $t \to 0^{\pm}$,

$$f_s^{(\text{surface})} \approx q_\pm^{(\text{surface})} |at|^{(d-1)\nu}, \qquad (d = 2, 3), \tag{2.41}$$

$$f_s^{(\text{edge})} \approx q_\pm^{(\text{edge})} |at|^{(d-2)\nu}, \qquad (d = 3), \tag{2.42}$$

$$f_s^{(\text{corner})} \approx -\nu y^{(\text{corner})} \ln |t|, \qquad (d = 2, 3), \tag{2.43}$$

where the universal coefficients $q_\pm^{(\text{surface})}$, $q_\pm^{(\text{edge})}$ and $y^{(\text{corner})}$ depend only "locally" on the type of the surface boundary conditions, edge and corner shape and angles, *etc.*, but they do not depend on the global geometry. Examination of the series (2.36)-(2.37) suggests that the corner free energy, for each corner, can be written as

$$f^{(\text{corner})}(t) \approx -\nu y^{(\text{corner})} \ln |t| + \phi_0 + q_\pm^{(\text{corner})}, \qquad \text{for} \quad t \to 0^\pm, \tag{2.44}$$

where the subleading constant terms are different for $t > 0$ and $t < 0$, with nonuniversal ϕ_0, but universal difference $q_+^{(\text{corner})} - q_-^{(\text{corner})}$. Relation (2.40) can be extended similarly because for a given flat-surface sample,

$$f^{(c)}(t) = \sum_{\text{corners}} f^{(\text{corner})}(t) \quad . \tag{2.45}$$

One can *define* [33] surface, edge, corner free energies by postulating that expansions (2.28)-(2.29) can be formally applied in the critical region. Consider, for illustration [33], $3d$ boxes of size $L \times \bar{L}^2$, with periodic boundary conditions in the $\bar{L}$ directions. In the direction of L we impose periodic (p) or free (f) boundary conditions. We then employ (2.28) to write

$$f(t;L) \simeq f^{(b)}(t), \qquad (p), \tag{2.46}$$

$$f(t;L) \simeq f^{(b)}(t) + \frac{1}{L} f^{(s)}, \qquad (f), \tag{2.47}$$

and we further employ the relation $f^{(s)} = 2f^{(\text{surface})}$, appropriate for the ($f$) geometry, to write

$$f(t;L) \simeq f^{(b)}(t) + \frac{2}{L} f^{(\text{surface})}, \qquad (f). \tag{2.48}$$

Thus, we can define

$$f^{(\text{surface})}(t;L) \equiv \frac{L}{2}\left[f_{(f)}(t;L) - f_{(p)}(t;L)\right], \tag{2.49}$$

and use this definition even at T_c. The singular behavior of this quantity follows from those of $f_{(f)}$ and $f_{(p)}$,

$$f_s^{(\text{surface})}(t;L) \approx L^{-(d-1)} Y^{(\text{surface})}\left(atL^{1/\nu}\right), \tag{2.50}$$

where

$$2Y^{(\text{surface})}(\tau) \equiv \left[Y_{(f)}(\tau,0) - Y_{(p)}(\tau,0)\right]. \tag{2.51}$$

Note, however, that $f^{(\text{surface})}(t;L)$ thus defined [and similarly, for $f^{(\text{edge})}(t;L)$, *etc.*] will depend on the original reference geometries, except in the strict $L \to \infty$ limit (*i.e.*, $|\tau| >> 1$), when they reduce to $f^{(\text{surface})}(t)$, *etc.*, considered before.

2.7. Finite-Size Properties of d > 4 Systems

For $d > 4$, finite-size scaling in its simple form described in Sect. 1.3 breaks down, as first noted in [14]. As with bulk critical behavior, the basic mechanism has been identified [4,34] as due to a dangerous irrelevant variable [35]. The RG scaling relations, compare (2.11) for ξ, apply:

$$f_s(t,H;L) \approx L^{-d}\bar{Y}\left(atL^2, bHL^{(d+2)/2}; uL^{4-d}\right), \tag{2.52}$$

$$\xi(t,H;L) \approx L\bar{X}\left(atL^2, bHL^{(d+2)/2}; uL^{4-d}\right), \tag{2.53}$$

where we used the RG eigenexponents

$$\lambda_t = 2, \quad \lambda_H = \frac{d+2}{2}, \quad \lambda_u = 4-d, \tag{2.54}$$

at the Gaussian fixed point, and the irrelevant variable $u > 0$ is proportional to the coefficient of the fourth-order term in the Ginzburg-Landau Hamiltonian. However, the scaling functions $\bar{Y}(\tau,\omega;\rho)$ and $\bar{X}(\tau,\omega;\rho)$ are singular as $\rho \to 0$ and cannot be expanded in powers of ρ.

The most studied case is that of a nearly-hypercubic sample with periodic boundary conditions. Sum-rule [34] and heuristic [4] arguments, as well as constant order parameter mean-field [5,16] calculations, suggest that as $\rho \to 0$,

$$\bar{Y}(\tau, \omega; \rho) \approx \mathcal{P}\left(\tau\rho^{-\frac{1}{2}}, \omega\rho^{-\frac{1}{4}}\right). \tag{2.55}$$

Introducing nonuniversal constants

$$\bar{a} = au^{-\frac{1}{2}} \qquad \text{and} \qquad \bar{b} = bu^{-\frac{1}{4}}, \tag{2.56}$$

we then have

$$f_s(t, H; L) \approx L^{-d}\mathcal{P}\left(\bar{a}tL^{d/2}, \bar{b}HL^{3d/4}\right). \tag{2.57}$$

Thus, a universal two-variable scaling function suffices in this case (periodic cubes) and, in fact, $\mathcal{P}$ can be obtained as a quadrature from the constant order parameter mean-field theory [5,16]. The correlation length scaling has been investigated in lesser detail [34]. Moment definitions of the correlation length [34] for finite samples, usually have a built in bound $\xi(t, H; L) \leq$ (const L), so that it is natural to conjecture that

$$\xi(t, H; L) \approx L\mathcal{R}\left(\bar{a}tL^{d/2}, \bar{b}HL^{3d/4}\right). \tag{2.58}$$

However, this conjecture has not been generally substantiated. In fact, the correlation length scaling is expected to be very sensitive to the precise definition of ξ.

For near-cubic systems with free boundary conditions, finite-size behavior of the free energy for $d > 4$ has been investigated within the mean-field type Gaussian approach [36]. While the general relation (2.52) seems to apply, one no longer has a simple pattern of a singular dependence on ρ, as in (2.55).

In the periodic cylinder geometry, $L^{d-1} \times \infty$, there are indications [4-5] that (2.55) is replaced by

$$\bar{Y}(\tau, \omega; \rho) \approx \rho^{1/3} \tilde{\mathcal{P}}\left(\tau \rho^{-2/3}, \omega \rho^{-1/2}\right), \tag{2.59}$$

$$\bar{X}(\tau, \omega; \rho) \approx \rho^{-1/3} \tilde{\mathcal{R}}\left(\tau \rho^{-2/3}, \omega \rho^{-1/2}\right), \tag{2.60}$$

where the correlation length is defined by the leading transfer matrix spectral gap. Among the various predictions that follow, we note that

$$\xi(0,0;L) \approx u^{-1/3} \tilde{\mathcal{R}}(0,0) L^{(d-1)/3}, \tag{2.61}$$

where $\tilde{\mathcal{R}}(0,0)$ is universal, and $d > 4$. Additional results for $d > 4$ spherical models in various geometries will be mentioned in Sect. 3.

<u>For $d = 4$</u>, *i.e.*, at the upper marginal dimensionality, one anticipates [5,14,16] logarithmic correction factors to the ordinary ($d < 4$) scaling laws. Explicit results are generally limited, and one has to evaluate finite-size quantities numerically [16].

2.8. *Other Research Topics*

In the preceding subsections, we described the theoretical foundations of the finite-size scaling description of static, isotropic critical behavior in prototype short-range n-vector models for $d < 4$, and we also outlined some results for $d > 4$. Here, for completeness, we list several other recently active research topics. We do not provide a detailed survey of these developments, but mostly list the relevant literature.

2.8.1. Microcanonical and Fixed-M Ensembles

Finite-size scaling in the *microcanonical* ensemble, where the system energy is the controlled variable, instead of the temperature, has been developed in [37], with an emphasis on applications in MC data analyses.

The magnetization density, $M(t, H; L)$, is obtained from the free energy via

$$M \equiv -k_B T \frac{\partial f}{\partial H}. \tag{2.62}$$

Study of finite-size effects in the fixed-M ensemble, *i.e.*, with system properties considered as functions of $(t, M; L)$, has been reported in [38-39]. Hyperuniversality of the scaling functions has been emphasized in [39].

2.8.2. Anisotropic Models

Finite-size behavior of anisotropic (directed) models has some interesting new aspects due to different correlation length divergence along different sample dimensions, and also due to the emergence, in some cases, of characteristic size-dependent length scales which diverge in the bulk limit even for $t \not\equiv 0$. A comprehensive summary of various results and concepts is given in [40-41].

2.8.3. Dynamics

Finite-size effects in critical dynamics and in some nonequilibrium systems have been investigated recently [42]. Results were also obtained on finite-size effects on nucleation processes [43].

2.8.4. Random Systems

Finite-size scaling analysis of numerical simulation results for random systems, such as spin glasses or random-field models, has new interesting aspects associated with the use of two averages, one over randomness and the usual configurational ensemble average, as well as slow equilibration effects. These topics are reviewed in Young's article in this volume.

3. CRITICAL-POINT SCALING: SURVEY OF RESULTS

3.1. *Definitions and notation*

In this section we review recent numerical and analytical results for selected $2d$ and $3d$ models with an emphasis on scaling function universality and finite-size amplitudes. However, we first introduce notational conventions for thermodynamic and other quantities of interest.

3.1.1. Thermodynamic Quantities

Relation (2.62) for the magnetization contains an inconvenient factor k_BT. It is customary to define the reduced field

$$h = \frac{H}{k_BT}, \tag{3.1}$$

so that

$$M = -\frac{\partial f}{\partial h}. \tag{3.2}$$

Similarly, we define the susceptibility,

$$\chi = \frac{\partial M}{\partial h}, \tag{3.3}$$

i.e., with an extra factor of k_BT as compared to the usual thermodynamic definition $\frac{\partial M}{\partial H}$. Note that no claim is made that h is dimensionless. In fact, we do not pay much attention to the particular choice of units of various quantities, as long as it is consistent.

We also introduce nonlinear "magnetization",

$$M^{(nl)} = \frac{\partial \chi}{\partial h} \quad , \tag{3.4}$$

and "susceptibility",

$$\chi^{(nl)} = \frac{\partial M^{(nl)}}{\partial h} \quad . \tag{3.5}$$

For bulk amplitudes in the limit $t \to 0^{\pm}$, we denote

$$M(t, 0; \infty) \approx B_{-}|t|^{\beta}, \qquad t \to 0^{-} \quad , \tag{3.6}$$

$$\chi(t, 0; \infty) \approx \Gamma_{\pm}|t|^{-\gamma}, \qquad t \to 0^{\pm} \quad , \tag{3.7}$$

$$M^{(nl)}(t, 0; \infty) \approx B_{-}^{(nl)}|t|^{-(\gamma+\Delta)}, \qquad t \to 0^{-} \quad , \tag{3.8}$$

$$\chi^{(nl)}(t, 0; \infty) \approx \Gamma_{\pm}^{(nl)}|t|^{-(\gamma+2\Delta)}, \qquad t \to 0^{\pm} \quad . \tag{3.9}$$

3.1.2. Binder's Cumulant Ratio

Binder [38] proposed to define the quantity

$$G(t; L) = -\left[\frac{\chi^{(nl)}}{L^d \chi^2}\right]_{H=0} \quad . \tag{3.10}$$

By differentiating (with respect to h) the finite-size scaling form (1.10) for f, we conclude that G scales according to

$$G(t;L) \approx \mathcal{G}\left(atL^{1/\nu}\right), \qquad (3.11)$$

and $G(0;L)$ approaches a universal value, $\mathcal{G}(0)$, as $L \to \infty$. More precisely, we have

$$\mathcal{G}(\tau) = \left[(\partial^4 Y/\partial\omega^4) \Big/ (\partial^2 Y/\partial\omega^2)^2\right]_{\omega=0}, \qquad Y \equiv Y(\tau,\omega). \quad (3.12)$$

If we denote by Ψ the fluctuating order parameter density, *i.e.*, $\langle\Psi\rangle = M$, then for hypercubic or near-hypercubic samples of volume L^d, we have

$$G(t;L) = \frac{3\langle\Psi^2\rangle^2 - \langle\Psi^4\rangle}{\langle\Psi^2\rangle^2} \quad . \qquad (3.13)$$

This formula suggests the term "cumulant ratio" for G. Such "fluctuation" relations usually can be written down only for fully finite systems. For example, for the cylindrical geometry $L^{d-1} \times \bar{L}$, we have

$$G(t;L,\bar{L}) = \frac{\bar{L}}{L}\frac{3\langle\Psi^2\rangle^2 - \langle\Psi^4\rangle}{\langle\Psi^2\rangle^2} \quad . \qquad (3.14)$$

This quantity has a finite limit, which we denote $G(t;L)$, as $\bar{L} \to \infty$.

3.1.3. Types of Surfaces (Boundary Conditions)

Surface interactions may be characterized by J_{wall}, different from the bulk coupling J. As long as J_{wall}/J is not too large, the surface critical behavior is driven by the bulk. This is called the ordinary (O) surface transition. The most familiar case is $J_{wall} = J$, *i.e.*, *free boundary conditions.* It is believed that a continuous field-theoretical description near T_c with the constraint that the order parameter vanishes at the wall, corresponds to the O-type surface transition.

When J_{wall}/J increases, the surface may actually order above T_c. However, at T_c, singularities still develop in surface properties. This is termed the extraordinary (E) transition. The borderline case is termed the special (Sp) surface transition. Finally, wall interactions may result in a positive (+) or negative (−) ordering field H_{wall} acting at the surface. This case is believed to be universal also with a continuous field-theoretical order parameter fixed at positive (+) or negative (−) value at the surface.

The above surface types are, strictly speaking, valid only for the short range Ising models in $3d$. In $2d$, only O and $\pm$ surfaces are possible. For models other than Ising, different classifications of boundary conditions may be appropriate.

For the finite-size scaling description, only the surface type (O, E, Sp, +, −, *etc.*) is important. The finite-size scaling functions are different for various types of boundary conditions.

3.2. Results for Cumulant and Other Ratios

3.2.1. Binder's Cumulant Ratio at T_c

For *hypercubic Ising* models, the value of $\mathcal{G}(0)$ in (3.11) has been estimated by MC [34,38,44-47], transfer matrix combined with MC [48], Wilson's [49] approximate RG [50], and ϵ-expansion [5] methods, mostly with *periodic* boundary conditions. For periodic systems, estimates in $2d$ [44,48] suggest values in the range 1.830-1.835; estimates in $3d$ [5,38,46-47] suggest $\mathcal{G}(0) = 1.2$-1.4. Numerical studies in $5d$ [34] suggest the value $\mathcal{G}(0) \sim 1.0$, barely consistent with the mean-field result $0.81156\ldots$ [5]. Further results for *free* boundary conditions can be found in [38].

Calculations for finite systems which are subblocks of larger systems (this can be regarded as another type of boundary conditions) yield values in the range 1.53-1.74 in $2d$ [38,50] and ~ 0.6 to ~ 0.9 in $3d$ [38,45,50]. Results in $4d$ [38], suggesting $\mathcal{G}(0) \sim 0$, should be considered preliminary, especially in view of the fact that the modified scaling relation (2.57), which in turn suggests

$$G(t;L) \approx \mathcal{G}\left(\bar{a}tL^{d/2}\right), \tag{3.15}$$

has been established only for $d > 4$, while the behavior of $G(t;L)$ in $4d$ has not been investigated theoretically.

For hypercubic, periodic n-vector models with $n > 1$, the only available result [5] is the $O(\epsilon)$ expansion in powers of $\sqrt{\epsilon}$, which is rather cumbersome and is not reproduced here. Note also that higher-order cumulant ratios, involving $\langle \Psi^6 \rangle$, *etc.*, have been considered [5,38,51], but no definitive numerical results are available. Finally, we quote an estimate $\mathcal{G}(0) = 1.934 \pm 0.001$ for $2d$ percolation [52].

Turning now to the *periodic cylinder* geometry, $L^{d-1} \times \infty$, we first quote the $3d$ results for the Ising model, $\mathcal{G}(0) \sim 3$, and for percolation, 5.0 ± 0.2, obtained by the transfer matrix MC method [52]. More results are available in $2d$. Conformal invariance predictions [53-54] for the $2d$ Ising correlation functions on the $L \times \infty$ strip, have been integrated [48] to yield

$$\mathcal{G}(0) = 7.38132 \pm 0.00006, \tag{3.16}$$

which has been confirmed by transfer matrix [48] and transfer matrix MC [52] estimates to about 1%. Numerical results are also available for $2d$ percolation, 9.90 ± 0.06 [52], and 3-state Potts model, 2.49 ± 0.09 [55]. For $2d$ models with small exponent η, a general conformal invariance prediction [7] is

$$\mathcal{G}(0) \simeq \frac{6}{\pi\eta}, \qquad \text{for} \qquad \eta << 1. \tag{3.17}$$

For disordered $2d$ Ising and 3-state Potts models, the estimates ~ 7.38 and 4.5 ± 0.9, respectively, [55] illustrate a clear change of the universal $\mathcal{G}(0)$ value as compared to "pure" models, only in the 3-state Potts case ($\alpha = \frac{1}{3} > 0$), but not in the Ising case [compare (3.16)] which is borderline ($\alpha = 0$) according to the Harris [56] criterion.

<u>3.2.2.</u> Cumulant Scaling Asymptotics

The scaling function $\mathcal{G}(\tau)$ in (3.11) is universal for any value of τ. Numerical evaluation of this function for $\tau \neq 0$ (*e.g.*, [34,38]) has been limited, as compared to the results at $\tau = 0$ described above. However, one can derive the asymptotic behavior in the limits $\tau \to \pm\infty$. For $\tau \to +\infty$, one can show [38,51], by considering large-τ behavior of the finite-size scaling functions, that

$$\mathcal{G}(\tau \to +\infty) \approx -\Gamma_+^{(nl)} \Gamma_+^{-2} (\tau/a)^{\alpha-2} . \tag{3.18}$$

This result applies for hypercubic systems, for all n, although the corrections to the power low decay (3.18), which are probably exponentially small in τ for periodic boundary conditions, have been considered carefully only for $n = 1$ [51]. Note that the coefficient of $1/\tau^{2-\alpha}$ is generally positive.

The detailed asymptotic prediction for $\tau \to -\infty$ has been derived [51] for periodic hypercubes, for $n = 1$ only, based on certain results on finite-size rounding of Ising first-order transitions [4], see Sect. 5. With

$$\bar{\tau} \equiv -\tau/a , \tag{3.19}$$

we have

$$\mathcal{G}(\tau \to -\infty) \approx \frac{2B_-^4 \bar{\tau}^{3(2-\alpha)} - 4B_- B_-^{(nl)} \bar{\tau}^{2-\alpha} - \Gamma_-^{(nl)}}{\bar{\tau}^{2-\alpha} \left(B_-^2 \bar{\tau}^{2-\alpha} + \Gamma_- \right)^2} , \tag{3.20}$$

where again the corrections are believed to be exponentially small in $\bar{\tau}$. Note that $\mathcal{G}(-\infty) = 2$.

3.2.3. High-Order Free Energy Derivative Ratios at T_c

Due to the $H \leftrightarrow -H$ symmetry which is not spontaneously broken even below T_c in fully finite, or effectively $1d$ cylindrically shaped systems, odd h-derivatives of the free energy vanish when $H \to 0^{\pm}$, as long as $L < \infty$. However, one can introduce [57] functions

$$f_{(+)}(T, H; L) = f_{(-)}(T, -H; L), \tag{3.21}$$

calculated from the two largest Ising model transfer matrix eigenvalues in the periodic cylinder geometry, which break the $H \leftrightarrow -H$ symmetry for $L < \infty$, and also approximate the $T < T_c$ "spontaneous" values of the free energy derivatives at $H = 0$, for $L < \infty$: see also Sect. 5.

Let $Y_{(+)}(\tau, \omega)$ denote the finite-size scaling function for $f_{(+)}$. The function $Y_{(+)}(0, \omega)$ was studied in [4], for the square and triangular lattice Ising model strips $L \times \infty$, with L up to ten lattice spacings. Universal ratios

$$\left. \frac{\left[\partial^{k+1} Y_{(+)}(\tau,\omega)/\partial\omega^{k+1}\right]\left[\partial^{k-1} Y_{(+)}(\tau,\omega)/\partial\omega^{k-1}\right]}{\left[\partial^{k} Y_{(+)}(\tau,\omega)/\partial\omega^{k}\right]^2} \right|_{\tau,\omega=0} \tag{3.22}$$

were estimated for $k = 2, 3, 4, 5, 6$. The values for two lattices agree to within 4% for $k = 2, 6$, but to better than 1% for $k = 3, 4, 5$.

3.3. *Free Energy and Correlation Length Scaling*

Most of this subsection is devoted to a survey of numerical and analytical tests of the finite-size scaling relations (1.10)-(1.11) and (1.13)-(1.14) in those geometries where these relations apply (see Sect. 2). Results for free-boundary-condition cubes and squares, where additional logarithmic scaling contributions enter (Sect. 2.5), are presented in the last Sect. 3.3.8.

3.3.1. Spherical Model Results

The $H = 0$ free energy scaling relation (1.10) was checked for spherical models in several geometries, with continuously varying $2 < d < 4$, and also with formally continuous number of finite dimensions, d^*, *i.e.*, for sample shapes $L_1 \times L_2 \times \ldots \times L_{d^*} \times \infty \times_{\ldots} \times \infty$. Rather complicated implicit equations for $Y(\tau, 0)$ were derived, confirming its universality, for systems with periodic [58] and antiperiodic [59] boundary conditions, as well as for a class of Bose gas models [60] which are in the same universality class. Some scaling predictions for $d > 4$, see (2.52)-(2.53), *etc.*, were tested by spherical model calculations, reported in [61].

Tests of the correlation length scaling have been focussed mainly on calculating the universal amplitude $X(0,0)$ in (1.14), in the periodic cylinder geometry $L^{d-1} \times \infty$, with varying $2 < d < 4$. Large-n limit results (equivalent to the spherical model [62]) are $X(0,0) \simeq 0.6614$ in $3d$ [14] and

$$X(0,0) \approx (4\pi^2\epsilon)^{-1/3} \tag{3.23}$$

for small $\epsilon = 4 - d$ [14],

$$X(0,0) \approx (\pi\bar{\epsilon})^{-1} \tag{3.24}$$

for small $\bar{\epsilon} = d - 2$ [63], *etc.* Results for antiperiodic boundary conditions are also available [64], and furthermore it is known [64] that the appropriate amplitudes for the energy-energy correlation length are twice the spin-spin amplitudes, for both periodic and antiperiodic boundary conditions. Detailed results for correlation length scaling for $d \geq 4$ were also obtained, only at T_c, see, *e.g.*, [14]. Specifically, the logarithmically modified L-dependence,

$$\xi(0,0;L) \approx L\left[\left(4\pi^2\right)^{-1}\ln L\right]^{1/3}, \qquad (d=4), \tag{3.25}$$

is among the very few explicit results available in $d = 4$ [14].

3.3.2. ϵ-Expansion Results

As mentioned in Sect. 2, ϵ-expansions for finite-size systems [5,16-18,30-32,39] generally confirm the phenomenological scaling predictions. However, explicit results are difficult to derive. For periodic cylinders, an $O(\epsilon)$ closed-form expression is available [5] for $X(0,0)$, extending (3.23) to all $n \leq \infty$.

3.3.3. Numerical Estimates of 3d Amplitudes

For *periodic cubes*, L^3, MC estimates are available for the SC and BCC Ising models [65], suggesting $Y(0,0) = -(0.64\text{-}0.69)$. For both periodic and antiperiodic Ising cylinders, $L^2 \times \infty$, in the Hamiltonian (anisotropic) limit, numerical results are available [66-69] supporting the universality of $Y(0,0)$ and $X(0,0)$, both spin-spin and energy-energy.

For 3d slabs, $L \times \infty^2$, with various boundary conditions at each of the two surfaces, the singular part of the free energy scales according to (1.10) because this nonperiodic geometry has no corners (see Sect. 2.5). However, $Y(0,0)$ depends on both boundary conditions which, as described in Sect. 3.1.3, can be any pair selected from $O, E, Sp, +, -$.

Results for 3d Ising amplitudes $Y(0,0)$ have been summarized in [70]. Their Migdal-Kadanoff real-space RG values are 0.279, $\sim 0^-$, 0.017, 0.051, 0.019, 0.017, and -0.015, for walls $+-$, $++$, $Sp+$, $O+$, $SpSp$, OSp, OO, respectively. Note also obvious symmetries $Y_{O+} = Y_{O-}$, *etc.* The OO value in the range $-(0.10\text{-}0.02)$ is also suggested by ϵ-expansion results [30]. For type E walls, it has been conjectured [70] that

$$Y_{AE}(0,0) = Y_{A+}(0,0), \tag{3.26}$$

for $A = O,\ E,\ Sp,\ +,\ -$. There are also certain mean-field relations among amplitudes which may hold approximately in 3d, see [70]. Numerical results for $n > 1$ are limited to some estimates for the case $n = 2$: see [71] for details.

Numerical MC results are also available [33] for surface free energy amplitudes: see discussion leading to relation (2.51), in Sect. 2.6. Universality of $Y^{(\mathrm{surface})}(0)$ has been checked [33] for various aspect ratios of $3d$ boxes (Sect. 2.6), for SC and BCC Ising models.

3.3.4. Conformal Invariance in $2d$

Conformal invariance (*e.g.*, [7, 72]) yields exact values of critical exponents and universal finite-size amplitudes at T_c, for a large number of undirected $2d$ models. We do not review this rapidly growing field here. However, in the following subsections (Sect. 3.3.5-3.3.8), when surveying results for $2d$ amplitudes, we quote various conformal invariance predictions. We limit our discussion to the most familiar models (Ising, Potts, *etc.*).

3.3.5. Correlation Lengths for $L \times \infty$ Strips

For the spin-spin correlation length of the exponential decay of correlations along $L \times \infty$ strips in $2d$, we have [53] conformal invariance results

$$X(0,0) = \frac{1}{\pi\eta}, \qquad \text{and} \qquad X(0,0) = \frac{2}{\pi\eta_{\|}}, \tag{3.27}$$

for *periodic* and *free* boundary conditions, respectively, where η and $\eta_{\|}$ are the standard bulk and surface exponents. Results for other correlations (*e.g.*, energy-energy) are similar, with appropriate exponents. For example, for the Ising model $\eta = \eta_{\text{spin-spin}} = \frac{1}{4}$, while

$\eta_{\text{energy-energy}} = 2.$

Another line of modification is to fix boundary conditions other than free (type O): (3.27) then applies with modified $\eta_{\|}$. For example, we have $\eta_{\|} = \eta_{\|}(OO) = 1$, but $\eta_{\|}(++) = \eta_{\|}(--) = 4$, for the Ising model. These results have been derived and reviewed in [7,73]. Various exponents entering (3.27) and generalizations, have been determined exactly for many $2d$ models, including n-vector models with $|n| \leq 2$, and q-state Potts models with $0 \leq q \leq 4$, see [7,73]. For some cases (Ising, 3,4-state Potts), results have been derived also for antiperiodic and mixed boundary conditions [73].

More general tests of the finite-size scaling relation (1.11) have been mostly limited to the $2d$ Ising model in zero field. Thus $X(\tau, 0)$ has been calculated for periodic, antiperiodic, free, fixed, and some mixed-type boundary conditions [67-69,74-75]. Numerical transfer-matrix tests of universality of quantities derivable from $X(\tau, \omega)$ were also reported [75] for the 3-state Potts, spin-$\frac{1}{2}$ and spin-1 Ising, and some other models.

3.3.6. Second-Moment Correlation Length Amplitudes in 2d

For a correlation length defined via the second moment of the correlation function of $L \times \infty$ strips, instead of the exponential decay definition (as in Sect. 3.3.5), the conformal invariance expression replacing (3.27) (assuming *periodic* boundary conditions), is [76]

$$X(0,0) = \frac{1}{4\pi}\sqrt{2\psi'\left(\frac{\eta}{4}\right) - \frac{\pi^2}{\sin^2(\pi\eta/4)}} \quad , \qquad (3.28)$$

where

$$\psi'(x) = \sum_{k=0}^{\infty} \frac{1}{(x+k)^2} \qquad (3.29)$$

is the derivative of the standard digamma function $\psi(x)$. MC tests of this result for the self-avoiding walk model [76], for which $\eta = \frac{5}{24}$ and (3.29) yields $X(0,0) = 1.527\ldots$, confirm the predicted value to about 2%, for strips up to 15 lattice spacings wide. Numerical MC results were also obtained [76] for strips with *free* boundary conditions, and compared with some approximate conformal invariance predictions (see [7,76]).

<u>*3.3.7.*</u> Tests of Free Energy Scaling in 2*d*

For *periodic* strips, $L \times \infty$, we have [77]

$$Y(0,0) = -\frac{\pi c}{6}, \qquad (3.30)$$

where c is the conformal anomaly number. For example, $c = \frac{1}{2}$ for the 2*d* Ising universality class. For *free* or *fixed* (same on both sides of the strip) boundary conditions, we have [77]

$$Y(0,0) = -\frac{\pi c}{24}. \qquad (3.31)$$

A point of caution is in order for the case of the 2*d* Ising model. When $\alpha = 0$, results like (3.30) refer to the "regular" scaling functions of the type $W(\tau)$ in (2.22), with logarithmic size dependence separated out (and vanishing at T_c, since logarithmic terms are proportional to t^2), not to functions $\bar{W}_{\pm}(\tau)$, as in (2.21). Thus, $W(0) = -\pi/12$. For some models, conformal invariance results are also available for more complicated boundary conditions [73]. For instance, we have [73] $W(0)/\pi = \frac{1}{6}, \frac{23}{48}, \frac{1}{24}$, for Ising strips with antiperiodic, mixed $+-$, and mixed $O+$ (or $O-$) boundary conditions, respectively.

Numerical tests of the above predictions have been largely limited to systems with periodic boundary conditions and were carried out for several models, *e.g.*, $0 \le q \le 4$ Potts, *etc.* [67-69,75,77].

Results on the free energy scaling away from T_c are limited to the zero-field Ising model with various boundary conditions, in rectangle geometry $L_1 \times L_2$, see [13,78-79]. However, at T_c, results for $Y(0,0)$ are available for $0 \le q \le 4$ Potts model rectangles with several types of boundary conditions [80]. The expressions are rather complicated, involving θ-functions, *etc.* In some limits, explicit answers were given [78,80]. For example, for periodic Ising squares,

$$W(0) = -\ln\left(2^{1/4} + 2^{-1/2}\right) = -0.6399\ldots \quad . \tag{3.32}$$

3.3.8. Free-Energy Amplitudes for Systems with Corners

Thus far, the only numerical and analytical results testing predictions (2.33)-(2.34) were obtained at T_c, where the characteristic universal size dependence $yL^{-d}\ln L$ can be isolated ($d = 2, 3$). In $3d$, MC estimates

$$y = 0.009 \pm 0.005 \qquad \text{and} \qquad 0.012 \pm 0.003 \tag{3.33}$$

were reported [29] for, respectively, the SC and BCC-lattice Ising cubes, L^3, with free boundary conditions. These studies also yielded universal surface and edge free energy amplitudes, see [29].

In $2d$, conformal invariance predictions are available for y [28], in rather general geometries with curved boundaries, corners, and in curved space. The amplitude y is then a global property [28]. We only survey the case of flat boundaries where we can attribute $y^{(\text{corner})}$ to each corner, see (2.35). With same (free or fixed) boundary conditions all over the polygon, each corner "geometry" is fully specified by the opening angle δ, and we have [28] the remarkable result

$$y^{(\text{corner})} = \frac{c\delta}{24\pi}\left[1 - (\pi/\delta)^2\right] . \tag{3.34}$$

4. SIZE EFFECTS ON INTERFACIAL PROPERTIES

4.1. *Interfacial Free Energy Near Criticality*

4.1.1. Periodic Interfaces

Similar to the surface free energy (Sect. 2.6), interfacial free energy for finite-size systems is defined by a difference in the free energies of two samples with different boundary conditions. Consider for example $L^{d-1} \times \bar{L}$ Ising boxes with periodic boundary conditions in all $d-1$ dimensions L. A "floating" periodic interface of hyperarea L^{d-1} can be introduced by imposing antiperiodic or $+-$ boundary conditions along $\bar{L}$. The appropriate boxes with no interfaces are with periodic and $++$ (or $--$) boundary conditions along $\bar{L}$, respectively. Note that we have to use $++$ for $+-$ to avoid any excess surface free energy. The excess interfacial free energy is then defined per $k_B T$ and per unit area,

$$\sigma\left(t; L, \bar{L}\right) = \bar{L}\left[f_{+-}\left(t; L, \bar{L}\right) - f_{++}\left(t; L, \bar{L}\right)\right]. \tag{4.1}$$

A similar difference can be defined for antiperiodic *vs.* periodic boundary conditions along $\bar{L}$. The resulting function $\sigma(t; L, \bar{L})$ depends on the boundary conditions and other geometrical features such as the aspect ratio $\bar{L}/L$. In the limit $L, \bar{L} \to \infty$ with fixed t, the quantity $\sigma(t; L, \bar{L})$ approaches the bulk surface tension

$$\sigma^{(b)}(t) = \sigma(t; \infty, \infty), \tag{4.2}$$

where $\sigma^{(b)}(t) = 0$ for $t \geq 0$, and

$$\sigma^{(b)}(t) \sim (-t)^{\mu}, \quad \text{for small} \quad t < 0. \tag{4.3}$$

The finite-size free energy difference (4.1) can in principle be considered for any model, in nonzero field, *etc.* However, it is customary to restrict the consideration to discrete spin systems (*e.g.*, Ising, Potts) at the phase boundaries ($H = 0$, $t \leq 0$ for the Ising case) where phases coexist and are separated by sharp interfaces.

Thus, most of our discussion refers to the $H = 0$ Ising models. However, we keep general t. Indeed, the hyperuniversal scaling form [4,81]

$$\sigma(t; L, \bar{L}) \approx L^{1-d} \Sigma \left(atL^{1/\nu}; \bar{L}/L \right) \tag{4.4}$$

involves a scaling function which is, as usual, smooth (analytic) at $\tau = atL^{1/\nu} = 0$. Note that we have not used "singular part" in (4.4). Indeed, for periodic "floating" interfaces relation (4.1) suggests that all the "nonsingular" terms in the free energies cancel out.

4.1.2. Exact and Numerical Results for Periodic Interfaces

All the available data on interfacial finite-size properties near T_c is limited to $2d$ models. For the $2d$ Ising model, the function $\Sigma(\tau; \bar{L}/L)$ can be evaluated exactly [82] in the $+-$ geometry described in Sect. 4.1.1. For example,

$$\Sigma(0; 1) = 0.86837\ldots \quad . \tag{4.5}$$

For the 3-state Potts model, with $+-$ replaced by fixing Potts variables at different values at two walls, a MC study [83] of the scaling relation (4.4) yields, *e.g.*,

$$\Sigma(0; 1) \simeq 2.1. \tag{4.6}$$

For the antiperiodic geometry (Sect. 4.1.1), results are available for $0 \le q \le 4$ Potts models at T_c [80] (consult [80] for the definition of "antiperiodic" for $q \neq 2$). For Ising ($q = 2$) squares, we have [80]

$$\Sigma(0;1) = \ln(1 + 2^{3/4}) \simeq 0.9865, \tag{4.7}$$

which, when compared with (4.5), illustrates dependence on boundary conditions.

Note that singling out the size L to "scale" (4.4) leads to a singular dependence on $\bar{L}/L$ in the limit of a long strip $L \to \infty$. For instance in the Ising case [84], we have

$$\Sigma(0;z) \approx \frac{\pi}{4z}, \quad \text{as} \quad z \to 0, \tag{4.8}$$

which simply means

$$\sigma(0;\infty,\bar{L}) \approx \pi/(4\bar{L}). \tag{4.9}$$

Obviously, one can redefine (4.4), scaled with $\bar{L}$ or any other combination of L and $\bar{L}$.

<u>*4.1.3.*</u> Pinned Interfaces

One can define interfacial free energy in geometries with interfaces with pinned boundaries (pinned ends in $2d$). An Ising $L \times \bar{L}$ system in $2d$, with an interface inclined by an angle δ, is illustrated in Fig. 1. Let Z_δ denote the appropriate partition function, and Z_+ denote the partition function with all + boundary conditions. Then we have

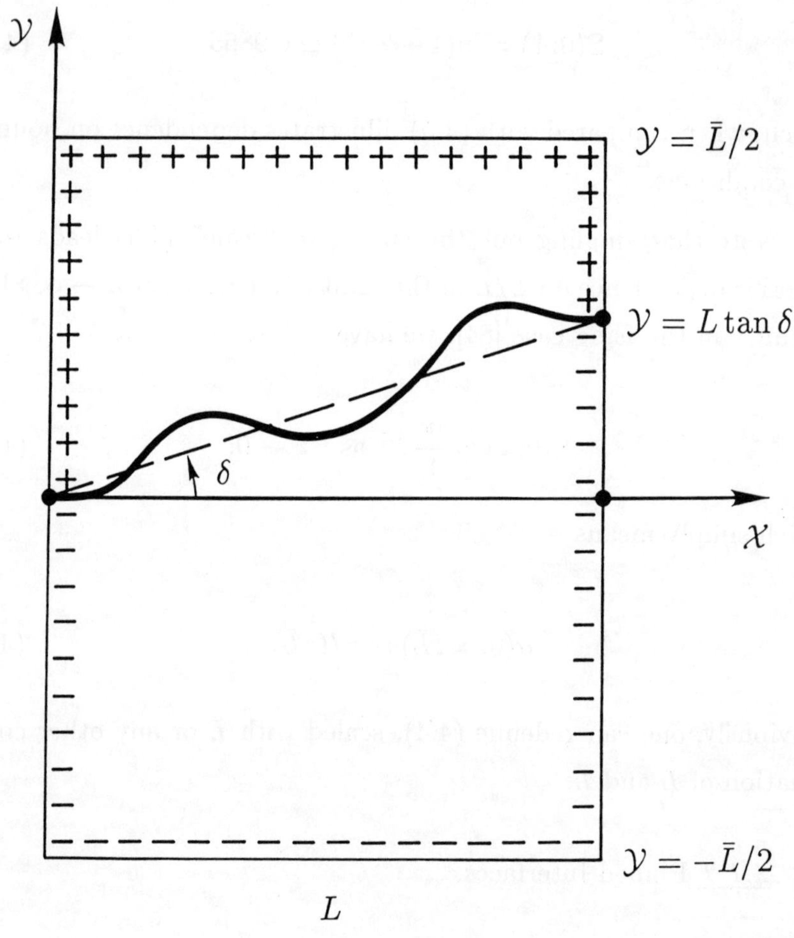

Figure 1: Pinned-end $2d$ interface inclined by an angle δ with respect to the $\mathcal{X}$-axis.

$$\sigma(t; L, \bar{L}; \delta) = -\frac{\cos\delta}{L} \ln \frac{Z_\delta}{Z_+}, \tag{4.10}$$

where $(L/\cos\delta)$ is the interfacial length (Fig. 1).

Quite generally, viewing interfacial free energy as a properly "normalized" free energy difference of two d-dimensional systems, we note that pinned-boundary interfaces have additional geometrical features of relative size $\sim L^{-2}, L^{-3}, \ldots, L^{-d}$, per unit volume, as compared to the reference system with no interface. Thus, the nonsingular background terms will not cancel out in these orders. When considered per unit area, we thus anticipate background terms of order $L^{-1}, L^{-2}, \ldots, L^{-(d-1)}$, in σ, to play the role of similar terms in the free energy scaling. Specifically, the $L^{-(d-1)}$ contribution will "resonate" with the scaling part (in integer d) to yield universal terms $\sim L^{1-d} \ln L$. Thus, for nonperiodic interfaces of characteristic size L, one can propose [85] the equivalents of (2.33)-(2.34), with integer d $(d = 2, 3)$,

$$\sigma(t; L) \approx \frac{1}{L}\sigma_{ns}^{(1)}(t) + \ldots + \frac{1}{L^{d-1}}\sigma_{ns}^{(d-1)}(t) \\ + L^{1-d}\left[\tilde{\Sigma}\left(atL^{1/\nu}\right) + s\ln\left(L/l\right)\right]. \tag{4.11}$$

Here s and $\tilde{\Sigma}(\tau)$ are universal, although they may depend on the geometries used to define $\sigma(t; L)$, *i.e.*, shape, boundary conditions, *etc.* Note that as for the free energy, the background terms of higher order and correction-to-scaling terms must be allowed for a fuller description.

4.1.4. Exact 2*d* Ising Results for Pinned Interfaces

For the 2*d* Ising geometry of Fig. 1 with $\bar{L} \to \infty$, the partition function ratio in (4.10) was calculated exactly for $t < 0$ [86] and recently analyzed in the limit $t \to 0^-$ [87]. For this 2*d* geometry, relation (4.11) is indeed confirmed. One finds [85,87]

$$s = 1, \tag{4.12}$$

and an explicit quadrature for $\tilde{\Sigma}(\tau \leq 0)$, as well as background terms of order $1/L$ [see (4.11)] and higher. Interestingly, $\tilde{\Sigma}(\tau)$, but not s, depends on the angle δ which is a geometrical feature here.

4.2. *Interfacial Free Energy Below Criticality*

In this subsection we consider finite-size free energy contributions due to Gaussian "capillary" interfacial fluctuations above the roughening temperature, which are a critical-type soft mode phenomenon. Thus, we take fixed $T_R < T < T_c$, where $T_R = 0$ in 2*d*, but $T_R > 0$ for 3*d* lattice models and, in fact, may depend on interface orientation. Since we are now considering behavior away from T_c, there is no critical part. However, capillary fluctuations contribute [88-89] a universal (geometry-dependent) term $vL^{1-d} \ln L$. Thus far, only integer d were studied; for an interface of a characteristic size L, we expect [88-89], in integer d,

$$\sigma(t;L) = \sigma(t;\infty) + \frac{1}{L}\sigma^{(1)}(t) + \frac{1}{L^2}\sigma^{(2)}(t) + \dots$$
$$+ \frac{1}{L^{d-1}}\sigma^{(d-1)}(t) + \frac{v}{L^{d-1}} \ln\left(\frac{L}{l}\right) + o\left(\frac{1}{L^{d-1}}\right). \tag{4.13}$$

The association of terms in (4.13) with geometrical characteristics is not straightforward. Indeed, for periodic interfaces, the contributions $\sigma^{(j)}(t)/L^j$ vanish identically for $j = 1, \dots, d-2$, however, the logarithmic and $O(L^{1-d})$ terms are still present [89], which is a special feature of Gaussian models. One finds [89] that

$$v = 1 \tag{4.14}$$

for "floating" periodic interfaces in any integer d, provided L is defined so that L^{d-1} is the interface area. For other boundary conditions on the interface (pinned, "floating" with free ends, *etc.*), v is explicitly geometry-dependent [88-89].

Turning now to the $2d$ geometry of Fig. 1 for which detailed results are available [88] provided δ is small, and $\bar{L} >> O(\sqrt{L})$, we note two interesting features. Firstly, size dependence no longer enters in ratios. The L-dependence is as in (4.13), with $d = 2$. However, finite-$\bar{L}$ effects are of order $\exp\left(-\kappa\bar{L}^2/L\right)$, where $\kappa(T)$ is the surface stiffness coefficient [88]. Secondly, the universal amplitude v is a function of the inclination angle δ. For small δ, we have [85,88]

$$v = \frac{1}{2} - \frac{1}{4}\phi^2 + O(\phi^4). \tag{4.15}$$

Finally, we point out that for the geometry of Fig. 1 with $\bar{L} \to \infty$, exact $2d$ Ising model results [86] confirm the Gaussian model predictions.

4.3. *Other Research Topics*

The purpose of this subsection is to provide a brief reference to three additional research topics of growing interest, where both finite-size and interfacial properties are involved. No comprehensive review is attempted here. However, for completeness, we list recent literature. The subjects discussed are wetting transitions in finite-size geometries, fluctuations of unpinned interfaces including the relation of interfacial properties and the transfer matrix spectrum in cylindrical geometries, and recent MC results for step free energies near roughening transitions.

4.3.1. Size Effects on Wetting

Recently, several mostly mean-field type calculations were reported for $3d$ wetting transitions in the slab geometry $L \times \infty^2$ [90], on cylinders and spheres (in $3d$) [91], and also for the case where the interface itself is finite-size [92].

In $2d$, critical wetting in the strip geometry $L \times \infty$ was studied within the solid-on-solid model [93]. Another project [94] involved the exact $2d$ Ising solution for a finite-size rounded first-order interface unbinding transition in a $2d$ surface-plus-defect geometry, as well as some solid-on-solid model results. Generally, finite-size effects on wetting are a developing field, detailed survey of which is outside the scope of this review.

4.3.2. Fluctuating Interfaces in Cylindrical and Slab Geometries

In Sect. 4.2 we discussed size effects on the free energy of interfaces of finite extent in some or all $d-1$ directions. In some other geometries, interfacial fluctuations are cut off due to finite size in the dimension transverse to the interface. For example, interfaces induced by $+-$ or antiperiodic boundary conditions in $2d$ strips ($L \times \infty$) or $3d$ slabs ($L \times \infty^2$) have infinite area (length in $2d$). Much progress has been made on fluctuations of line interfaces in $2d$, specifically, in the strip geometry, by numerical and exact $2d$ Ising [84,95] calculations and by solid-on-solid type modeling [93,96-98], equivalent to diffusion. Available results for slabs [98] are more limited. The issues investigated are how the free energy and correlations are modified by the transverse size. Specific results are, however, not reviewed here.

Recently, several authors considered transfer matrix spectrum and in particular, the leading spectral gap of Ising type models, with boundary conditions inducing an interface along a $2d$ strip [99]. For the $+-$ or antiperiodic boundary conditions, the correlation lengths defined by the transfer matrix spectral gaps have been related [97] to the bulk surface stiffness coefficient $\kappa(T)$. For example, the leading-gap correlation length is given by [97]

$$\xi(T < T_c; L) \approx \frac{2\kappa(T)L^2}{\pi^2} \quad \text{and} \quad \frac{2\kappa(T)L^2}{3\pi^2}, \tag{4.16}$$

for antiperiodic and $+-$ strips $L \times \infty$, with $L \to \infty$ at fixed $0 < T < T_c$. For $3d$ cylinders, $L \times \bar{L} \times \infty$, with antiperiodic or $+-$ boundary conditions along L, and with, *e.g.*, periodic boundary conditions along $\bar{L}$ (so that the interface has width $\bar{L}$), the appropriate results have been derived for $T_R < T < T_c$ and separately, for $0 < T < T_R$. Above T_R, the spectrum is determined by $\kappa(T)$ and relations similar to (4.16) apply, with an extra factor of $\bar{L}$ on the right hand sides. Below T_R, however, the spectrum is determined by the *step free energy*, $s(T)$, measured per $k_B T$ and per unit step length in a "terrace-shaped" rigid interface. The full expressions [97] require introducing further notation. Thus, we only quote that for large L,

$$\xi(T < T_R; L, \bar{L}) \propto L^2 \sqrt{\bar{L}}\, \mathrm{e}^{s(T)\bar{L}}, \tag{4.17}$$

without specifying the form of the coefficient (which is T-dependent). For free, instead of periodic, boundary conditions along $\bar{L}$, the factor $\sqrt{\bar{L}}$ should be omitted. The behavior at T_R has not been investigated, see also Sect. 4.3.3 below.

For cylindrical geometries, $L^{d-1} \times \infty$, with fully periodic, free, or other boundary conditions such that no *longitudinal* interfaces will be introduced, there occurs an interesting phenomenon of spontaneous generation of *transverse* interfaces [4,100] (for $T < T_c$ and $H = 0$). Their exponential separation defines a correlation length associated with the leading transfer matrix spectral gap,

$$\xi(T < T_c; L) \sim \mathrm{e}^{\sigma(T) L^{d-1}}. \tag{4.18}$$

This form has been discussed recently [4-5,97,101] in connection with the finite size contributions to the difference

$$\tilde{\sigma}(T;L) - \sigma(T) \equiv \tilde{\sigma}(T;L) - \tilde{\sigma}(T;\infty), \tag{4.19}$$

where $\tilde{\sigma}$ is *defined* via

$$\xi(T;L) = \exp\left[\tilde{\sigma}(T;L)L^{d-1}\right]. \tag{4.20}$$

We will not review this topic further here (see [97,101] and Sect. 5.4.4 for details).

4.3.3. Step Free Energy Near Roughening Temperature

Recently, MC results were obtained for the step free energy of the cubic-shaped $3d$ Ising model with a stepped interface induced by periodic-antiperiodic-helical boundary conditions [102]. The results as $T \to T_R$ from below seem to be described [102] by a scaling form with L scaled by $[s(T)]^{-1}$. This single scaling variable description cannot, however, be fully satisfactory in general due to the Gaussian nature of the rough phase above T_R. In fact, finite-size behavior at the Kosterlitz-Thouless transitions in XY systems is also not fully understood theoretically, see, *e.g.*, recent accounts [103] and references quoted therein.

5. FIRST-ORDER TRANSITIONS

5.1. *Opening Remarks*

This section is devoted to finite-size effects at first-order transitions. We consider cubic or nearly cubic systems of volume $V = L^d$, as well as a crossover to long cylinders, $A \times L_{\|}$, of cross sectional area A and volume $V = AL_{\|}$, with, eventually, $L_{\|} \to \infty$. Most theoretical results available to date (see below) pertain to these geometries and furthermore, assume *periodic boundary conditions* in all finite dimensions.

It is generally believed that first-order transitions are controlled by the so-called *discontinuity fixed points* [104] of the RG transformations, which are characterized by having one or more RG eigenexponents equal the space dimensionality d. Results on finite-size rounding of first-order transitions can be generally cast in the appropriate RG scaling forms, and furthermore, one can discuss a crossover to critical point scaling near T_c: all these issues will be considered in the following subsections.

Reviews and comprehensive expository articles on finite-size effects at first-order transitions, both theory and numerical MC simulations, include references [4-6,105-107]. Certain numerical transfer matrix results are also available (see Sect. 5.4.5).

5.2. *Hypercubic Ising Models at the Phase Boundary*

Let us denote by M_0 the bulk spontaneous magnetization,

$$M_0 \equiv M\left(T < T_c, H \to 0^+; L = \infty\right). \tag{5.1}$$

In this subsection we consider *symmetric* first-order transitions exemplified by hypercubic or near-hypercubic ferromagnetic Ising models at fixed $T < T_c$, with the bulk magnetization approaching $\pm M_0$ as $H \to 0^{\pm}$. We assume periodic boundary conditions, so that the full $H \leftrightarrow -H$ symmetry is preserved in a finite system.

When such a system with $H \simeq 0$ is evolving in time under some model dynamics, it spends most of the time in the states with the fluctuating order parameter Ψ near $\pm M_0$, as is indeed observed in MC simulations [108]. The probability distribution for Ψ is then sharply peaked at $\pm M_0$ [38]. Thus, the "singular part" of the H-dependence of the free energy can be estimated [4,108-109] from the mimic partition function

$$\mathrm{e}^{hM_0V} + \mathrm{e}^{-hM_0V}, \tag{5.2}$$

where we used (3.1). For the magnetization, this yields

$$M_s(T, H; L) \approx M_0 \tanh(hM_0V), \tag{5.3}$$

which suggests that the transition (discontinuity $\pm M_0$ in the magnetization) is rounded on the scale

$$H_V = \frac{k_B T}{M_0 V} \sim \frac{1}{V}. \tag{5.4}$$

The susceptibility of finite system develops a sharp peak of width $\sim V^{-1}$ and area $2M_0$, representing the rounded delta-function $2M_0\delta(h)$. The shape of this peak is described by

$$\chi_s \approx M_0^2 V / \cosh^2(hM_0V) \quad . \tag{5.5}$$

Substantiation of the phenomenological arguments leading to (5.4)-(5.5) can be achieved in part [4] by considering sum rules like

$$\chi\left(T < T_c, H \equiv 0; L < \infty\right) \simeq M_0^2 V + \chi\left(T < T_c, H \to 0^+; L \equiv \infty\right), \tag{5.6}$$

[compare (5.5) with $h = 0$], the validity of which relies on the fact that for finite Ising cubes, odd correlation functions vanish by symmetry ($H \leftrightarrow -H$) so that "connected" finite-size correlation functions are similar to bulk "full" correlation functions, see [4] for details. Better control of correction to, *e.g.*, (5.5), can be gained in the transfer matrix formulation [4] described in Sect. 5.4 below.

5.3. *Cubic n-Vector Models in the Giant-Spin Approximation*

5.3.1. Néel's Mean-Field Theory of Superparamagnetism

We now extend our consideration to n-vector models with $n = 1, 2, 3, \ldots$, still assuming periodic boundary conditions and near-cubic shapes. Néel [110] proposed a phenomenological mean-field type description of the development of the susceptibility contribution $\sim V$ in small single-domain magnetic particles, typically with

Heisenberg ($n = 3$) symmetry, see a review [111]. This phenomenon is termed *superparamagnetism.* In Néel's approach, the only fluctuations considered are rotations of the global order parameter $\vec{\Psi}$, which is an n-component vector for $n = 2, 3, 4, \ldots$. Denoting unit vectors by overhats, as usual, the mimic partition function generalizing (5.2) to $n > 1$ is given by [107]

$$\int e^{hM_0V(\hat{\Psi}\cdot\hat{H})}\, d^{n-1}\hat{\Psi}, \tag{5.7}$$

where the integration is over the $n - 1$ orientational (angular) coordinates of the vector $\vec{\Psi} = M_0\hat{\Psi}$, with respect to the direction of the applied field $\vec{H} = H\hat{H}$. For the (longitudinal) magnetization, this yields

$$M_s \approx M_0 m(hM_0V), \tag{5.8}$$

where the scaling function $m(x)$ for general n can be expressed [107] in terms of the standard Bessel functions as follows,

$$m(x) = I_{\frac{n}{2}}(x) \Big/ I_{\frac{n-2}{2}}(x). \tag{5.9}$$

For $n = 3$, $m(x)$ reduces [111] to the well known Langevin function,

$$m_{n=3} = \operatorname{cotanh}(x) - x^{-1}, \tag{5.10}$$

while for $n = 1$, relation (5.3) is formally reproduced. For the leading order longitudinal susceptibility peak at $H = 0$, we get

$$\chi_s(H{=}0) = M_0^2 V/n \,. \tag{5.11}$$

Corrections to the leading asymptotic results (5.8), (5.11), *etc.*, for $n = 2, 3, \ldots$, turn out to be much more significant than for the case $n = 1$, due to spin-wave fluctuation contributions [107]. We will return to this issue in Sect. 5.5 below.

5.3.2. Giant-Spin Concept and Discontinuity Fixed Point Scaling

The scaling combination in (5.8) can be rewritten as

$$hM_0 V = \frac{M_0(T)H}{k_B T} L^d , \tag{5.12}$$

where all the T-dependence has been made explicit. As $T \to 0$, $M_0(T)$ approaches a constant value, so that the scaling combination involves a *linear scaling field* h, at the discontinuity fixed point at $H, T = 0$, with eigenexponent $\lambda_h^{(\text{discontinuity})} = d$, as anticipated [104]. Furthermore, it is natural to conjecture that the full T-dependent scaling field is $M_0(T)h$, where T enters as an irrelevant variable which in turn has negative eigenexponent $\lambda_T^{(\text{discontinuity})} = 1 - d$ or $2 - d$, for $n = 1$ and $n > 1$, respectively.

When taken seriously, this RG scaling interpretation can be used to calculate the scaling functions like $m(x)$ in (5.8), see [112]. Indeed, let l denote a microscopic lattice spacing, and iterate the RG transformation L/l times to obtain

$$h' = h(L/l)^d, \quad T' = T(l/L)^{d-d_0} \quad , \tag{5.13}$$

where $d_0 = 1$ for $n = 1$, $d_0 = 2$ for $n = 2, 3, \ldots$. Thus, we obtained one spin with a "giant" magnetic coupling $\sim hL^d$, at a very low temperature $\sim TL^{d_0-d}$. By solving for the free energy of this "giant" spin, and using the standard RG free energy recursion

$$f_s(T, h; L) \approx (L/l)^{-d} f_s(T', h'; 1), \tag{5.14}$$

one can actually obtain information on the form of the scaling functions. In the cubic geometry, this method yields, *e.g.*, the result (5.9). Applications to cylindrical geometries [112] sometimes run into difficulties [4] associated with dangerous irrelevant variables, see Sect. 5.4.4 below.

5.3.3. Upper Critical Dimension for First-Order Transitions

The fact that RG iteration down to a single "giant" spin, and Néel's mean-field theory which also, in a sense, assumes a rotating "giant" overall spin variable $\vec{\Psi}$, with $|\vec{\Psi}| = M_0$, both give the same results, at least in the leading order, suggests that first-order transitions are always technically above their upper critical dimension d_0. The dimensionality d_0 is also the lower critical dimension for the critical point at $T_c > 0$. As mentioned, $d_0 = 1$ for Ising models ($n = 1$), and $d_0 = 2$ for n-vector models with $n = 2, 3, 4, \ldots, \infty$. Exact calculations [113] for the long-range spherical models ($n = \infty$) for which d_0 varies continuously ($0 < d_0 < 2$) confirm this observation.

5.4. *Crossover to Cylindrical Geometry: Ising Case*

5.4.1. Longitudinal Correlation Length

For Ising models in the cylindrical geometry $A \times L_{\|}$, with periodic boundary conditions, it is useful to consider the transfer matrix evaluation of the partition function. If the transfer matrix $\mathcal{T}$ adds up a slice of size $A \times l$, where l is of the order of lattice spacing, then the partition function is given by $\mathrm{Tr}(\mathcal{T}^{L_{\|}/l})$. In terms of the ordered set of eigenvalues of $\mathcal{T}$,

$$\Lambda_0 > \Lambda_1 \geq \Lambda_2 \geq \ldots, \tag{5.15}$$

we have for the partition function,

$$\Lambda_0^{L_{\|}/l} + \Lambda_1^{L_{\|}/l} + \ldots \quad . \tag{5.16}$$

The free energy of the infinite cylinder, $A \times \infty$, is given by f_0, where

$$f_j \equiv -\frac{1}{lA} \ln \Lambda_j \,, \qquad (j = 0, 1, \ldots), \tag{5.17}$$

denote the so-called free energy levels [57]. One can also introduce correlation lengths associated with the transfer matrix spectral gaps,

$$\xi_j \equiv l/\left[\ln\left(\Lambda_0/\Lambda_j\right)\right], \qquad (j = 1, 2, \ldots). \tag{5.18}$$

For sufficiently long cylinders, one can no longer argue that the first order transition will be rounded according to (5.3) on the scale H_V defined in (5.4). Indeed, when $L_{\|} \to \infty$, the total volume V diverges and so $H_V \to 0$. However, the transition must remain rounded in a one-dimensional cylinder. One then argues [4] that (5.3) is essentially a single-domain result. On the other hand, as mentioned in Sect. 4.3.2, long cylinders in zero field spontaneously form $\pm$ domains of length [4,100] $\sim e^{\sigma A}$, associated with entropy gain [114]. This longest length scale $\xi_{\|}$ in the $H = 0$ cylinder, describing the persistence of the order parameter values $+M_0$ or $-M_0$, must correspond to ξ_1,

$$\xi_{\|}(T;L) = \xi_1(T, H=0; L), \tag{5.19}$$

as is indeed confirmed by exact $2d$ Ising results [84] and other general considerations [4,100].

For the rounded first-order transition, one then anticipates that

$$M_s(T,H;L) \approx M_0 \tilde{m}(hM_0V, L_{\|}/\xi_{\|}), \tag{5.20}$$

with

$$\tilde{m}(x, z \to 0) = m(x), \quad \text{for} \quad L_{\|} << \xi_{\|}, \tag{5.21}$$

$$\tilde{m}(x, z \to \infty) = \bar{m}(x/z), \quad \text{for} \quad L_{\|} >> \xi_{\|}. \tag{5.22}$$

The correlation length $\xi_{\|}$ is generally described by relations (4.18)-(4.20). We will return to its precise form in Sect. 5.4.4 below.

For the moment, the important feature is its exponential leading-order L-dependence, (4.18), which has been confirmed by exact $2d$ Ising results [84,100], and by numerical transfer matrix studies in $3d$ [115]. Note that for cylinders it is convenient to use the convention $A = L^{d-1}$ (Sect. 4), which should not be confused with $V = L^d$ used in subsections on cubes.

5.4.2. Calculation of the Scaling Functions

Before considering the implications of the scaling relation (5.20), it is helpful to outline calculations [4] yielding the exact result

$$\tilde{m}(x,z) = \frac{x}{\sqrt{x^2 + \frac{1}{4}z^2}} \tanh \sqrt{x^2 + \frac{1}{4}z^2} \,, \tag{5.23}$$

which satisfies (5.21) and also gives [see (5.22)]

$$\bar{m}(\bar{x}) = \frac{\bar{x}}{\sqrt{\bar{x}^2 + \frac{1}{4}}}, \qquad (\bar{x} = x/z). \tag{5.24}$$

The theory leading to these explicit Ising model results is rather complicated [4,116]. Only the main steps will be presented here. First, we note that the two largest eigenvalues of the transfer matrix are asymptotically degenerate at $H = 0$ (for $T < T_c$), as is implied by the general relation

$$f_j - f_0 = (\xi_j A)^{-1}. \tag{5.25}$$

Indeed, for $j = 1$ the free energy values are split $\sim \mathrm{e}^{-\sigma A}$. However, this pair is isolated from the rest of the spectrum because the next gap, $f_2 - f_0$, defines the bulk correlation length $\xi^{(b)}$ which is finite at $H = 0$. Thus, $f_2 - f_0 \sim 1/A$. It turns out [4] that the two lowest free energies are well separated from the rest of the spectrum for

$$|H| < H_V\left(L_{\|}\Big/\xi^{(b)}\right) \sim \frac{1}{A}, \tag{5.26}$$

compare (5.4).

The range (5.26) is wider than the finite-size rounding, both for $L_{\|} = O(L)$ and for $L_{\|} \to \infty$ (see below). One can show [4] that in this regime the mimic partition function describing the rounded transition is given by (5.16) with only the first two terms kept. The next step is to use the two-level model [57] description of the near degeneracy of f_1 and f_0 at $H = 0$. This allows a representation of f_0 and f_1 in terms of the *bulk* quantities, and $\xi_{\|}$, by using the concept of the single-phase free energy functions $f_{(\pm)}$, mentioned in Sect. 3.2.3, see (3.21). We only present the final expressions valid up to corrections of order H^3 and $\mathrm{e}^{-L/\xi^{(b)}}$ (for periodic boundary conditions). Higher orders in h can be also evaluated, see [4,51] for details. We have

$$f_{\substack{0\\1}} \approx f^{(b)} \mp \sqrt{M_0^2 h^2 + \frac{1}{4A^2\xi_{\|}^2}} - \frac{1}{2}\chi^{(b)} h^2, \tag{5.27}$$

where (b) denotes evaluation in the bulk limit, with $h \to 0^+$ *after* setting $A, L_{\|} = \infty$. Finally, the expressions (5.27) are substituted in the mimic partition function (5.16), with (5.17).

The transfer matrix approach yields the leading-order scaling results for cubes, cylinders, and in the crossover regime $L_{\|} \simeq \xi_{\|}$. Thus, (5.23) is obtained from the leading free-energy expression [4]

$$f_s \approx -\frac{1}{V} \ln \left[2\cosh \left(V \sqrt{M_0^2 h^2 + \frac{1}{4A^2 \xi_{\|}^2}} \right) \right] , \tag{5.28}$$

where the size dependence enters not only via V, A, but also $\xi_{\|}$. Furthermore, as mentioned, control of correction terms is possible, although derivation of explicit results is rather cumbersome [51]. For illustrations, we mention the leading order correction to the "cubic" result (5.5),

$$\chi_s \approx \frac{M_0^2 V}{\cosh^2(hM_0 V)} + \chi^{(b)} \quad . \tag{5.29}$$

This relation extends the sum rule (5.6) to fields $|H| \leq O(H_V)$.

5.4.3. Cylindrical Limit Scaling

In the limit $L_{\|} \to \infty$, obtained for $L_{\|} >> \xi_{\|} \sim e^{\sigma A}$, relations (5.20) with (5.22) and (5.24) should be used. One has

$$M_s \approx M_0 \bar{m}(hM_0 A\xi_{\|}), \tag{5.30}$$

so that the scaling combination involves the "correlated" volume $A\xi_{\|}$ instead of the total volume $AL_{\|}$. For the susceptibility peak, of area $2M_0$, we get

$$\chi_s \approx \frac{M_0^2 A\xi_{\|}}{4\left[\left(hM_0 A\xi_{\|}\right)^2 + \frac{1}{4}\right]^{3/2}}, \tag{5.31}$$

which is exponentially large (in A) at $h = 0$, where we have $\chi_s(h=0) \approx 2M_0^2\xi_{\|}A$.

5.4.4. Size Dependence of $\xi_{\|}$ and RG Scaling

As already alluded to in the discussion of relations (4.18)-(4.20), the cross-section size (and shape, *etc.*, or what we have generally called *geometry*, in Sect. 1-4) dependence of the quantity $\tilde{\sigma}(T; L)$ is of interest for several reasons [97,101]. Its general pattern (at least, for rough interfaces) should be similar to that of the Gaussian prediction (4.13). We will denote coefficients in such a relation for $\tilde{\sigma}$, by $\tilde{\sigma}^{(j)}(t)$ and $\tilde{v}$, compare (4.13).

Exact results for periodic [4], free [99,117], and some other [99] boundary conditions are available for $2d$ strips $L \times \infty$. The form (4.13) then applies, with $\tilde{\sigma}^{(1)}$ and $\tilde{v}$ terms present in $2d$. For the universal amplitude $\tilde{v}$, one finds

$$\tilde{v}_{\text{periodic},2d} = \frac{1}{2}, \qquad \tilde{v}_{\text{free},2d} = 0. \tag{5.32}$$

However, the appropriate single-interface Gaussian model results (Sect. 4.2) were [89]

$$v_{\text{periodic}} = 1, \qquad v_{\text{free},2d} = \frac{1}{2}. \tag{5.33}$$

It has been suggested that the difference $v - \tilde{v}$ is due to fluctuations of the dilute gas of interfaces [5,101,118]. This effect warrants further study.

The size dependence of $\tilde{\sigma}(T; L)$ complicates identification of (5.20) or (5.30) as RG scaling results. Indeed, in RG scaling we expect dependence on the *relevant* scaling combination hM_0V, already identified from the results for cubes. Shape ratio dependence is also possible, and allows us to isolate the $L_{\|}$ dependence in (5.20),

$$L_{\|}/\xi_{\|} = (L_{\|}/L)(L/\xi_{\|}). \tag{5.34}$$

The remaining scaling variable is conveniently rewritten as

$$\frac{L}{\xi_{\|}} = \mathrm{e}^{-\bar{\sigma}L^{d-1}}, \tag{5.35}$$

where

$$\bar{\sigma}(T; L) = \tilde{\sigma}(T; L) - L^{1-d} \ln L. \tag{5.36}$$

Since we define the interfacial free energies *per* k_BT, we have $\sigma(T) \sim T^{-1}$ as $T \to 0$. Thus, in the leading order in L, as $L \to \infty$, the scaling combination $L/\xi_{\|}$ indeed depends on the leading irrelevant variable $\sim T$ at the discontinuity fixed point. The first-order transition scaling functions in the cylindrical limit $L_{\|} >> \xi_{\|}$, and also for $L_{\|} = O(\xi_{\|})$ [relations (5.30) and (5.20)] are highly singular in their TL^{1-d} dependence, so that the irrelevant variable is dangerous.

The interpretation of the subleading L-dependence of the product $\bar{\sigma}^{-1}L^{1-d}$ is, however, less clear. Indeed, if we identify $\bar{\sigma}^{-1}$ as the full nonlinear irrelevant RG scaling field with eigenexponent $\lambda_T = 1 - d$, then it has explicit L-dependence. In the critical-point finite size scaling (Sect. 2) it has been emphasized that the scaling fields g_t, g_H, g_u [see (2.6)-(2.7), *etc.*] do not acquire L-dependence for $L < \infty$, and L enters as a "macroscopic" length. The origin of the breakdown of this observation at the Ising discontinuity fixed points is not clear. Ambiguities noted in [4], with the RG rescaling [112] of Ising cylinders down to $1d$ giant-spin chains, are also related to the importance of the L-dependent corrections to the irrelevant-variable scaling combination $\sim TL^{1-d}$.

5.4.5. Exact and Numerical Tests of Scaling for Ising Cylinders and Cubes

Exact calculations for finite-size rounded Ising first-order transitions have been reported for two models. First, studies of the infinite-range Ising model, which is usually used as a prototype model of mean-field critical points, yield results [4] on the phase boundary which in the leading order are consistent with the periodic cube predictions (5.3), *etc.* However, the pattern of corrections is generally different, see [4] for details. The other exact calculation is for the first-order *wetting transition* [94] in a certain $2d$ Ising geometry, mentioned in Sect. 4.3.1. In this case the mechanism of the rounding is similar to cylinders and follows the pattern of the two-level free-energy model.

Numerical transfer-matrix tests of the rounded magnetization scaling in $2d$ Ising-type model ***strips*** were reported in [119]. MC tests were also reported, for ***square*** Ising-model systems [105-106,108], including detection of the correction term in (5.29).

5.5. *Spin-Wave Effects for $n = 2, 3, \ldots, \infty$*

5.5.1. Cylinder-Limit Scaling

Instead of breaking into ± domains, the vector-spin models in cylindrical geometries have their order parameter slowly rotate. This random drift of the order parameter orientation, which may also be viewed as an ensemble of diffuse Bloch walls, or $\vec{k}_\perp = 0$, small-$k_\parallel$ spin waves, leads to the absence of ferromagnetic n-vector ordering in $1d$. Analysis of this phenomenon within an effective longest-mode transfer matrix formulation was carried out in [107,120]. We do not review here this rather complicated formalism, but only summarize some results.

For the zero-field correlation length defining the persistence distance of the order parameter orientation, one finds [107,121],

$$\xi_\parallel(T; L) \approx \frac{2\Upsilon(T)A}{(n-1)k_B T} \quad . \tag{5.37}$$

Here $\Upsilon(T)$ is the helicity modulus which some readers may have encountered in the form of the so-called spin-wave stiffness coefficient,

$$\rho_s \equiv \Upsilon(T)/k_B T \quad . \tag{5.38}$$

The transfer matrix formulation [107] with nonzero applied field was used to establish the validity of the scaling description (5.20)-(5.22), for $|H| \leq O(H_V)$, with $m(x)$ in (5.21) given by (5.9). Exact closed form expressions for $\tilde{m}$ and $\bar{m}$ are not known. However, $\bar{m}(\bar{x})$ can be obtained from the ground state of a certain quantum-mechanical Hamiltonian [107]. Thus, numerical and approximate analytical results for $\bar{m}(\bar{x})$ can be derived, although results available to date are very limited.

By considering the scaling combination $L_{\|}/\xi_{\|}$, similar to the Ising case, we can conclude that the irrelevant RG field $\sim T$, with eigenexponent $\lambda_T = 2 - d$, is given by

$$\frac{(n-1)k_B T}{\Upsilon(T)} \,. \tag{5.39}$$

Since this expression is not dependent on L, the ambiguities of the Ising case are not present here. Indeed, the scaling relation (5.20), reformulated in terms of the free energy, can be written in the standard RG form

$$f_s \approx L^{-d}\mathcal{F}\left[M_0(T)hL^d, \frac{(n-1)k_B T}{\Upsilon(T)}L^{2-d}; r_{ij}\right], \tag{5.40}$$

where r_{ij} denote collectively various shape ratios. In this subsection we only specify $L_{\|}/L$ explicitly. Recall, however, that even in the long cylinder geometry, there may also be some dependence on the *shape* of the cross-section, not only its area $A \equiv L^{d-1}$. Furthermore, $\mathcal{F}$ is also n-dependent, beyond the factor $n-1$ shown explicitly.

Note that $M_0(T)$ and $\Upsilon^{-1}(T)$ play here the role of the "nonuniversal metric factors" discussed extensively in connection with the critical-point finite-size scaling. As in the Ising case, here also the irrelevant variable is dangerous. Due to a more conventional way of emergence of the RG scaling for $n > 1$ (*i.e.*, no L-dependence in the irrelevant scaling field), procedures of renormalizing down to a 1d chain of "giant" spins [112] do not have the ambiguities noted for the Ising ($n = 1$) case (Sect. 5.4.4).

5.5.2. Spin-Waves in Cubic Geometries

Scaling forms considered until now for n-vector models with $n = 1, 2, 3, \ldots$ describe finite size rounding of the $\pm M_0$ jump in the (longitudinal) magnetization. For cubes, the jump is rounded on the H-scale $H_V \sim V^{-1}$, given by (5.4), while for long cylinders the rounding occurs on the scale $k_B T/M_0 A\xi_{\parallel} \sim A^{-2}$. However, additional finite-size effects will be present in both geometries on much larger H-scales, of order $\sim L^{-2}$. Indeed, for *bulk* vector-spin models below T_c, there is a divergent contribution, due to spin-waves, to the susceptibility in the $H \to 0^{\pm}$ limit, see, *e.g.*, [122]. The hydrodynamic (Gaussian) theory of spin wave fluctuations which, loosely, represent local rotations of the order parameter, has been reviewed in [107]. We only quote some results. The bulk correlation length behaves for small H according to

$$\xi^{(b)} \approx \sqrt{\Upsilon/M_0|H|} \quad , \tag{5.41}$$

while the susceptibility has a singularity of the form

$$\chi_s^{(b)} \approx \text{const}(n-1)k_BT(M_0/\Upsilon)^{d/2}|H|^{-(4-d)/2}, \qquad (d>2), \quad (5.42)$$

where const depends on d only, and an extra factor of $\ln|H|$ must be added in (5.42) for $d = 4, 6, 8, \ldots$. This divergence (for $2 < d \leq 4$) results in the infinite slope of the magnetization as $H \to 0^{\pm}$.

Descriptions of the combined finite-size rounded first-order transition discontinuity and spin-wave singularities turn out to be rather complicated and have been considered in some detail only for (periodic) hypercubic or near hypercubic-shaped samples [107].

It was argued in [107] that the RG scaling expression (5.40) should apply in all regimes, with a more complicated dependence on the shape ratios r_{ij} and irrelevant-field argument $\sim TL^{2-d}$, see (5.40), than the form required to describe the leading-order ("jump") rounding for cubes and cylinders. For "cubic" samples, three regimes can be identified.

a. Néel's regime

As long as $|H| \leq O(H_V)$, so that the relevant scaling combination in (5.40) is not too large (it can be zero or up to $O(1)$ in magnitude), the dominant fluctuations are the "giant-spin" rotations. This can be viewed as the $\vec{k} = 0$ spin-wave mode (with no spatial orientation gradients) which is highly anharmonic for small hM_0V. In this regime the scaling functions for the magnetization, susceptibility, *etc.*, can be expanded in the irrelevant variable to yield Néel's leading order results, *with spin-wave corrections* of relative magnitude L^{2-d}, with coefficients proportional to $(n-1)k_BT/\Upsilon$. For the susceptibility, this correction at $H = 0$ has been considered in some detail [107]. Thus, (5.11) is replaced (for $n > 1$) by

$$\chi(H=0) = \frac{M_0^2}{n}V + \frac{(n-1)M_0^2 k_B T}{(n+2)\Upsilon}\text{const}L^2 + \ldots, \tag{5.43}$$

where the (positive) <u>const</u> is geometry and possibly also n-dependent. (This and related results were established within a self-consistent scheme which becomes exact for $n = \infty$ [107].)

<u>b. Spin-Wave Regime</u>

For positive $hM_0V >> 1$, the leading order expression (5.8) gives $M_s \approx M_0$. Expansion in the irrelevant variable acquires, however, a "dangerous" character. It turns out that to the first order in TL^{2-d}, one can write generally [107]

$$M_s \approx M_0 \left[m(hM_0V) + \frac{(n-1)k_B T L^{2-d}}{\Upsilon} m_1 \left(\frac{(n-1)HM_0L^2}{\Upsilon}; r_{ij} \right) \right]. \tag{5.44}$$

Thus, the function m_1 depends on the *product* ($\sim L^2$) of two scaling combinations and cannot be further expanded in the irrelevant combination $\sim TL^{2-d}$ in the regime where the relevant combination $\sim hL^d$ is large.

By (5.41), the new "product" scaling combination is proportional to $\left[L/\xi^{(b)}\right]^2$. Thus, for $hM_0V >> 1$, when the first term in (5.44) becomes M_0 (*i.e.*, $m(+\infty) = 1$), the second term will vary for fields of order $\sim L^{-2}$, and it represents the rounding of the bulk spin-wave singularities. This effect was observed in a MC study [123]. In this regime the $\vec{k} = 0$ spin-wave mode ("giant spin") is frozen along the applied field. Fluctuations of the $\vec{k} \neq 0$ modes can be regarded harmonic and treated within a finite-size Gaussian model. Explicit results for the scaling functions like m_1 in (5.44) can then be obtained [107]. They are strongly geometry-dependent.

c. Intermediate Regime

When $|H| << O(L^{-2})$, and $|H|$ approaches values of order L^{-d}, the $\vec{k} = 0$ spin-wave mode is no longer frozen, and it interacts anharmonically with the $\vec{k} \neq 0$ modes. The resulting many-body problem cannot be analyzed except in the large-n limit. It has been conjectured [107] that (5.44) applies generally in the first order in TL^{2-d}, and a form of m_1 was proposed, valid to this order in all regimes.

Note that the rounded spin-wave susceptibility peak, of width $\sim L^{-2}$, reaches values of order $\sim L^{4-d}$ in this regime (see [107] for more detailed expressions), and smoothly crosses over to the "central" peak of width $\sim L^{-d}$ and maximal value given by (5.43).

5.5.3. Spherical Model Results

Spherical model calculations serve to test the general n-vector model predictions [107] outlined above (with properly rescaled factors of n), in the limit $n \to \infty$. Results have been obtained [113] for general cubic-like and cylindrical geometries in $d > 2$, as well as for long-range interactions decaying $\sim 1/R^{d+\sigma}$, with $0 < \sigma < 2$, for general $d > \sigma$. Note that $\sigma = 2$ corresponds to the short-range case, and that for $\sigma < 2$, spin-wave concepts have not been formulated (for $n \leq \infty$) so that the scaling expressions are less physically transparent (see [113] for details).

Another extension achieved to date only for spherical models [124], are finite-size results for geometries other than cubic and cylindrical. General-geometry results obey the RG scaling, *e.g.*, (5.40); consult [124] for details.

5.4.4. Quantum Models

Several authors studied finite-size effects on first-order transitions in models where the order parameter is a quantum-mechanical vector operator. This recent development will not be reviewed here; see, *e.g.*, [125].

5.6. *Nonsymmetric $n = 1$ Transitions*

5.6.1. General Formulation

If $\bar{k}$ distinct phases coexist, their (bulk) free energies are equal at the transition. Let us denote these free energies by $f_j^{(b)}(K_i)$, where $j = 1, 2, \ldots, \bar{k}$, while K_i represent various thermodynamic fields (such as H, T) taking values $K_i^{(0)}$ at the transition. The mimic partition function for periodic cubes (we consider only this geometry here) is simply

$$\sum_{j=1}^{\bar{k}} e^{-Vf_j^{(b)}(K_i)} \quad . \tag{5.45}$$

While this result is easy to state, it is not that easily substantiated. There are actually two separate issues: (1) the meaning of $f_j^{(b)}\left(K_i \simeq K_i^{(0)}\right)$ outside the part of the phase diagram where j is the stable phase (*i.e.*, where $f_j^{(b)}$ is minimal among the set $f_i^{(b)}$, at least, within mean-field theory); (2) the exponentials in (5.45) are added up with no coefficients. In order to avoid introduction of excessive notation, we will explore the above issues in the case $\bar{k} = 2$.

5.6.2. Two-Phase Coexistence

We turn again to magnetic type notation. Thus, we assume that the transition occurs when a certain "field" h equals zero. For *small* h, we assume that it couples to the order parameter Ψ, normalized in such a way that the Boltzmann factor due to this interaction (which is only a part of the total interaction energy density) is given by

$$\exp(h\Psi V). \tag{5.46}$$

At the phase transition the free energies of the coexisting phases can be expanded as

$$f_1^{(b)}(h) = f_0^{(b)} + M_1^{(b)}h + \frac{1}{2}\chi_1^{(b)}h^2 + \ldots, \tag{5.47}$$

$$f_2^{(b)}(h) = f_0^{(b)} + M_2^{(b)} h + \frac{1}{2}\chi_2^{(b)} h^2 + \dots \ . \tag{5.48}$$

These series are asymptotic [126]. However, in finite-size analyses we usually need only a few terms and the resulting scaling expressions only involve the "spontaneous" values of $M_1^{(b)} > M_2^{(b)}$, *etc.* Thus, the issue of nonanalyticity of thermodynamic quantities as coexistence ($h = 0$) is approached, and their analytic continuation to the "metastable region", are irrelevant.

When series (5.47)-(5.48) are substituted in (5.45), with $\bar{k} = 2$, the leading order finite-size scaling results can be easily calculated. For instance, for the magnetization we get

$$M_s \approx \frac{M_1^{(b)} + M_2^{(b)}}{2} + M_0 \tanh(hM_0V), \quad (\bar{k} = 2), \tag{5.49}$$

$$M_0 \equiv \left[M_1^{(b)} - M_2^{(b)} \right] / 2. \tag{5.50}$$

This simple result, reminiscent of the symmetric case, with no shift in the location of the transition, obviously relies strongly on the assumption (2) above, which was that the exponentials in (5.45) add up with no relative coefficients. Results (5.45) and (5.49)-(5.50) can be substantiated by exact and transfer matrix considerations which are briefly discussed in Sect. 5.6.4. A more detailed discussion will be based on the formalism of the probability density of the order parameter Ψ [108,127]. (We emphasize that the no-shift property in valid only for $\bar{k} = 2$.)

5.6.3. Probability Distribution of the Order Parameter

For convenience we assume the Ψ is a continuous scalar variable, taking values in $(-\infty, +\infty)$, which can be done without any loss of generality. The probability density distribution for Ψ, $P_L(\Psi) \equiv P_L(\Psi; T, h)$, is sharply peaked at the equilibrium "magnetization" M, and several studies [38,108] suggest that near M, it is Gaussian. The form usually assumed for $\Psi \simeq M$ is then

$$P_L(\Psi) \propto \exp\left[-\frac{(\Psi - M)^2}{2\chi} V\right], \tag{5.51}$$

where the width $\sim \sqrt{\chi/V}$ is a standard thermodynamic fluctuation result. Note that (5.51) is not normalized.

When $P_L(\Psi)$ is peaked at *two values*, $M_{1,2} = M_{1,2}^{(b)} + \chi_{1,2}^{(b)} h + \ldots$, as happens at $h \simeq 0$, it is important to decide how to add up the peaks at M_1 and M_2. The argument of [108] was essentially that one should add up *normalized* peaks so that away from phase coexistence, the dominant peak will yield normalized $P_L(\Psi)$. This line of argument leads to results equivalent to (5.45).

Another argument [127] is that $P_L(\Psi)$ must follow from the constrained fixed-Ψ zero-field free energy $p(\Psi; L)$, as

$$P_L(\Psi) = \frac{e^{-V[p(\Psi;L) - h\Psi]}}{\int_{-\infty}^{\infty} e^{-V[p(\psi;L) - h\psi]}\, d\psi} \; . \tag{5.52}$$

It is usually assumed that (5.51) follows from a minimum of $[p(\Psi; L) - h\Psi]$ at $\Psi = M$,

$$p - h\Psi \approx \frac{1}{2\chi}(\Psi - M)^2. \tag{5.53}$$

If there are two such minima in $p(\Psi; L)$ at $M_{1,2}^{(b)}$, it suggests that the correct approximation is to add up the exponentials *before* normalizing [127], which leads to results somewhat different from, *e.g.*, (5.49). However, note that this is a typical mean-field consideration. Usually, $p(\Psi; L)$ for $M_2^{(b)} < \Psi < M_1^{(b)}$ is corrected for multiphase effects [38] which are beyond mean-field theory. However, the normalization problem can be resolved if we also allow for rather mild finite-size modifications of $p(\Psi; L)$ beyond the strict mean-field type form. We can put, *e.g.*,

$$p - h\Psi \approx \frac{1}{2\chi}(\Psi - M)^2 - \frac{1}{2V}\ln\frac{V}{2\pi\chi}, \quad \text{for} \quad \Psi \simeq M, \tag{5.54}$$

to get a normalized peak near $\Psi = M$ *in the numerator* of (5.52). Corrections of orders L^{-d} and $L^{-d}\ln L$ are not unusual for free energy quantities, although here they are not clearly associated with geometrical features. Thus, we can "normalize" the peaks at $M \simeq M_{1,2}^{(b)}$ by making the minima of p slightly unequal,

$$|\Delta p_{\min}| = \left|\frac{1}{2V}\ln\left(\chi_2^{(b)}/\chi_1^{(b)}\right)\right|, \tag{5.55}$$

resolving all the discrepancies.

5.6.4. Exact and Transfer Matrix Results

Recently, the general relation (5.45) has been established *rigorously* [128] within the Pirogov-Sinai theory of phase coexistence, which applies to a wide class of models at sufficiently low temperatures, see a review [129]. The $\bar{k} = 2$ result (5.49)-(5.50) has been also confirmed by transfer matrix considerations [130]. In fact, the original transfer matrix formulation [4] used the $H \leftrightarrow -H$ Ising symmetry only in relations like (5.47)-(5.48). Extension of the transfer matrix analysis to $\bar{k} > 2$ should also be possible.

5.7. *Miscellaneous Topics*

5.7.1. Matching of Scaling Forms

Several authors [4,34,51,107,131] considered a crossover from the critical-point scaling near T_c to the first-order transition scaling which sets up for $t < 0$ and $L >> \xi^{(b)}$, *i.e.*, when $tL^{1/\nu}$ is large and negative. Thus, one can derive the asymptotic form of the critical-point scaling function $Y(\tau, \omega)$ for $\tau \to -\infty$, whenever explicit "first-order" scaling results are available (*e.g.*, [4,51]). We will not elaborate on specific results one of which [51] is relation (3.20). Instead, we comment of some points of general interest.

First, we note [4] that "first-order" scaling combinations depend on d. Thus, they can be put in the standard hyperuniversal scaling form only for $d < 4$. Specifically,

$$hM_0L^d \approx \frac{H}{k_BT_c}B_-|t|^\beta L^d \propto \left(HL^{\Delta/\nu}\right)|tL^{1/\nu}|^\beta L^{(d\nu-2+\alpha)/\nu}, \quad (5.56)$$

$$\sigma L^{d-1} \sim |t|^\mu L^{d-1} = |tL^{1/\nu}|^\mu L^{[(d-1)\nu-\mu]/\nu}, \quad (5.57)$$

$$(k_BT)^{-1}\Upsilon L^{d-2} \sim |t|^\upsilon L^{d-2} = |tL^{1/\nu}|^\upsilon L^{[(d-2)\nu-\upsilon]/\nu}, \quad (5.58)$$

where we introduced the helicity modulus exponent υ, and used *scaling* relations $\Delta = \beta + \gamma$ and $\alpha + 2\beta + \gamma = 2$. Thus, (5.56)-(5.58) reduce to critical-point scaling combinations only if the standard hyperscaling relations can be used, $2-\alpha = d\nu$ for (5.56), $\mu = (d-1)\nu$ for (5.57), and $\upsilon = (d-2)\nu$ for (5.58). In fact, matching of scaling forms has led to understanding of some of the aspects of modified finite-size behavior for $d > 4$ [4,34], see Sect. 2.7.

Another interesting property was found for spherical models [131] where it turns out that not only the scaling combinations near T_c and below T_c are compatible, but that a single *scaling function* can be used, see [131] for details. It has been speculated [131] that this property applies for all $n = 2, 3, \ldots, \infty$. However, it is most likely restricted to models whose critical behavior is intrinsically Gaussian. These include the spherical model, but also in a certain sense the conformally invariant $2d$ models, where a similar effect (in the strip geometry) is the so-called Zamolodchikov's [132] c-theorem, see [133].

5.7.2. Extension to nonperiodic boundary conditions

Incorporation of surface effects in a systematic theory of finite-size rounded first-order transitions would be highly desirable, but no such development has been reported to date. Leading-order results (but not the form of the corrections) will remain unchanged for boundary conditions which (1) introduce no additional asymmetries, and (2) introduce no interfaces. For example, the Ising result (5.3) should apply also for free boundary conditions but not for, say, all + boundary conditions. In the latter case, a shift in h of order $1/L$ towards negative values will be needed. More qualitative changes should be expected for boundary conditions which induce interfaces [134].

5.7.3. Loci of Partition Function Zeros

In the complex-T or H planes, partition function zeros typically fall on line loci (with nonzero density). It has been pointed out [26,135] that these lines are reminiscent of first-order transitions in that in the bulk, the imaginary part of the free energy corresponds to discontinuous "order parameter" [135]. Furthermore, one can use relations of the type (5.45) to study the distribution of zeros in finite-size systems [26].

6. CONCLUSION

Overall, the theory of finite-size effects has been well developed. However, many topics still require theoretical work and numerical tests, notably surface and interfacial effects, applications to quantum systems, dynamics, and nonequilibrium phase transitions. The author hopes that this review has provided both an introduction and a survey of available results. However, he wishes to point out that many topics and results were left out or mentioned only in outline.

Acknowledgements

The author thanks his colleagues and collaborators, D.B. Abraham, M.C. Bartelt, K. Binder, J.L. Cardy, C.R. Doering, M.E. Fisher, G. Forgacs, G. Gaspari, M.L. Glasser, M. Nauenberg, M.P. Nightingale, R.K. Pathria, S. Redner, J. Rudnick, L.S. Schulman, N.M. Švrakić, A.M. Szpilka and A.P. Young for rewarding interactions. He also acknowledges support of his work by the United States National Science Foundation, and by the Donors of the Petroleum Research Fund, as well as a Graham research award.

REFERENCES

1. M.E. Fisher, in: *Critical Phenomena, Proc. 1970 E. Fermi Int. School of Physics,* M.S. Green *ed.* (Academic, NY, 1971), Vol. **51**, p. 1.
2. M.N. Barber, in: *Phase Transitions and Critical Phenomena,*

C. Domb and J.L. Lebowitz *eds.* (Academic, NY, 1983), Vol. **8**, p. 145.

3. M.P. Nightingale, J. Appl. Phys. **53**, 7927 (1982).
4. V. Privman and M.E. Fisher, J. Stat. Phys. **33**, 385 (1983).
5. E. Brézin and J. Zinn-Justin, Nucl. Phys. B**257**[FS**14**], 867 (1985).
6. K. Binder, Ferroelectrics **73**, 43 (1987) and J. Comput. Phys. **59**, 1 (1985).
7. J.L. Cardy, in: *Phase Transitions and Critical Phenomena*, C. Domb and J.L. Lebowitz *eds.* (Academic, NY, 1987), Vol. **11**, p. 55.
8. Reprint volume: *Finite-Size Scaling*, J.L. Cardy *ed.* (North-Holland, Amsterdam, 1988).
9. V. Privman and M.E. Fisher, Phys. Rev. B**30**, 322 (1984).
10. V. Privman, P.C. Hohenberg and A. Aharony, in: *Phase Transitions and Critical Phenomena*, C. Domb and J.L. Lebowitz *eds.* (Academic, NY, *in print*), see Ch. 10 there.
11. M.E. Fisher, Rev. Mod. Phys. **46**, 597 (1974).
12. M.E. Fisher and M.N. Barber, Phys. Rev. Lett. **28**, 1516 (1972).
13. See also M.E. Fisher and A.E. Ferdinand, Phys. Rev. Lett. **19**, 169 (1967).
14. E. Brézin, J. Physique **43**, 15 (1982).
15. M.P. Nightingale and H.W.J. Blöte, J. Phys. A**16**, L657 (1983).
16. J. Rudnick, H. Guo and D. Jasnow, J. Stat. Phys. **41**, 353

(1985).

17. A.M. Nemirovsky and K.F. Freed, J. Phys. A**18**, L319 (1985) and A**19**, 591 (1986).

18. H. Guo and D. Jasnow, Phys. Rev. B**35**, 1846 (1987), see also Jasnow's article in this volume.

19. B. Widom, J. Chem. Phys. **43**, 3898 (1965).

20. D. Stauffer, M. Ferer and M. Wortis, Phys. Rev. Lett. **29**, 345 (1972).

21. F. Wegner, Phys. Rev. B**5**, 4529 (1972).

22. V. Privman and M.E. Fisher, J. Phys. A**16**, L295 (1983); V. Privman, Physica **123**A, 428 (1984); H.J. Hermann and D. Stauffer, Phys. Lett. **100**A, 366 (1984); B. Derrida and D. Stauffer, J. Physique **46**, 1623 (1985).

23. A. Aharony and M.E. Fisher, Phys. Rev. B**27**, 4394 (1983).

24. S.I. Chase and M. Kaufman, Phys. Rev. B**33**, 239 (1986).

25. V. Privman and J. Rudnick, J. Phys. A**19**, L1215 (1986).

26. M.L. Glasser, V. Privman and L.S. Schulman, Phys. Rev. B**35**, 1841 (1987).

27. V. Privman, Phys. Rev. B**38**, 9261 (1988).

28. J.L. Cardy and I. Peschel, Nucl. Phys. B**300** [FS**22**], 377 (1988).

29. P.-Y. Lai and K.K. Mon, Phys. Rev. B, *in print*.

30. K. Symanzik, Nucl. Phys. B**190**[FS**3**], 1 (1981).

31. E. Eisenriegler, Z. Physik B**61**, 299 (1985).

32. V. Dohm, Z. Physik B**75**, 109 (1989).

33. K.K. Mon and M.P. Nightingale, Phys. Rev. B**31**, 6137 (1985), and references therein. See also Mon's article in this volume.

34. K. Binder, M. Nauenberg, V. Privman and A.P. Young, Phys. Rev. B**31**, 1498 (1985).

35. M.E. Fisher, In: *Renormalization Group in Critical Phenomena and Quantum Field Theory*, J.D. Gunton and M.S. Green *eds.* (Temple University, Philadelphia), p. 65.

36. J. Rudnick, G. Gaspari and V. Privman, Phys. Rev. B**32**, 7594 (1985).

37. R.C. Desai, D.W. Heermann and K. Binder, J. Stat. Phys. **53**, 795 (1988).

38. K. Binder, Z. Physik B**43**, 119 (1981) and Phys. Rev. Lett. **47**, 693 (1981).

39. E. Eisenriegler and R. Tomaschitz, Phys. Rev. B**35**, 4876 (1987).

40. K. Binder and J.-S. Wang, J. Stat. Phys., *in print*, see also Binder's article in this volume.

41. V. Privman and N.M. Švrakić, *Directed Models of Polymers, Interfaces, and Clusters: Scaling and Finite-Size Properties*, Lecture Notes in Physics **338** (Springer-Verlag, Berlin, 1989).

42. Y.Y. Goldschmidt, Nucl. Phys. B**280**[FS**18**], 340 (1987) and B**285**[FS**19**], 519 (1987); H.W. Diehl, Z. Physik B**66**, 211 (1987); C.R. Doering and M.A. Burschka, *preprint.*

43. W. Vogelsberger, J. Colloid Interface Sci. **103**, 448 (1985); K.K. Mon and D. Jasnow, Phys. Rev. Lett. **59**, 2983 (1987); J. Viñals and D. Jasnow, Phys. Rev. B**37**, 9582 (1987).

44. A.D. Bruce, J. Phys. A**18**, L873 (1985).
45. K. Kaski, K. Binder and J.D. Gunton, Phys. Rev. B**29**, 3996 (1984).
46. M.N. Barber, R.B. Pearson, D. Toussaint and J.L. Richardson, Phys. Rev. B**32**, 1720 (1985).
47. A. Sariban and K. Binder, J. Chem. Phys. **86**, 5859 (1987).
48. T.W. Burkhardt and B. Derrida, Phys. Rev. B**32**, 7273 (1985).
49. K.G. Wilson, Phys. Rev. B**4**, 3174 and 3184 (1971).
50. A.D. Bruce, J. Phys. C**14**, 3667 (1981).
51. V. Privman, Physica **129**A, 220 (1984).
52. H. Saleur and B. Derrida, J. Physique **46**, 1043 (1985).
53. J.L. Cardy, J. Phys. A**17**, L385 (1984).
54. J.L. Cardy, Nucl. Phys. B**240**[FS**12**], 514 (1984).
55. B. Derrida, B.W. Southern and D. Stauffer, J. Physique **48**, 335.
56. A.B. Harris, J. Phys. C**7**, 1671 (1974).
57. V. Privman and L.S. Schulman, J. Stat. Phys. **29**, 205 (1982) and J. Phys. A**15**, L231 (1982).
58. S. Singh and R.K. Pathria, Phys. Rev. B**31**, 4483 (1985). See also J. Shapiro and J. Rudnick, J. Stat. Phys. **43**, 51 (1986). For earlier finite-size studies of the spherical models, see M.N. Barber and M.E. Fisher, Ann. Phys. (NY) **77**, 1 (1973).
59. S. Singh and R.K. Pathria, Phys. Rev. B**32**, 4618 (1985).
60. S. Singh and R.K. Pathria, Phys. Rev. B**31**, 1816 (1985) and B**35**, 4814 (1987).

61. S. Singh and R.K. Pathria, Phys. Rev. B**38**, 2740 (1988), and references therein.

62. H.E. Stanley, Phys. Rev. **176**, 718 (1968).

63. J.M. Luck, Phys. Rev. B**31**, 3069 (1985).

64. M. Henkel, J. Phys. A**21**, L227 (1988).

65. K.K. Mon. Phys. Rev. Lett. **54**, 2672 (1985).

66. M. Henkel, J. Phys. A**19**, L247 (1986).

67. M. Henkel, J. Phys. A**20**, L769 (1987).

68. M. Henkel, J. Phys. A**20**, 3669 (1987).

69. M. Henkel, J. Phys. A**20**, 995 (1987).

70. J.O. Indekeu, M.P. Nightingale and W.V. Wang, Phys. Rev. B**34**, 330 (1986).

71. J.O. Indekeu, J. Chem. Soc. *Faraday Trans.* **II, 82**, 1835 (1986); K.K. Mon and M.P. Nightingale, Phys. Rev. B**35**, 3560 (1987).

72. Reprint volume: *Conformal Invariance and Applications to Statistical Mechanics*, C. Itzykson, H. Saleur and J.-B. Zuber *eds.* (World Scientific, Singapore, 1988).

73. J.L. Cardy, Nucl. Phys. B**275**[FS**17**], 200 (1986).

74. T.W. Burkhardt and I. Guim, Phys. Rev. B**35**, 1799 (1987).

75. J.-M. Debierre and L. Turban, J. Phys. A**20**, 1819 (1987), and references therein.

76. V. Privman and S. Redner, J. Phys. A**18**, L781 (1985).

77. H.W.J. Blöte, J.L. Cardy and M.P. Nightingale, Phys. Rev. Lett. **56**, 742 (1986). See also I. Affleck, Phys. Rev. Lett. **56**,

746 (1986).

78. A.E. Ferdinand and M.E. Fisher, Phys. Rev. **185**, 832 (1969).
79. M.E. Fisher and H. Au-Yang, Physica **101**A, 255 (1980); H. Au-Yang and M.E. Fisher, Phys. Rev. B**11**, 3469 (1975) and B**21**, 3956 (1980).
80. H. Park and M. den Nijs, Phys. Rev. B**38**, 565 (1988).
81. K.K. Mon and D. Jasnow, Phys. Rev. A**30**, 670 (1985).
82. D.B. Abraham and N.M. Švrakić, Phys. Rev. Lett. **56**, 1172 (1986).
83. D. Jasnow, K.K. Mon and M. Ferer, Phys. Rev. B**33**, 3349 (1986).
84. L. Onsager, Phys. Rev. **65**, 117 (1944).
85. V. Privman, *preprint.*
86. N.M. Švrakić, V. Privman and D.B. Abraham, J. Stat. Phys. **53**, 1041 (1988).
87. D.B. Abraham, *preprint.*
88. V. Privman, Phys. Rev. Lett. **61**, 183 (1988).
89. M.P. Gelfand and M.E. Fisher, Int. J. Thermophysics **9**, 713 (1988).
90. R. Lipowsky and G. Gompper, Phys. Rev. B**29**, 5213 (1984). For related results, see B. Legait and P.G. de Gennes, J. Physique Lett. **45**, L-647 (1984); R. Evans and U.M.B. Marconi, J. Chem. Soc. *Faraday Trans.* **II, 82**, 1763 (1986).
91. M.P. Gefand and R. Lipowsky, Phys. Rev. B**36**, 8725 (1987); J.O. Indekeu, P. Upton and J. Yeomans, Phys. Rev. Lett.

61, 2221 (1988) and Phys. Rev. B**40**, 666 (1989). For related results, see also R.J. Creswick, H.A. Farach and C.P. Poole, Jr., Phys. Rev. B**35**, 8467 (1987).

92. D.M. Kroll and G. Gompper, Phys. Rev. B**39**, 433 (1989).
93. V. Privman and N.M. Švrakić, Phys. Rev. B**37**, 3713 (1988).
94. G. Forgacs, N.M. Švrakić and V. Privman, Phys. Rev. B**37**, 3818 (1988) and B**38**, 8996 (1988).
95. J. Stecki and J. Dudowicz, Proc. R. Soc. Lond. A**400**, 263 (1985).
96. M.E. Fisher and D.S. Fisher, Phys. Rev. B**25**, 3192 (1982).
97. V. Privman and N.M. Švrakić, J. Stat. Phys. **54**, 735 (1989) and Phys. Rev. Lett. **62**, 633 (1989).
98. J. Bricmont, A. El Mellouki and J. Fröhlich, J. Stat. Phys. **42**, 743 (1986).
99. G.G. Cabrera and R. Jullien, Phys. Rev. Lett. **57**, 393 and 3297 (1986), and Phys. Rev. B**35**, 7062 (1987); J. Zinn-Justin, Phys. Rev. Lett. **57**, 3296 (1986); M.N. Barber and M.E. Cates, Phys. Rev. B**36**, 2024 (1987); D.B. Abraham, L.F. Ko and N.M. Švrakić, Phys. Rev. Lett. **61**, 2393 (1988) and J. Stat. Phys. *in print.*
100. M.E. Fisher, J. Phys. Soc. Japan Suppl. **26**, 87 (1969).
101. G. Münster, *preprint* (DESY-89-011).
102. K.K. Mon, S. Wansleben, D.P. Landau and K. Binder, Phys. Rev. Lett. **60**, 708 (1988) and Phys. Rev. B**39**, 7089 (1989).
103. G.A. Williams, Japan. J. Appl. Phys. **26**, Suppl. **26-3**, 305

(1987); K.Y. Szeto and G. Dresselhaus, Phys. Rev. B**32**, 3142 (1985).

104. B. Nienhuis and M. Nauenberg, Phys. Rev. Lett. **35**, 477 (1975); M.E. Fisher and A.N. Berker, Phys. Rev. B**26**, 2507 (1982).
105. K. Binder, Rep. Prog. Phys. **50**, 783 (1987); see also Bhanot's and Landau's articles in this volume.
106. M.S.S. Challa, D.P. Landau and K. Binder, *preprint.*
107. M.E. Fisher and V. Privman, Phys. Rev. B**32**, 447 (1985).
108. K. Binder and D.P. Landau, Phys. Rev. B**30**, 1477 (1984).
109. C. Dasgupta and B.I. Halperin, Phys. Rev. Lett. **47**, 1556 (1981).
110. L. Néel, C. R. Acad. Sci. **228**, 664 (1949); Ann. Géophys. **5**, 99 (1949).
111. I.S. Jacobs and C.P. Bean, in *Magnetism*, G.T. Rado and H. Suhl *eds.* (Academic, NY, 1963), Vol. **III**, p. 271.
112. J.L. Cardy and P. Nightingale, Phys. Rev. B**27**, 4256 (1983); H.W.J. Blöte and M.P. Nightingale, Physica **112**A, 405 (1982); see also Barber's review [2].
113. M.E. Fisher and V. Privman, Com. Math. Phys. **103**, 527 (1986).
114. This mechanism of absence of ordering in $1d$ systems is credited to L.D. Landau: see L.D. Landau and E.M. Lifshitz, *Statistical Physics*, Theoretical Physics, Vol. **V** (Science, Moscow, 1964, *in Russian*), Sect. 152.

115. C.J. Hamer and C.H.J. Johnson, J. Phys. A**19**, 423 (1986).

116. V. Privman and M.E. Fisher, J. Appl. Phys. **57**, 3327 (1985).

117. D.B. Abraham, Stud. Appl. Math. **50**, 71 (1971).

118. Numerical results in $4d$, were reported by K. Jansen, J. Jersák, I. Montvay, G. Münster, T. Trappenberg and U. Wolff, Phys. Lett. **213**B, 203 (1988); K. Jansen, I. Montvay, G. Münster, T. Trappenberg and U. Wolff, *preprint* (DESY-88-169).

119. G.G. Cabrera, R. Jullien, E. Brézin and J. Zinn-Justin, J. Physique **47**, 1305 (1986).

120. V. Privman and M.E. Fisher, J. Magn. Magnet. Mater. **54-57**, 663 (1986).

121. M.E. Fisher, *unpublished* (1973).

122. M.E. Fisher, M.N. Barber and D. Jasnow, Phys. Rev. A**8**, 1111 (1973); A.L. Patashinskii and V.L. Pokrovskii, Zh. Eksp. Teor. Fiz. **64**, 1445 (1973) [Sov. Phys. JETP **37**, 733 (1974)].

123. H. Müller-Krumbhaar, Z. Physik **267**, 261 (1974), studied $3d$ Heisenberg ($n = 3$) systems of up to 16^3 spins.

124. S. Singh and R.K. Pathria, Phys. Rev. B**37**, 7806 (1988).

125. H. Neuberger and T. Ziman, Phys. Rev. B**39**, 2608 (1989); D.S. Fisher, Phys. Rev. B**39**, 11783 (1989).

126. S.N. Isakov, Com. Math. Phys. **95**, 427 (1984). For additional literature, see [57].

127. M.S.S. Challa, D.P. Landau and K. Binder, Phys. Rev. B**34**, 1841 (1986).

128. C. Borgs and R. Kotecký, *work in progress.*

129. C. Borgs and J. Imbrie, Com. Math. Phys. **123**, 305 (1989).

130. M.E. Fisher, *unpublished notes.*

131. S. Singh and R.K. Pathria, Phys. Rev. B**34**, 2045 (1986), see also [124].

132. A.B. Zamolodchikov, Pisma Zh. Eksp. Teor. Fiz. **43**, 565 (1986) [JETP Lett. **43**, 730 (1986)].

133. D. Boyanovsky and C.M. Naon, *preprint.*

134. J.H. Mayer and W.W.Wood, J. Chem. Phys. **42**, 4268 (1965); M.C. Bartelt and V. Privman, *work in progress.*

135. C. Itzykson, R.B. Pearson and J.-B. Zuber, Nucl. Phys. B**220**, 415 (1983).

II. FINITE-SIZE SCALING, HYPERSCALING AND THE RENORMALIZATION GROUP

David Jasnow

TABLE of CONTENTS

Finite-Size Scaling, Hyperscaling and the Renormalization Group

by

David Jasnow

Department of Physics and Astronomy

University of Pittsburgh

Pittsburgh, Pennsylvania 15260

I. FORWARD

In this article I will review finite-size scaling within the context of the field-theoretic renormalization group. The review will concentrate on work done in collaboration with colleagues Joseph Rudnick and Hong Guo, who deserve much of the credit for what will be presented here.

Specifically, I will review renormalization group calculations for systems in box geometry in $d = 4$ and $d = 4 - \epsilon$ dimensions. To do this, a form of mean-field theory, which yields a rounded transition when applied to a finite system, will be introduced. This development allows an appropriate expansion in fluctuations. Explicit numerical results for the specific heat, both above and below the critical temperature, will be presented along with an evaluation of the universal scaling function to $O(\epsilon)$.

The structure of finite-size scaling within the framework of the renormalization group will be examined in further detail. The implications of two-scale factor (or hyper-) universality on finite-size scaling will be discussed. In that context it will be shown that the scaling is of the form discussed by Privman and Fisher, which has the consequence that the singular part of finite-system free energy is universal at the bulk critical point.

II. INTRODUCTION

Simulations have been playing an ever increasing part in studies of equilibrium and more recently nonequilibrium problems in condensed matter physics. Clearly this is due to the widening availability of high-speed general purpose computers and specially designed processors. But it is also due to the appreciation, particularly evident, for example, within the framework of nonequilibrium phenomena and critical phenomena, that analytic progress is especially hard to come by and that properly conceived numerical attacks can yield a great deal of useful qualitative and quantitative information.

Nonetheless, even with the best available technology, simulations are often restricted to what are essentially microscopic systems. This produces an immediate problem in numerical studies of phase transitions since the mathematical singularities in thermodynamic functions which characterize phase transitions only occur in the thermodynamic limit. More to the point, continuous transitions are associated with divergent correlation lengths; any finite-system simulation will eventually run aground. The major task of interpreting, say, Monte Carlo results in studies of phase transitions, is to extrapolate to the infinite-system limit results for relatively small systems. Finite-size scaling allows such extrapolation to be made in a variety of cases. When one knows something about the nature of the extrapolations, in a sense one turns the system-size parameter into a weapon instead of a serious liability.

The phenomenology and application of finite-size scaling will be discussed amply elsewhere in this volume. For the moment a brief encapsulation will suffice. The simplest and most traditional form holds that the singular part of a thermodynamic function, denoted quite generally by A, which is a function of the reduced temperature $t = (T - T_c)/T_c$, the system size L, and ordering field h, is of the form

$$A(t, h, L) \sim t^{-\tau} X(tL^{1/\nu}, \frac{h}{t^{\Delta}}). \tag{1}$$

Here τ is the usual bulk critical exponent describing the particular function, A, and ν and Δ are the standard critical exponents for the correlation length $\xi \sim t^{-\nu}$ and 'gap' respectively. When finite-size scaling obtains, it is expected to hold in the regime $L >> a$, $\xi >> a$, where a is the microscopic length (or lattice spacing) in the problem. In general, the scaling function $X(x, y = 0)$ is expected to have a singularity as $x \rightarrow 0$, which allows the function A to behave perfectly smoothly as $T \rightarrow T_c$ for fixed L. Conversely the proper bulk scaling must be obtained in the alternate limit $x \rightarrow \infty$. The form of Eq. (1) can be easily generalized to include correlation functions.

To obtain, for example, an estimate of the critical exponent τ, or more specifically τ/ν, from numerical data obtained in a simulation, one might typically evaluate $A(t = 0, h = 0, L)$, assuming the bulk transition temperature is sufficiently well known or has been estimated. From Eq. (1) and the limits discussed above, one expects $A(t = 0, h = 0, L) \sim L^{\tau/\nu}$. Many successful evaluations of bulk critical exponents[1] and even 'beyond-the-bulk' exponents such as the surface tension[2] have exploited this technique.

The scaling form (1) has been shown to arise from a number of exactly solvable models. Fisher and Barber,[3] Singh and Pathria,[4] and Rudnick (see also the article contained in this volume) have carried out calculations on spherical models, and Brézin[5] considered the N-vector model. Brézin also indicated how the scaling form could evolve from the structure of the field-theoretic renormalization group. The first attempt to consider finite-size effects within the framework of renormalization group was due to Suzuki,[6] who used the more intuitive momentum shell integration techniques.

It is reasonable to inquire what can be expected from a renormalization group treatment of the statistical mechanics of finite-size systems. First, the manner in which the structure of finite-size scaling fits into the general renormalization group framework can provide useful insights. Here one has in mind, for example, the implications of hyper- or two-scale factor universality. Practically speaking, a proper renormalization group treatment holds out the possibility of calculating the universal scaling functions analogous to $X(x, y)$ in Eq. (1) in a manner which can be realistically compared with simulations in three and four dimensions. Finally, renormalization group treatments are particularly effective in dealing with logarithmic corrections. It is known that special subtleties are to be expected for critical phenomena in $d \geq 4$, associated with the breakdown of hyperscaling; indeed some of these features are seen within the context of spherical models. At the upper critical dimension ($d_c = 4$, or $d_c = 3$ for tricriticality) one finds logarithmic modifications to mean-field behavior, but these are generally difficult to observe experimentally or via simulation. In the special dimensions renormalization group calculations are presumably asymptotically exact. In principle these may be compared to standard Ising simulations with the adjustment of three parameters (two nonuniversal amplitudes and a coupling constant). One can hope to use such renormalization group results as a guide for the reliable extraction of logarithmic corrections, but again one requires sufficient knowledge of the scaling functions.

The remainder of this review is laid out as follows. In Section III a mean-field theory appropriate to a finite system is introduced. Section IV deals with the evaluation of a fluctuation integral in $d = 4$. In Section V the renormalization group approach is examined. Numerical results for $d = 4$ are discussed in Section VI, while the following section deals with modifications for $d = 4 - \epsilon$. At that point there is a brief recapitulation. Sections IX through XI address the consequences of hype runiversality. Section XII contains a few concluding remarks and two appendices are reserved for some additional details. Most of the results quoted have appeared previously in the work of Rudnick *et al.*[7] and Guo and Jasnow.[8,9] Additional details appear in Guo's thesis.[10]

III. MEAN-FIELD THEORY FOR A FINITE SYSTEM

When dealing with renormalization group treatments for a finite system, one is immediately faced with a conceptual problem, one that has caused some confusion. RG methods, at least those employing field theoretic approaches, generally require a perturbative expansion in fluctuations. Furthermore, the fluctuations are, loosely speaking, around the mean-field or classical solution. The first task is to develop an appropriate mean-field theory for a finite system, one that yields the expected rounded transition. Traditional mean-field theories do not properly consider system-size and tend to yield a sharp transition even for a finite system. A major contribution in developing a proper mean-field theory was made by Rudnick and employed by Rudnick *et al.*[7]

One begins with the standard Ginzburg-Landau description of the finite system in which the reduced Hamiltonian is given by

$$H[s] = \int_\Omega d^d x \, [\frac{1}{2}(\nabla s)^2 + \frac{1}{2}rs^2 + \frac{1}{4!}us^4 - hs] \tag{2}$$

where the spatial integrations are over a cubical domain of linear extent L in each of its $d-$dimensions. In this work the system is always considered to have such 'box' geometry; generalizations to other shapes and to an $N-$ component order parameter are straightforward. The parameters r and $u > 0$ are phenomenological, with the usual interpretation that $r \propto T - T_0$, where T_0 is a reference temperature whose significance will appear momentarily. The thermodynamics of the system follow from knowledge of the partition function

$$Z = \int Ds \, e^{-H[s]}. \tag{3}$$

It is assumed that the system is confined to the finite box with periodic boundary conditions applied. One expands $s = M + \psi$, with the understanding that $M = const$ and that ψ contains no $k = 0$ part. Then in the usual way one finds

$$H = L^d(\frac{1}{2}rM^2 + \frac{1}{4!}uM^4 - hM)$$

$$+ \int_\Omega d^d x [\frac{1}{2}(\nabla\psi)^2 + \frac{1}{2}(r + \frac{uM^2}{2})\psi^2 + \frac{uM\psi^3}{6} + \frac{1}{4!}u\psi^4]. \tag{4}$$

Notice that the remaining term which involves $\int \psi \, d^d x$ vanishes identically since, by construction, ψ has no $k = 0$ part. Schematically the partition function becomes

$$Z = \int dM e^{-H_{mf}(M)} e^{-\Gamma_1(M)} \tag{5}$$

where

$$\exp(-\Gamma_1(M)) \equiv \int D\psi\, e^{-H(\psi;M)}. \tag{6}$$

Here $H(\psi; M)$ is the fluctuation or ψ-dependent part of the Hamiltonian (4), and $H_{mf}(M)$ is the (mean-field) ψ-independent part of the Hamiltonian. In the usual way $\Gamma_1(M)$ may be evaluated via a diagram expansion (i.e., loop expansion; see, e.g., Ref.[11]). Formally the terms begin at $O(u^0)$ with

$$\Gamma_1(M) = \frac{1}{2}\sum_{\mathbf{k}\neq 0} \ln(k^2 + r + \frac{1}{2}uM^2) + \cdots \tag{7}$$

where it has been explicitly indicated that the $\mathbf{k} = 0$ term is absent; the dots indicate higher terms in u from the loop expansion.

The mean-field contribution to the free energy in zero field is obtained by neglecting Γ_1. It is formally $O(u^{-1})$ and takes the form

$$\ln Z_{mf} = \ln\{\int dM \exp[-L^d(rM^2/2 + uM^4/4!)]\} \tag{8}$$

from which the entropy and specific heat may be found by differentiation with respect to the temperature parameter r. Looking ahead for the moment, notice that as $L \to \infty$ the integral may be evaluated by steepest descent methods; the dominant contribution comes for $rM + uM^3/6 = 0$, which can be recognized as the usual mean-field equation of state for the infinite system. (It also shows that in the thermodynamic limit $M \sim u^{-1/2}$ which accounts for the ordering of terms noted above in the loop expansion.) Notice also that from the bulk equation of state $r = 0$ is identified as the mean-field critical temperature. For finite size, $L < \infty$, moments of M, defined by

$$< M^{2j} > \equiv Z_{mf}^{-1} \int dM\, M^{2j} \exp[-H_{mf}(M)] \tag{9}$$

are strictly finite even at the bulk critical point. Odd moments vanish identically, but the finite value for even moments indicates that susceptibilities do not diverge for the finite system. Within mean-field theory the moments, which will play an important role in practical calculations to follow, have a scaling form

$$< M^{2j} >= (uL^d)^{-j/2} D_j(\frac{rL^{d/2}}{u^{1/2}}). \tag{10}$$

Note the characteristic appearance of the factor $u^{-1/2}$ which occurs naturally in this approach. The total heat capacity for the system then has the form

$$C \equiv \frac{\partial^2 \ln Z_{mf}}{\partial r^2} = \frac{L^{2d}}{4}[< M^4 > - < M^2 >^2]. \tag{11}$$

The scaling functions D_j are perfectly smooth through zero argument. From Eq. (8) it is seen again via steepest descents arguments that a singularity can emerge in the thermodynamic limit. The specific heat C/L^d is shown for several system sizes in Fig. 1 as a function of the temperature variable, r, within this finite-size mean-field theory. (Lengths are assumed to be in units of the microscopic length for the problem.)

Notice the sharpening of the curve and the appearance of the discontinuity in the specific heat as $L \to \infty$. The singularity appears in the thermodynamic limit for all $d > 0$, a well-known defect of mean-field theories, which also appears at 'zero order' in renormalization group treatments of bulk systems. However, this version of mean field theory specifically eliminates a sharp transition for a finite system regardless of the dimensionality. The message of how easily one might be misled on the nature or even the existence of a phase transition from restricted numerical information on finite systems is clear from Fig. 1 and the underlying the analytic calculations. That message will reappear in different ways in accompanying articles in this volume.

To make progress in the evaluation of fluctuation corrections, a basic fluctuation integral, which appears at one-loop order, must be evaluated.

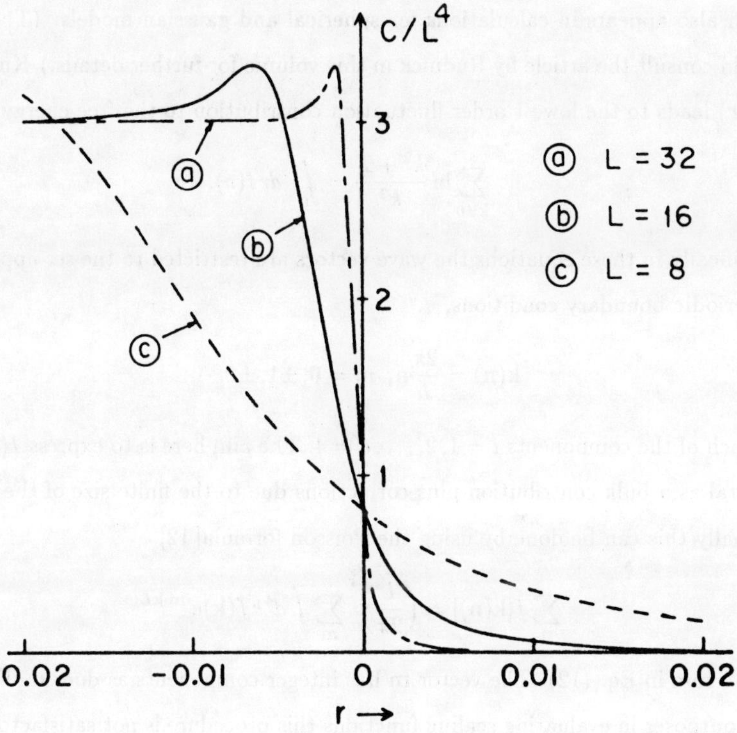

Figure 1: Specific heat as a function of reduced temperature from mean-field calculation for several system sizes, L. The coupling constant has been set as $u = 1$.

IV. ONE-LOOP TERM

An essential part of the finite-size calculation requires the evaluation of a sum of the form

$$I(r) = \sum_{\mathbf{k}\neq 0} \frac{1}{r + k^2} \tag{12}$$

which also appears in calculations for spherical and gaussian models. (The reader should consult the article by Rudnick in this volume for further details.) Knowledge of $I(r)$ leads to the lowest order fluctuation contribution to the free energy since

$$\sum_{\mathbf{k}\neq 0} \ln \frac{k^2 + T}{k^2} = \int_0^T dr I(r). \tag{13}$$

Specifically in these equations the wave vectors are restricted to the set appropriate to periodic boundary conditions,

$$\mathbf{k}(\mathbf{n}) = \frac{2\pi}{L}\mathbf{n}, \ n_i = 0, \pm 1, \pm 2, \cdots \tag{14}$$

for each of the components $i = 1, 2, \ldots, d = 4$. The aim here is to express $I(r)$ or its integral as a bulk contribution plus corrections due to the finite size of the system. Formally this can be done by using the Poisson formula[12]

$$\sum_{\mathbf{n}} f[\mathbf{k}(\mathbf{n})] = \left(\frac{L}{2\pi}\right)^4 \sum_{\mathbf{m}} \int d^4k f(\mathbf{k}) e^{i\mathbf{m}\cdot\mathbf{k}L} \tag{15}$$

on the sum in Eq. (12). The vector $\mathbf{m}$ has integer components as does $\mathbf{n}$. For practical purposes in evaluating scaling functions this procedure is not satisfactory since it leaves a slowly convergent sum. Convergence can be improved by reexpressing $I(r)$ as

$$I(r) = \sum_{\mathbf{k}\neq 0} \left[\int_0^B d\alpha\, e^{-\alpha(r+k^2)} + \int_B^\infty d\alpha\, e^{-\alpha(r+k^2)} \right] \tag{16}$$

where one may write the arbitrary constant B as $B = c(L/2\pi)^2$ and choose c for good convergence. The first term in Eq. (16) is evaluated using the Poisson formula. This contribution has important high wavenumber contributions and actually produces the ultraviolet divergence contained in $I(r)$. The second term of Eq. (16), which will be dominated by small $r + k^2$, can be evaluated in terms of the function

$$X(y) \equiv \sum_{n=-\infty}^{\infty} e^{-yn^2} \tag{17}$$

which converges rapidly for positive argument. Details of these computations can be found in the original reference[7] and in the article in this volume by J. Rudnick. The result is worth including for further comment. Combining the two parts of $I(r)$ which were separately evaluated one finds

$$\begin{aligned}\sum_{\mathbf{k}\neq 0} \frac{1}{r+k^2} = \left(\frac{L}{2\pi}\right)^4 \int \frac{d^4k}{r+k^2} + c\left(\frac{L}{2\pi}\right)^2 \int_1^\infty dx\, e^{-rc(L/2\pi)^2 x}[X^4(cx) - 1] \\ + \frac{\pi^2}{c}\left(\frac{L}{2\pi}\right)^2 \int_1^\infty dx\, e^{-rc(L/2\pi)^2 x^{-1}}[X^4(\frac{\pi^2 x}{c}) - 1] \\ + \pi^2\left(\frac{L}{2\pi}\right)^4 r e^{-rc(L/2\pi)^2} \int_0^\infty \frac{e^{-x}dx}{x + rc(L/2\pi)^2} \\ - \frac{\pi^2}{c}\left(\frac{L}{2\pi}\right)^2 e^{-rc(L/2\pi)^2} + \frac{1}{r}(e^{-rc(L/2\pi)^2} - 1).\end{aligned} \tag{18}$$

The first term is immediately identified with the bulk contribution. The remaining terms are finite-size corrections in which $r \to 0$ may be taken; in other words the corrections will not produce problems at bulk criticality. From the structure of Eq. (18) it is apparent that only the usual bulk renormalizations (or counter terms) are required to eliminate ultraviolet divergences and take the high momentum cutoff to infinity. There is a slight subtlety here: intuitively one expects that ultraviolet problems are associated with short-distance behavior having nothing to do with the finite size of the system (it is always assumed that the system size is much greater than microscopic lengths). The intuition is correct, but the result of Eq. (18) is not a good test. More general terms in the loop expansion for Γ_1 will introduce ultraviolet divergences in finite-size corrections. Nonetheless, the usual bulk renormalizations are sufficient to deal with all ultraviolet divergences; a two-loop calculation[8,9] which presumably captures the general spirit is sketched in the Appendices. Finally, note that formally the finite-size corrections come in at $O(u^0)$.

Performing the integral of $I(r)$ to obtain the free energy contribution is relatively straightforward, and the reader should consult the references for details. The main point to be made is that the lowest-order free energy contribution (13) has finite-size corrections which may be evaluated without numerical or convergence difficulties. The evaluation of the fluctuation integral forms a central part of the application of the full renormalization group treatment, which is now explored more thoroughly.

V. RENORMALIZATION GROUP

In this section the renormalization group approach will be examined. Since an asymptotically correct finite-size calculation can be constructed in four dimensions, the specific interest at first is in the case $d = 4$. The situation in $d = 4 - \epsilon$ will be considered later.

As noted above, standard bulk counterterms succeed in renormalizing the theory. (For a review of the field theoretic approach to bulk critical phenomena, the reader should consult, for example, the review of Brézin *et al.*[11] and the Amit's monograph.[13]) The renormalization constants are introduced into the Hamiltonian so that the complete (renormalized) exponential, now reduced by the system volume, $\Gamma = L^{-d}(H_{mf} + \Gamma_1)$, becomes

$$\Gamma_R(t, M, u, \kappa; L) = \Gamma_B(r, M, u, \Lambda; L) - a_1^\infty - a_2^\infty r - a_3^\infty r^2 \qquad (19)$$

where the subscripts R and B refer to the renormalized and bare functions, respectively. The following replacements have to be made on the right-hand side: $r \Rightarrow Z_{\phi^2} t + \delta t, u \Rightarrow \kappa^\epsilon Z_u u, M \Rightarrow Z_\phi^{1/2} M$, with M renormalizing like the field. It is understood henceforth that t, u, M and so on, are all *renormalized* quantities and $d = 4 - \epsilon$ (or $d = 4$ when explicitly stated). The (multiplicative) renormalization constants (the Z's) are chosen to remove ultraviolet divergences in the usual way. There are three *additive* renormalizations required on the right for the free energy; these are taken to be the same as for the bulk system, as indicated by the superscripts. Finally, Γ_B is computed diagramatically as noted above, the leading contributions being the mean-field part $L^{-d}H_{mf}(M)$, and the lowest fluctuation contribution, which is represented by the explicit term in Eq. (7). Space does not permit complete detail here; suffice it to say that formally the loop expansion is the same as for a bulk homogeneous system. The changes involve the treatment of the mean-field part and the fact that fluctuations are expressed as sums in momentum space rather than integrals. All loop corrections are evaluated with M as a parameter. At the end an integration over M must be performed as suggested in Eq. (5). All momentum integrals which appear may be regularized with a hard cutoff Λ which is eventually removed to infinity, and the constant κ introduced in the definition of the renormalized coupling constant sets the basic (inverse) length scale of the renormalized theory. Acting on Eq. (19) with $\kappa d/d\kappa$, one arrives at the inhomogeneous renormalization group equation,

$$[\kappa\frac{\partial}{\partial\kappa}+\beta(u)\frac{\partial}{\partial u}-\frac{1}{2}\eta(u)M\frac{\partial}{\partial M}+\gamma_{\phi^2}(u)t\frac{\partial}{\partial t}]\Gamma_R=G(t,u,\kappa) \tag{20}$$

[To avoid writing many definitions, the notation here is essentially that of Amit.[13]] The inhomogeneous term $G(t,u,\kappa)$ arises from the additive renormalizations required for the free energy. The inhomogeneous term is itself renormalized and is finite, meaning that it contains no divergences as the cutoff is taken to infinity. In a practical calculation it should be appropriate to two-loops, as this is necessary to get all $O(\epsilon^0)$ contributions to the free energy. There is some freedom in choosing the additive renormalizations, and specifics (see below) require a choice of renormalization scheme.

One formally solves Eq. (20) in the usual fashion[11,13] by introducing the flow equations (or running coupling constants) for general d

$$\begin{aligned}
\rho\frac{\partial t(\rho)}{\partial\rho} &= [-2+\gamma_{\phi^2}(u(\rho))]t(\rho)\\
\rho\frac{\partial u(\rho)}{\partial\rho} &= \beta(u(\rho))\\
\rho\frac{\partial M(\rho)}{\partial\rho} &= -\frac{1}{2}\eta(u(\rho))M(\rho).
\end{aligned} \tag{21}$$

(The particular definition of the running constant $t(\rho)$ may appear unnatural in view of the form of Eq. (20). The choice has been made for historical reasons.) In terms of these functions the renormalized free energy may be expressed for general d as

$$\Gamma_R = \Gamma_R[\rho^2 t(\rho), M(\rho), u(\rho), \kappa\rho; L] + \int_\rho^1 \frac{d\lambda}{\lambda} G[\lambda^2 t(\lambda), u(\lambda), \kappa\lambda] \tag{22}$$

in which the last term (the so-called 'trajectory integral') is formally $O(u^{-1})$. The inhomogeneous term $G(t, u, \kappa)$ leading to the trajectory term has been evaluated in a dimensionally regularized scheme; details can be found in the references.[8,10] For future discussion it is useful to note that the rescaled free energy in Eq. (22) may be written formally as

$$\begin{aligned}
&\Gamma_R[\rho^2 t(\rho), M(\rho), u(\rho), \kappa\rho; L] \\
&\quad = \frac{1}{2}\rho^2 t(\rho) M^2(\rho) + \frac{1}{4!}(\kappa\rho)^\epsilon u(\rho) M^4(\rho) \\
&\quad + L^{-d}\Gamma_1[L^2 T(\rho), L^{d-2} M^2(\rho), u(\rho); \kappa\rho L].
\end{aligned} \tag{23}$$

Here Γ_1 is the full loop correction to the free energy, and the useful combination

$$T(\rho) = \rho^2 t(\rho) + \frac{1}{2}(\kappa\rho)^\epsilon u(\rho) M^2(\rho) \tag{24}$$

has been defined.

Now specializing to $d = 4$ and using a standard renormalization scheme (of fixing the vertices of the theory at criticality[11,13]), one has $\beta(u) = \frac{3}{2}S_4 u^2$, $\gamma_{\phi^2}(u) = \frac{1}{2}S_4 u$, $\eta(u) = \frac{1}{24}(S_4 u)^2$. As usual S_d is the surface of a $d-$ dimensional unit sphere. The solution of the flow equations, in particular, $u(\rho) = u/(1 - \frac{3}{2}S_4 u \ln \rho)$, and so on, in $d = 4$ are used in Eq. (22), which in turn gives the partition function for the finite-size system

$$Z = \int_{-\infty}^{\infty} dM\, e^{-L^d \Gamma_R}. \tag{25}$$

Some comments are in order. First, the renormalization group equations are formally the same as for the corresponding bulk system. The size L enters as a parameter, so that the system-size plays a special role. In this treatment it is not analogous to a coupling field like the temperature or ordering field. Secondly, at this stage, ρ is a parameter in the formal solution; for an explicit evaluation of the finite-size free energy an appropriate choice of $\rho = \rho^*$ must be made. This choice actually involves some physical intuition on the nature of finite-size effects in critical phenomena.

The lowest order calculation is so-called renormalized mean-field theory (denoted here by 'Rmf'), which includes the effects of fluctuations in the flow equations but neglects explicit loop corrections to the correlation functions of interest. The loop contribution, Γ_1 is accordingly neglected. This approximation yields the correct free energy to leading logarithms in $d = 4$, to which the remainder of this section is specialized. As noted one neglects loop corrections to Γ_R, but one includes the trajectory integral contribution, which is formally $O(u^{-1})$, so that in $d = 4$

$$\Gamma_{Rmf}(t, M, u, \kappa; L) = \frac{1}{2}\rho^2 t(\rho) M^2(\rho) + \frac{1}{4!}(\kappa\rho)^{\epsilon} u(\rho) M^4(\rho)$$

$$+ \frac{t^2}{16\pi^2 u}[1 - (1 - \frac{3}{2}S_4 u \ln \rho)^{1/3}] \tag{26}$$

where the explicit contribution of the trajectory term has been inserted.[8] The specific heat follows by differentiation and has the form

$$\frac{C}{L^4} = L^{-4}\frac{\partial^2}{\partial t^2}\ln Z_{Rmf}$$

$$= L^4 < (\frac{\partial \Gamma_{Rmf}}{\partial t})^2 > -L^4 < \frac{\partial \Gamma_{Rmf}}{\partial t} >^2 - < \frac{\partial^2 \Gamma_{Rmf}}{\partial t^2} > \tag{27}$$

where averages are evaluated according to the distribution defined by H_{mf} as in Eq. (9), but with $r \to (\rho^*)^2 t(\rho^*), u \to u(\rho^*)$. (All averages will involve such a procedure.) Further technical details are supplied in the references.[7] Briefly, one notes that only explicit temperature derivatives are taken in Eq. (27); also the averages involve $< M^p(\rho) >$ for $p = 2, 4$, which have the scaled combination $wv^{-1/2}$ with $w = L^2(\rho^*)^2 t(\rho^*), v = u(\rho^*)$. A choice of ρ^* is required for a more explicit evaluation.

A natural choice[7] for ρ^* is to choose it as the solution of

$$(\rho^*)^2 t(\rho^*) + \frac{1}{2}u(\rho^*) < M^2(\rho^*) > + \frac{1}{L^2} = (\kappa\rho^*)^2. \tag{28}$$

The motivation is as follows. The first term on the left of Eq. (28) has the interpretation of the 'mass' entering the renormalized Hamiltonian. Furthermore, $(\kappa\rho^*)^{-1}$ has the interpretation of the effective 'Kadanoff' block size in the renormalization, i.e., the (inverse) block size begins at the characteristic scale κ and goes to $\kappa\rho^*$. When the reduced temperature $t \sim T - T_c$ is small so that the bulk correlation length is *large* compared to the system size L, renormalization must end when the effective block size $[\sim (\kappa\rho^*)^{-1}]$ is on the order of L. This is the *strongly finite-size* regime and $\rho^* \approx (\kappa L)^{-1}$. Conversely, when L is *much larger* than the bulk correlation length, renormalization transformations are carried out until the block size is on the order of the bulk correlation length. This is quite analogous to renormalization in bulk systems and corresponds to $(\rho^*)^2 t(\rho^*) + \frac{1}{2}u(\rho^*)M^2(\rho^*) = (\rho^*\kappa)^2$, which is indeed appropriate for bulk systems. Equation (28) interpolates between these two limits. Alternative interpolation schemes can be used, although the one proposed appears quite natural. The general structure of the theory allows freedom. To find $\rho^*(t, L)$ one must generally solve Eq. (28) numerically; the equation is implicit since ρ^* is itself required for evaluation of the moments. An iterative solution is therefore required. Before returning to explicit numerical results, the method of dealing with the loop corrections is addressed. For further details the reader should consult the references.

An iterative approach is also required for the loop corrections. The full renormalized vertex Γ_1 must be substituted into the integral over M to get the partition function. A complete expression for the renormalized Γ_1, (denoted by $\Gamma_{R,1}$) is provided by Rudnick *et al.*[7] It is possible to construct a cumulant expansion by writing

$$\Gamma_{R,1}(M^2(\rho)) = \Gamma_{R,1}(< M^2(\rho) >) + \Delta \tag{29}$$

where averages $< \cdot >$ are computed with respect to the renormalized mean-field theory, i.e., according to Eq. (10) with the proper substitutions, $r \to (\rho^*)^2 t(\rho^*)$, etc. (It is to be understood that an appropriate choice $\rho = \rho^*$ had been made.) Then one finds[7] an expansion of the form

$$\ln Z = \ln Z_{Rmf} - L^4 \Gamma_{R,1}(< M^2(\rho) >) + \sum_{n=1}^{\infty} \frac{(-1)^n}{n!} < (L^4 \Delta)^n >_c \qquad (30)$$

where $< \cdot >_c$ are cumulant averages. Recall that leading contributions to the specific heat are formally $O(u^{-1})$ so that one is looking for leading correction terms of order u^0. Such terms can be isolated and an explicit calculation has been performed to the first loop correction beyond renormalized mean-field theory.

It is now appropriate to consider the form of the scaling that emerges from this treatment. At renormalized mean-field theory level one sees directly from the nature of the moments that

$$\ln Z(t, u; \kappa, L) = \bar{F}[w, v] \quad [w = L^2 \rho^2 t(\rho), v = u(\rho)] \qquad (31)$$

where $\bar{F}$ is dimensionless (and a smooth background term has been omitted). Specifically, at renormalized mean-field level there is an additional simplification since $z = wv^{-1/2}$ is all that enters. In terms of this variable the free energy takes the simple form

$$\begin{aligned} \bar{F}_{Rmf}(w, v) &= f(z = wv^{-1/2}) \\ &= z^2 + \ln \int_{-\infty}^{\infty} dy \exp\{-(\frac{1}{2} z y^2 + \frac{1}{4!} y^4)\}. \end{aligned} \qquad (32)$$

This final result is correct to *leading logarithms* in $d = 4$.

The more general result, which includes the addition of higher order loop terms in $d = 4$ or, indeed, other dimensions consistent with use of the renormalization group equations, is

$$\ln Z(t, u; \kappa, L) = F[\frac{\rho^2 t(\rho)}{\kappa^2}, u(\rho), \rho \kappa L] \qquad (33)$$

which, of course, can reproduce (31). This is the general statement of finite-size scaling. Such scaling form has been discussed by Brézin.[5] In applications with

$2 < d < 4$, a fixed point is approached with $u(\rho) \to u^* \neq 0$ as $\rho \to 0$. Then one may show that a form like (33) reproduces the intuitive type of scaling described in the Introduction. However, in the special dimensionality, $d = 4$, $u(\rho)$ iterates slowly to zero as indicated above in the solution of the flow equations. This introduces additional L dependences and explicit dependence on the initial fourth order coupling constant, u. Such dependence on the coupling constant is associated with the breaking down of hyperscaling above $d = 4$, and arises in bulk phenomena as well. The apparent nonuniversality should not be construed as a breakdown of finite-size scaling, which indeed holds in the form of Eq. (33).

This is an appropriate place to consider the numerical results for $d = 4$.

VI. NUMERICAL RESULTS FOR d=4

In $d = 4$ loop corrrections are formally corrections to scaling, so the calculations outlined above are presumed to be asymptotically exact (to leading logarithms) and should be comparable to simulations. It is instructive to plot the specific heat following from the results of the last section; a more explicit expression in provided in the Appendix of Ref.[7] . First note in Fig. 1, for pure mean-field theory, that the specific heat peak sharpens but does not increase. The situation at the Rmf level is shown in Fig. 2. Note that the central ingredient to this calculation is the simple form of Eq. (26). It is straightforward to show that that $C(t = 0, L)$ diverges as $(\ln \kappa L)^{1/3}$ as $L \to \infty$, although such dependence isn't expected to be evident from the plots, which include only a limited variation in L. Note the peak is below $t = 0$. One may also demonstrate that the specific heat peak itself diverges with the same power of L. One knows by direct computation for the infinite system that the specific heat (heat capacity per unit volume) diverges as $C \sim |\ln|t/\kappa^2||^{1/3}$. Hence the mechanism by which the finite-system results go over to the bulk is that the specific heat values at the peak and at the bulk critical temperature diverge together. The peak position t_{max} approaches $t = 0$ rapidly as the system size increases; numerical

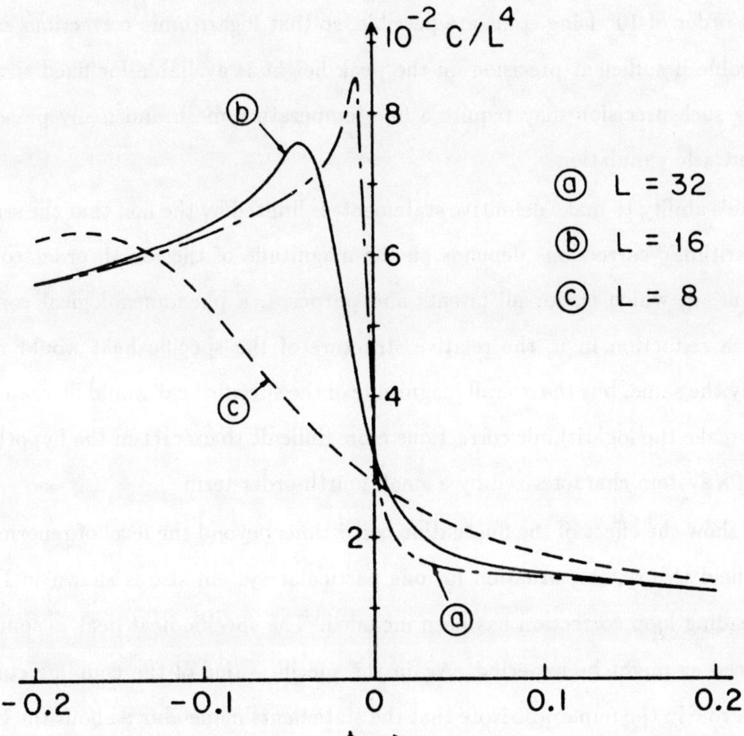

Figure 2: Specific heat at the level of renormalized mean-field theory at $d = 4$ for various sizes of the system, L. The parameter values $\kappa = 1$ and $uS_4 = 1$ have been chosen.

and analytic results suggest that $|t_{max}| \sim [L^2 \ln^{1/6}(\kappa L)]^{-1}$.

Some insight into the visibility of logarithmic corrections can be gained by considering $C(t_{max})/L^4$, the peak height, *versus* system-size, since this is precisely what a simulator might do to search for logarithmic behavior in $d = 4$. Simulations using on the order of 10^6 Ising spins are possible, so that logarithmic corrections may be observable if sufficient precision on the peak height is avaliable for fixed size. Obtaining such precision may require a fine temperature mesh and many passes in a Monte Carlo simulation.

One's ability to make definitive statements is limited by the fact that the strength of logarithmic corrections depends on the magnitude of the fourth order coupling constant, u, which is, for all intents and purposes, a phenomenological constant. Under a reduction in u, the relative structure of the specific heat would remain roughly the same, but the overall magnitude of the specific heat would increase. This would make the logarithmic corrections more difficult to extract in the hypothetical case of a system characterized by a small fourth-order term.

To show the effect of the fluctuation corrections beyond the level of renormalized mean-field theory, the situation for one particular system size is shown in Fig. 3. The leading loop correction has been included. The specific heat peak is somewhat enhanced as might be expected. Again the specific value of the coupling constant plays a role in the numerics. Note that the statements made above about the growth of the specific heat with system size continue to hold with the inclusion of the loop corrections.

VII. MODIFICATIONS FOR $d = 4 - \epsilon$

In Section VI the analysis was begun generally but quickly made specific to $d = 4$, since an asymptotically exact treatment is possible within renormalization group. To carry over the analysis to $d = 4 - \epsilon$ dimensions and compute systematically order-by-order, the steps are much the same as for bulk phenomena. One must take care to

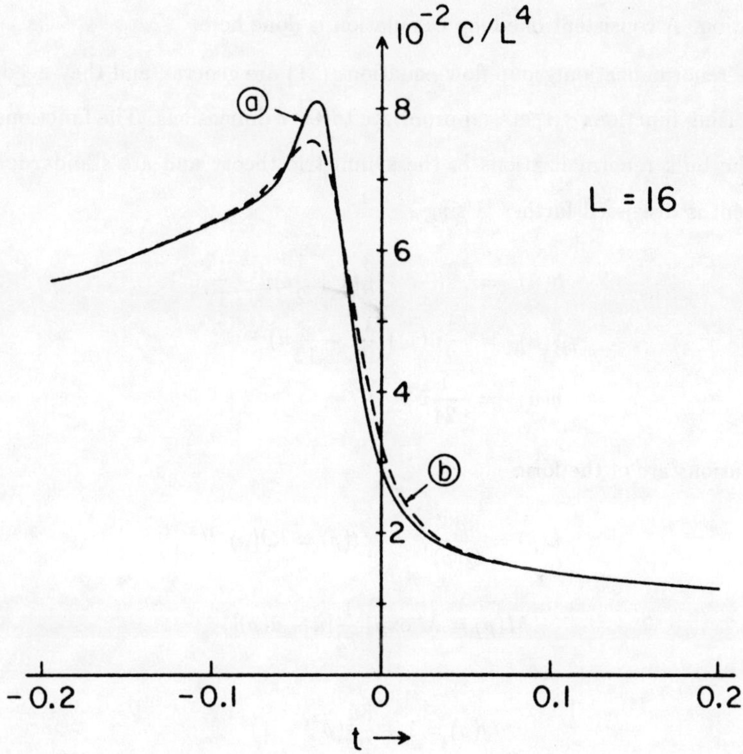

Figure 3: Specific heat for $L = 16$ in $d = 4$ showing (a) the result including the loop correction and (b) renormalized mean-field theory. The characteristic (inverse) length κ and coupling constant uS_4 have been set to unity.

do a consistent calculation. It has been remarked that the finite-size contributions begin at $O(u^0)$. Hence, bulk contributions must be included to a corresponding order; in particular, for leading corrections one must include two-loop contributions to the trajectory integral. This is quite in analogy with organizing a consistent bulk calculation. A consistent one-loop calculation is done here.

The renormalization group flow equations (21) are general, and they need to be solved using functions γ_{ϕ^2}, etc. appropriate to $4-\epsilon$ dimensions. The functions arise from the bulk renormalizations in the symmetric theory and are standard; below this point is discussed further. Using

$$\begin{aligned}\beta(\tilde{u}) &= \tilde{u}(-\epsilon+\frac{3}{2}(1+\frac{1}{2}\epsilon)\tilde{u}-\frac{17}{12}\tilde{u}^2)\\ \gamma_{\phi^2}(\tilde{u}) &= \frac{1}{2}\tilde{u}(1+\frac{1}{2}\epsilon-\frac{5}{12}\tilde{u})\\ \eta(\tilde{u}) &= \frac{1}{24}\tilde{u}^2\end{aligned}$$

the solutions are of the form

$$\tilde{u}(\rho)=\frac{\tilde{u}\rho^{-\epsilon}}{Q(\rho)},\qquad \rho^2 t(\rho)=tQ(\rho)^{-1/3}$$

$$M(\rho)=M\exp\{\frac{1}{72}[\tilde{u}-\tilde{u}(\rho)]\} \tag{34}$$

with

$$Q(\rho)=1+\frac{3\tilde{u}}{2\epsilon}(\rho^{-\epsilon}-1)$$

which are consistent to one-loop order. The geometrical factor S_d has been absorbed into $S_d u=\tilde{u}$. These solutions go over to the ones appropriate to four dimensions in the limit $\epsilon\to 0$. The contribution at the level of renormalized mean-field theory, including the trajectory integral contribution, is

$$\Gamma_{Rmf}=\frac{1}{2}\rho^{*2}t(\rho^*)m^2(\rho^*)+\frac{1}{4!}(\kappa\rho)^\epsilon u(\rho^*)m^4(\rho^*)-\frac{t^2\kappa^{-\epsilon}}{32\pi^2}\int_{\rho^*}^1\frac{dx}{x^{1+\epsilon}}[Q(x)]^{-2/3} \tag{35}$$

which once again reproduces the result for $d=4$ in the appropriate limit. There is a simplification to be noticed. The one-loop correction may be calculated consistently in $d = 4$, which means that the one-loop part, $\Gamma_{R,1}$, also goes over to the four-dimensional result. Quite by analogy with bulk calculations, it is possible to organize the expansion so that order-by-order one may recover the results for four dimensions in the limit $\epsilon \to 0$.

This procedure is conceptually nice (also for bulk calculations), since it recovers the proper limit and also includes corrections to scaling as $u(\rho)$ flows to the appropriate fixed point. The scaling is contained in Eq. (34). However, detailed expressions are more complicated, and considerable simplifications can be achieved if one agrees to fix $\epsilon > 0$ and to give up the notion of finding the limiting behavior in four dimensions. For then one may set $u = u^*$, the fixed point value, at the outset. One also has to this order $M(\rho) \approx M = const.$ As noted in Ref.[7] the one-loop correction to the finite-size free energy looks similar in structure to the four-dimensional result, a new factor of $(\rho\kappa L)^\epsilon$ appearing.

The essential point is that the final free energy has the form (within an additive constant and, for the present purposes, neglecting the trajectory term)

$$\ln Z = \ln \int dx\, e^{A(x)}.$$

Here $A(x)$ depends only on the combination $L^2(\rho^*)^2 t(\rho^*) + \frac{1}{2}u^* x^2$, so that the final free energy depends on the combination $L^2(\rho^*)^2 t(\rho^*)$. The scaling form appears most easily if one chooses ρ^* according to $\kappa\rho^* L = 1$. From the flow equation one has $t(\rho) = t\rho^{-1/\nu}$, since at the fixed point one identifies the critical exponent as $2 - \nu^{-1} = \gamma_{\phi^2}$.[13,11] Then $L^2(\rho^*)^2 t(\rho^*) = (t/\kappa^2)(\kappa L)^{1/\nu}$. The reduced free energy takes the form

$$f = -L^{-d} \ln Z = L^{-d} W[(t/\kappa^2)(\kappa L)^{1/\nu}] = \kappa^d (t/\kappa^2)^{d\nu} \tilde{W}[(t/\kappa^2)(\kappa L)^{1/\nu})].$$

Hyperscaling holds within the renrormalization group structure, and one identifies $d\nu = 2 - \alpha$ in the conventional notation. Two derivitives with respect to the temperature yields the leading part of the specific heat which then has the scaling form

$$\frac{\kappa^\epsilon C}{L^d} = (\kappa^{-2} t)^{-\alpha} X[\kappa^{-2} t(\kappa L)^{1/\nu}]. \tag{36}$$

As written, the function $X(y)$ is not itself universal. The limit $X(y \to \infty)$ yields the nonuniversal specific heat amplitude for the bulk system above T_c. Given an arbitrary overall amplitude, the function $X(y)$ is expected to be universal. Alternatively one may simply define $\tilde{X}(y) = X(y)/X(\infty)$, which is expected to be a universal function. The question of universality and the implications of hyperuniversality will be discussed in more detail below. For now just note that different systems in the same universality class would be represented by different starting values of the coupling constant u. On the critical manifold these different starting values flow to the fixed point value u^* which has been used in the calculations to construct the formal and explicit scaling forms.

Note that the form of the scaling became most quickly apparent with the above choice of ρ^*. More generally ρ^* is found from an equation analogous to Eq. (28), with $u(\rho^*)$ replaced by $u^*(\kappa\rho^*)^\epsilon$. Indeed there is a scaling solution of the form $\rho^* = (\kappa L)^{-1}R(tL^{1/\nu})$, which recovers the intuitive limits for the critical and strongly finite-size regimes as discussed above.

An approximate form for the scaling function for the specific heat in three dimensions (obtained simply by setting $\epsilon = 1$) is shown in Fig. 4. The more explicit forms appearing in the Appendix of Ref.[7] have been used. The argument y is constructed using the one-loop result $\nu = \frac{1}{2} + \frac{1}{12}\epsilon \approx 0.58$ at $\epsilon = 1$. An alternative scheme could employ the known values of ν and α for three dimensions. This might indeed produce better agreement with experiment or simulations, but from the point of view of a systematic development, there is no fundamental basis for using this approach.

The interesting structure which appears below T_c is a remnant of (renormalized) mean field theory. The cusp-like behavior is essential since the scaling function $X(y)$ must have a singularity at zero argument $X(y) \sim y^\alpha$ for the specific heat to remain finite for $L < \infty$. For comparison of $X(y)$ with simulations or experiments one could

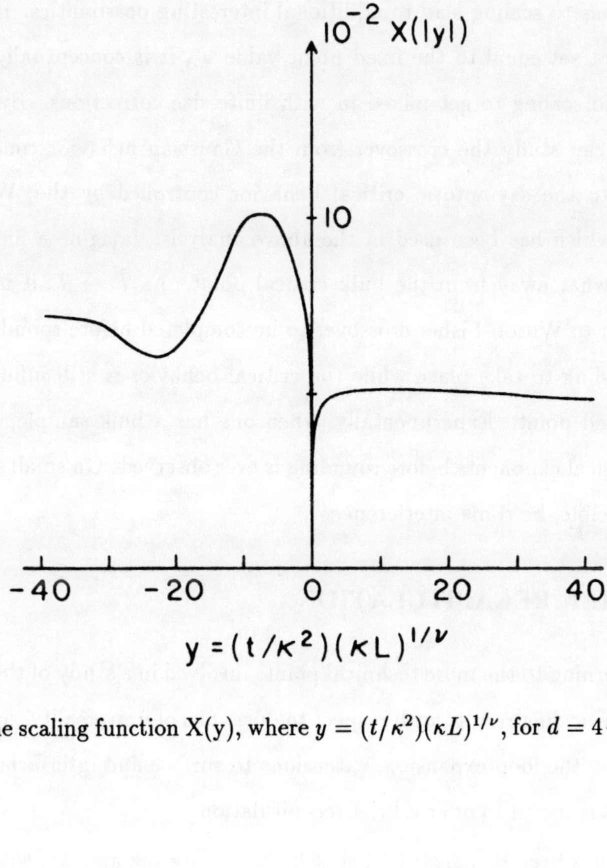

Figure 4: The scaling function X(y), where $y = (t/\kappa^2)(\kappa L)^{1/\nu}$, for $d = 4-\epsilon$ evaluated at $\epsilon = 1$.

plot

$$\frac{C}{L^d(\Delta T/T_c)^{-\alpha}} \quad vs. \quad L^{1/\nu}(\frac{\Delta T}{T_c})$$

with the understanding that one should allow for a rescaling of the magnitude of X (noted above) and also the magnitude of y associated with a nonuniversal metrical factor referring to the combination $tL^{1/\nu}$.

Corrections to scaling lead to additional interesting possibilities. If the starting value u is not set equal to the fixed point value u^*, it is conceptually possible for corrections to scaling to get mixed in with finite-size corrections. By allowing for $u \neq u^*$ one can study the crossover from the Gaussian behavior controlled by its fixed point to the asymptotic critical behavior controlled by the 'Wilson-Fisher' fixed point which has been used in the above analysis. Imagine a finite system of size L somewhat away from the bulk critical point. As $T \rightarrow T_c$ it is possible for the Gaussian to Wilson-Fisher crossover to be completed before rounding sets in or for the rounding to take place while the critical behavior is still influenced by the Gaussian fixed point. Experimentally, when one has a bulk sample, one typically sees nonclassical exponents before rounding is ever observed. On small systems there may, in principle, be some interference.

VIII. BRIEF RECAPITULATION

Before turning to the more technical points involved in a study of the structure of the finite-size scaling results with respect to questions of universality, size-dependent divergences in the loop expansion, extensions to surface and interfacial phenomena and so on, it is useful to offer a brief recapitulation.

So far it has been demonstrated that finite-size scaling appears within the basic formulation of the field theoretic renormalization group. The essential point (to be further discussed) is that the usual bulk counter terms are sufficient to renormalize the theory. The system size, L, has been shown to enter as a parameter, not as a scaling field. A practical method for producing the finite-size scaling functions, and results of explicit calculations of thermodynamic functions in the scaling regime $L >> a, ta^2 << 1$ with a the appropriate microscopic length, have been presented. In the development of the theory, various moments $< M^{2j} >$, evaluated with respect to a probability distribution determined by the Hamiltonian at the level of renormalized mean-field theory, play a crucial role. These lead to the characteristic appearance of factors $[u(\rho^*)]^{-1/2}$ as suggested in Eq. (10).

Since the results in $d = 4$ are ostensibly exact to leading logarithms, they provide a useful benchmark for comparisons with simulations. In principle by adjusting two nonuniversal constants, one corresponding to a dilation or contraction of the $T - T_c$ scale and the other, for example, corresponding to the overall amplitude of the specific heat, the theoretical results using these methods should be comparable in the scaling regime with results of simulations. In $d = 4$ one also has to fix the nonuniversal fourth-order coupling constant u; if this is sufficiently small, low order calculations (such as the one-loop calculations described by Rudnick *et al.*[7]) should give a reasonable representation of the region near criticality.

A major focus of the calculation specifically in $d = 4$ is the question of logarithmic corrections in the rounded regime near the bulk critical point. Information concerning the logarithmic corrections is contained in the increase of the height of the specific heat peak in Fig. 2. It has been estimated[7] that for $uS_4 \approx 1$ and sizes ranging from $L/a = 8 - 64$, the logarithmic corrections would be visible for specific heat peaks determined with an accuracy of about 5 %.

Below $d = 4$ one must work order by order in $\epsilon = 4 - d$. Explicit one-loop calculations of the scaling function for the specific heat have been produced.[7] Notice that now $u \rightarrow u^*$ so that only two metrical factors need adjustment. It would be interesting to see if simulations indicate structure in the scaling function below the bulk critical temperature, as indicated in Fig. 4. If so, it is a further suggestion that three dimensions is 'close to four.'

Some further properties of the scaling forms and the implications of hyperscaling (and two-scale-factor universality) are now discussed. It is shown from the general features of the loop expansion that the free energy of a finite system has a form suggested by Privman and Fisher,[14] and that the discussion can be extended to some 'beyond-the-bulk' properties.

IX. TWO SCALE FACTOR UNIVERSALITY: PHENOMENOLOGY

In this section the situation with respect to two-scale-factor or hyperuniversality is discussed. The reader should consult other articles in this volume for additional detail. In the next section the connection with the renormalization group treatment discussed above will be addressed.

The situation is restricted now to dimensions below the upper critical dimensionality, $d_c = 4$. Scaling asserts that there is one characteristic diverging length as bulk criticality is approached, say, the disordered-phase correlation length, $\xi = C_0|t|^{-\nu}$, and that all diverging lengths are proportional to it. Two diverging lengths will be related to one another by a universal ratio. Then the singular part of the free energy has the form noted in the Introduction, namely

$$(k_B T_c)^{-1} f_s(t, L) \approx A_1 |t|^{2-\alpha} W^{\pm}(L/\xi_\infty) \tag{37}$$

where the $\pm$ corresponds to temperatures above and below the critical temperature. Now, the amplitudes A_1 and C_0 are nonuniversal and will depend on details of the system described, but such details don't enter Eq. (37) in any additional way. Various phenomenological arguments[14,10] and, indeed, the renormalization group approach sketched above, so indicate. This will be discussed in more detail below.

When two-scale-factor or hyperuniversality[15–17] holds for the bulk system, one has an additional relation that $(k_B T)^{-1} f_s \xi_\infty^d$ is asymptotically universal, which forces a relation between the nonuniversal amplitudes for the free energy (or specific heat) and the correlation length. This has been shown to hold within the structure of the field-theoretic renormalization group for bulk homogeneous systems.[18] If one now demands that the scaling form of the free energy in Eq. (37) reproduce this relation among amplitudes in the thermodynamic limit $L \to \infty$, it imposes the condition that $A_1 C_0^d = U$, where U is universal. This in turn allows the free energy to be rewritten as

$$(k_B T_c)^{-1} f_s(t, L) \approx L^{-d} Y^{\pm}(C_1 t L^{1/\nu}) \tag{38}$$

which form was first suggested by Privman and Fisher.[14] The function $Y(y)$ is then universal implying that the *total* singular part of the free energy of the system (in units of $k_B T_c$) is universal at bulk criticality. This prediction can be checked via simulations and analytic methods.

Evidence for the form (38) has accumulated. For example, Singh and Pathria[4,19] have verified the form of the free energy within the framework of the spherical model and relativistic Bose gas. Mon[20] observed the universality of the total free energy in Ising simulations in three dimensions. Furthermore, Mon and Jasnow[21] found, again with three-dimensional Ising Monte Carlo simulations, that the total *interfacial* free energy for two coexisting phases at bulk criticality is universal. This would be the expectation from an interfacial version of Eq. (38), namely

$$(k_B T_c)^{-1} \sigma(t, L) \approx L^{-(d-1)} \Sigma(C_1 t L^{1/\nu}) \tag{39}$$

where one would also propose that the scaling function $\Sigma(x)$ is universal. Such form follows from phenomenological arguments based on scaling[22] and the structure of the renormalization group (see below). Finally, it is interesting to note that, in an additional extension of the Privman-Fisher hypothesis, Mon and Nightingale[23] concluded from Monte Carlo estimates, that the total incremental free energy of an Ising *free surface* is universal at bulk criticality.

The explicit renormalization group calculations of Rudnick et al.[7] as summarized above indeed yield a free energy consistent with the two-scale factor form given in Eq. (38) above. Here a general argument is sketched suggesting how a calculation to all orders would go. Some additional comments on interfacial properties are made.

X. HYPERUNIVERSALITY WITHIN RG

The general structure of the free energy has been displayed in Eqs. (22) and (23) above. Note the appearance of characteristic combinations in the loop corrections, as can be verified term by term. A test of hyperuniversality within the context of finite-size requires a similar approach as in bulk systems. One should include the possibility of the coupling constant u differing from the fixed point value in order to demonstrate 'system' independence. Nonuniversal amplitudes may then depend explicitly on u. One should also demonstrate that appropriate results are independent of renormalization scheme. This cannot be met fully, but the situation is the same as in calculations for a bulk, homogeneous system. This point will be considered in further detail.

The most explicit way to proceed is to linearize about the fixed point. A somewhat better formal treatment avoids linearization and involves the introduction of the metrical factors X and Y as in the references.[18,13] It is emphasized that there is nothing new here relating to the possible finiteness of the system.

In linearizing the flow equations one introduces the scaling fields, $g_t(\rho)$ and $g_u(\rho)$, which arise in the evaluation of ordinary bulk properties,

$$g_u(\rho) = g_u\rho^{\omega} = (\Delta u)\rho^{\omega}$$

$$g_t(\rho) = t(\rho)[1 - \frac{\nu' g_u(\rho)}{\nu^2\omega}] = g_t\rho^{-1/\nu} \tag{40}$$

with $\Delta u = u - u^*$. The exponent $\omega = \epsilon + O(\epsilon^2)$ is the correction-to-scaling exponent,[24] and the function $\nu(u)$, defined by $\gamma_{\phi^2}(u) = 2 - (1/\nu(u))$, enters the flow equations for $t(\rho)$ in the usual fashion. As usual, the correlation length exponent is $\nu = \nu(u^*)$, and $\nu' = (d\nu/du)_{u^*}$ has been defined.[11] The initial value at $\rho = 1$ is $g_t = t[1 - (\nu'/\nu^2\omega)\Delta u]$, which would reduce to t if the fixed point value of u were chosen.

Now scaling variables x, y and z may be defined as

$$x = \frac{g_t}{\kappa^2}(\kappa L)^{1/\nu}, \quad y = \frac{g_u}{(\kappa L)^{\omega}}, \quad z = \kappa\rho L \tag{41}$$

which may be substituted back into Eq. (23). The result is that $L^d\Gamma_R(t, M, u, \kappa; L)$ may be written as a function of scaling variables x, y, z and the combination $L^{d-2}M^2(\rho)$. One also finds that the trajectory integral contribution can also be expressed as a function of the scaling variables. Note that formally the trajectory contribution is the same as for the bulk system, but in Eq. (22) $\rho = \rho^*$ will generally depend on the system size. The renormalized vertex Γ_R must be substituted back into Eq. (25) in order to perform the integration over M. A change of integration variables can be made with the conclusion that the singular part of the reduced free energy is a function of the scaling variables x, y and z. All system dependence has been absorbed into the starting values of the scaling fields g_u and g_t. As noted above, it is possible to work more generally without linearizing these fields.

To get the final form of the results the ρ dependence must be removed. The total free energy is independent of this parameter; changing ρ merely trades contributions between the loop contributions and the trajectory integral. For explicit results a consistent choice has to be made. A rather cumbersome choice was made above to allow numerical analysis of the free energy or specific heat to span both the critical and finite-size regimes, but merely to demonstrate the scaling structure, the simpler choice $\rho\kappa L = 1$ may be made. This yields for the singular part of the free energy

$$(k_BT_c)^{-1}f_s(t, L) = L^{-d}[Y(x, y)]_{sing} \tag{42}$$

where the singular part must be extracted. [For the trajectory integral contribution, the singular contribution to the total free energy scales as $L^dt^{2-\alpha} = L^dt^{d\nu}$.] The final result is of the Privman-Fisher form,[14] in which the system dependences remain only in the scaling variables x, y. A specific representation for $Y(x, y)$ suitable for numerical evaluation from the critical to finite-size regimes has been discussed above, but is available only to one-loop order.[7]

To imagine comparing results of a renormalization group calculation, for example, the allegedly universal value of the total reduced free energy at bulk criticality, with those for an experimental system or a simulation, one should demonstrate that the results are independent of renormalization scheme. One knows that if the renormalization group scheme were changed, the functions $\gamma_\phi, \beta(u)$, and so on, would all change in general. The fixed point values would change, and so would the nonuniversal critical amplitudes, in general. So one is faced with essentially the same question as for ordinary bulk calculations, and one can demonstrate, by means of explicit low order calculations, independence of renormalization scheme (e.g., different regularization).

One can show formally that the final scaling functions, e.g. $Y(x,y)$, are independent of scheme in a limited sense. Within the general method of renormalizing in the symmetric theory at criticality, one can change the normalization conditions, for example, the definition of the renormalized coupling constant, u. The system dependence of the scaling variables x and y changes. Or, one may imagine computing counter terms at two different fixed temperatures above T_c. These types of demonstrations of independence of renormalization scheme can be done without explicit loop calculations, and the arguments are the same as for a bulk system. Explicit demonstrations of independence between hard cutoff and minimal subtraction methods have been carried out, but only to one-loop.[25]

XI. ADDITIONAL IMPLICATIONS OF HYPERUNIVERSALITY

It was noted above that hyperuniversality has additional implications and that the Fisher-Privman form can be extended to surface and interfacial phenomena. Surface and interfacial properties may be computed within the renormalization group framework as well. The reader is referred to the reviews by Diehl[26] and Jasnow[22,27] and references cited therein. The case of interfaces is considered here. Instead of considering 'box' geometry as above, one may imagine computing the free energy *difference* for a system in 'slab' geometry, having infinite extent in $(d-1)-$ dimensions and 'thickness' L, when it subjected to periodic (p) *vs.* antiperiodic (a) boundary conditions. For an Ising-like system confined to such a domain Ω, the interfacial free energy or ***surface tension*** may be defined by (see the reviews[22,27] for further discussion and references)

$$\sigma = \lim_{L\to\infty} L[f^a(t,L) - f^p(t,L)]. \tag{43}$$

In principle the individual free energies, and hence the difference, may be computed within the framework of the field theoretic renormalization group. Each of the free energies *densities* should have the form

$$(k_B T_c)^{-1} f^{p,a} = L^{-d} Y^{p,a}(C_1|t|L^{1/\nu})$$

where p, a correspond to the periodic and antiperiodic cases respectively. This form must reproduce the same bulk free energy density $\sim |t|^{d\nu}$ in the limit $L \to \infty$ for either type of boundary conditions. Now, the difference of free energy densities is related to the surface tension according to the definition Eq. (43) above. The thermodynamic surface tension is obtained in the limit of infinite L. The difference in free energy densities scales with the usual argument so that putting the pieces together one has

$$(k_B T_c)^{-1}\sigma = L^{-(d-1)} \Delta Y(C_1|t|L^{1/\nu})$$

where ΔY is the difference of the scaling functions. This is an extension of the Privman-Fisher form to interfacial phenomena and has been confirmed[21] via simulations. Singh and Pathria[28] have shown that the spherical model under antiperiodic boundary conditions in 'box geometry' has the hyperscaling form.

Some of the ground work for explicitly treating the slab geometry as outlined above has been worked out,[29] but a full calculation has not been reported. This is actually an interesting conceptual point since a renormalization group calculation of the surface tension according to the above prescription, *i.e.*, considering a finite system and taking the thermodynamic limit, places such calculations in closer analogy with rigorous statistical mechanical methods.

XII. CONCLUDING REMARKS

The use of finite-size scaling methods has become standard in a variety of problems in which the behavior of a large system must be inferred from the known behavior of a small system. An extrapolation scheme of this sort is essential in critical phenomena, for example, in which case it is known that true thermodynamic singularities cannot appear except in the thermodynamic limit. Using methods of finite-size scaling a simulator can hope to extract the behavior of the infinite system from 'rounded' data. The conceptual base of finite-size scaling has taken on additional significance through the application of the ideas to nonequilibrium growth problems[30–32] and to the study of finite-temperature field theories and conformal invariance (see, e.g., the references included in[33]) . Some related topics are discussed elsewhere in this volume.

The phenomenological form of finite-size scaling discussed in the Introduction is contained within the structure of the renormalization group equations. The essential point from a formal perspective is that the renormalizations required for a finite system of size L are the same as for an infinite system. From a more intuitive viewpoint, renormalization transformations must effectively be stopped when the size of the Kadanoff blocks is on the order of the system size, or equivalently, as degrees of freedom are successively integrated out, the process must end at a lower cutoff $k \sim L^{-1}$.[6]

In these pages a method has been reviewed for explicitly calculating thermodynamic functions in the scaling regime, $L >> a$, $ta^2 << 1$, where a represents the microscopic length. Attention was first paid to the situation in four dimensions, since a calculation correct to leading logarithms is possible. Such a calculation can provide a useful benchmerk for simulations. To compare with experiment or simulations two 'metrical' factors must be adjusted; these correspond to a rescaling of the $T - T_c$ axis, and to the nonuniversal amplitude of, for example, the specific heat. In addition the nonuniversal coupling constant, u, must be set for the system at hand.

Then the scaling forms (the RG result written more explicitly in the Appendix of the article by Rudnick *et al.*[7]) are expected to agree. There is a proviso implicit in the whole calculation, of course. Only if the value of u describing the physical system is not too large, can the one loop computation be expected to provide adequate numerical agreement.

The calculation in $d = 4$ also can address the question of visibility of logarithmic corrections. As discussed above, the specific forms derived allow a semiquantitative evaluation of the relative size of such corrections. This information is contained in the increase of the peak height of the specific heat as shown in Fig. 2. Recall the predicted slow increase of the specific heat peak proportional to $\ln^{1/3}(L/a)$. Here one is facing a window. The logarithmic corrections are due to fluctuations. To be visible the value of u describing the system must not be too small; on the other hand it cannot be too large or results based on low-order loop contributions will be unreliable.

Below four dimensions the methods presented here require one to work systematically in orders of $\epsilon = 4 - d$. It may be possible to develop summation techniques which allow working in fixed dimension, say $d = 3$, in analogy with what has been done for bulk properties.[34] The justifications require analysis of the structure of the perturbation expansion. A one-loop result within the $\epsilon-$ expansion has been produced yielding the universal scaling function for the specific heat. For $d < 4$ one is required to fix only two metrical factors, since $u \rightarrow u^*$ and requires no adjustment. A comparison of the $\epsilon-$ expansion result (at $\epsilon = 1$) with Ising simulations in three dimensions would be interesting. It would also be interesting to compare both to a one-loop calculation, as suggested above, in which the loop integrals and renormalizations are carried out in fixed $d = 3$. Such a calculation could be thought of as the lowest order result in a systematic expansion, but complete justification of this approach remains to be worked out.

It was also shown that the Privman-Fisher[14] form for the singular part of the free energy of a finite system is contained within the structure of the field-theoretic renormalization group. The essential point is that the finite-size free energy must go over to the bulk free energy in the thermodynamic limit. The bulk free energy, however, satisfies hyperscaling, which eliminates one critical amplitude as discussed in Section IX. The reflection of hyperscaling within the context of finite-size effects is the Privman-Fisher form.

Additional insights have been obtained as well. The size of the system, L, while entering the scaled combinations does not mix with the formation of the scaling fields g_u and g_t. From the outset (see Section V), in the construction of the renormalization group equations for the free energy (or other vertices), it is observed that L plays the role of a parameter[5] (much the same as the wave number in equations for the vertices) and is not renormalized. As emphasized above, no new counter terms are required for the finite-size calculations.

The implications of renormalization group and hyperscaling on 'beyond-the-bulk' properties like the surface tension were also discussed. Surface and interfacial properties can be calculated within the renormalization group framework. In the later case one can imagine calculating the free energy under periodic and antiperiodic boundary conditions for the 'slab' geometry discussed in Section XI. This line of argument leads to an extension of the Privman-Fisher form to the interfacial free energy, and this is consistent with the observed universality of the *total* interfacial free energy at the bulk T_c as observed in Monte Carlo simulations.[21]

ACKNOWLEDGMENTS

The author would like to thank colleagues Joseph Rudnick and Hong Guo, with whom most of the work reviewed here was performed. The author also thanks Hong Guo for his assistance with the preparation of this manuscript. The author is grateful to the National Science Foundation through the Division of Materials Research (Grant No. DMR86-13030) for support of this work.

APPENDIX A

In this appendix the calculation of the inhomogeneous term in the renormalization group equation is sketched. Essentially this involves a computation of the bulk free energy to two-loop order. Details appear in Guo's work.[10] Recall that

$$G(t,u,\kappa) = -\kappa \frac{d}{d\kappa}(a_1^\infty + a_2^\infty r + a_3^\infty r^2)$$

where bare parameters are held fixed. Furthermore, bare quantities are renormalized by the substitutions discussed after Eq. (19), at which point the renormalized quantities were introduced. The bare free energy per unit volume for the bulk homogeneous system is given by

$$\Gamma_B^\infty = \frac{1}{2} r M^2 + \frac{1}{4!} u M^4 + \frac{1}{2}$$ 

$$+\frac{1}{8}u$$  $$-\frac{1}{12}(uM)^2$$ 

where the diagrams are given by

 $$= \int dq \ln(q^2 + T_0)$$  $$= (\int dq \frac{1}{q^2+T_0})^2$$

 $$= \int dq dk \frac{1}{(q^2+T_0)(k^2+T_0)((\mathbf{q}+\mathbf{k})^2+T_0)}$$

and $T_0 = r + \frac{1}{2}uM^2$. For brevity $dq \equiv d^d q/(2\pi)^d$, etc. The renormalized coupling constants defined after Eq. (19) are to be substituted in the inhomogeneous term. The first and second additive constants, a_1^∞ and a_2^∞ vanish in a dimensionally regularized scheme. (The 'mass shift' δt vanishes as do other terms for $r = M = 0$.) One must then evaluate the third additive constant

$$-\frac{1}{2} r^2 \left. \frac{\partial^2 \Gamma_B^\infty}{\partial r^2} \right|_{r=Z_{\phi^2}\kappa^2, M=0}$$

which can be expressed schematically as

$-\frac{1}{2}r^2\Big(-\frac{1}{2}$  $+\frac{1}{4}u\ \Big[$  $\Big]^2$

$+\frac{1}{2}u$  $\Big)_{r=Z_{\phi^2}\kappa^2, M=0}$

The bare coupling constants are replaced by their renormalized counterparts and the renormalization constants (the Z' s) are substituted in to a consistent order in u. In the last two terms one can set $Z_{\phi^2} = Z_u = 1$, whereas in the first term one uses Z_{ϕ^2} to $O(u)$. Space does not permit further detail, for which the reader can consult Guo's work.[10] The trajectory integral follows directly by application of $\kappa d/d\kappa$ to the above expression.

APPENDIX B

In this appendix a few additional details on the nature of the finite-size corrections in the loop expansion are provided. The reader is referred to the references[8,9] for additional remarks.

It was stated above that it is reasonable to expect that bulk renormalizations are sufficient to regularize the theory for a finite system. This is due to the general understanding that the renormalizations are designed to handle *ultraviolet* divergences, while finite-size effects normally become apparent at distances long compared to any microscopic lengths. This statement is correct, and an example has already appeared in the evaluation of the one-loop fluctuation in Section IV. Observe in Eq. (18) that the only ultraviolet problem is in the bulk term, which the ordinary bulk counter-terms can deal with. However, the one-loop term is a poor example since it gives the (false) impression that ultraviolet-divergent finite-size terms never appear. This impression is corrected by going to two-loop order or beyond. The two-loop calculation is sketched here.

To that order the bare 'Boltzmann' factor $\Gamma(M)$ with a factor of L^d removed is

$$\Gamma(M) = \frac{1}{2}rM^2 + \frac{1}{4!}uM^4 + \frac{1}{2}\left(\frac{2\pi}{L}\right)^d \sum_{\mathbf{q}}{}' \ln(q^2 + T_0)$$

$$+\frac{1}{8}u\left[\left(\frac{2\pi}{L}\right)^d \sum_{\mathbf{q}}{}' \frac{1}{q^2+T_0}\right]^2 - \frac{1}{12}(uM)^2\left(\frac{2\pi}{L}\right)^{2d} \sum_{\mathbf{q},\mathbf{k}}{}' \frac{1}{(q^2+T_0)(k^2+T_0)[(\mathbf{q}+\mathbf{k})^2+T_0]} \tag{44}$$

where, as usual, the wave vectors are restricted to the set appropriate to periodic boundary conditions [see Eq. (14)], the primes are a reminder that there is no $\mathbf{q} = \mathbf{0}$ mode, and for convenience $T_0 = r + uM^2/2$ is used. One substitutes in the renormalized coupling constants using results from a purely bulk calculation employing, for example, minimal subtractions.[35] In the perturbation expansion the renormalization constants are expanded to consistent order so that Γ is calculated to $O(u)$. (Recall that formally one has $uM^2 \sim 1$.)

Now each sum in Eq. (44) is transformed using the Poisson formula Eq. (15) into the corresponding term for an infinite system along with size-dependent correction terms. In Section IV this procedure was demonstrated for the one-loop fluctuation term. The most interesting part of Eq. (44) is is the last term, which contains the kinds of complications one is likely to encounter in higher orders. Application of the Poisson summation technique yields[8,9]

$$\left(\frac{2\pi}{L}\right)^{2d} \sum_{\mathbf{q},\mathbf{k}}{}' \frac{1}{(q^2+T_0)(k^2+T_0)[(\mathbf{q}+\mathbf{k})^2+T_0]} = \int \frac{d^dq\, d^dk}{(q^2+T_0)(k^2+T_0)[(\mathbf{q}+\mathbf{k})^2+T_0]}$$

$$+3\sum_{\mathbf{n}}{}' \int d^dq\, d^dk \frac{e^{i\mathbf{q}\cdot\mathbf{n}L}}{(q^2+T_0)(k^2+T_0)[(\mathbf{q}+\mathbf{k})^2+T_0]}$$

$$-\frac{3}{T_0}\left(\frac{2\pi}{L}\right)^d \int \frac{d^dq}{(q^2+T_0)^2} - \frac{3}{T_0}\left(\frac{2\pi}{L}\right)^d \sum_{\mathbf{n}}{}' \int d^dq \frac{e^{i\mathbf{q}\cdot\mathbf{n}L}}{(q^2+T_0)^2}$$

$$+\frac{2}{T_0^3}\left(\frac{2\pi}{L}\right)^{2d} + \sum_{\mathbf{m}\neq\mathbf{0},\mathbf{n}\neq\mathbf{0},\mathbf{m}\neq\mathbf{n}} \int d^dq\, d^dk \frac{\exp(i\mathbf{q}\cdot\mathbf{m}L + \mathbf{k}\cdot\mathbf{n}L)}{(q^2+T_0)(k^2+T_0)[(\mathbf{q}+\mathbf{k})^2+T_0]} \tag{45}$$

The first term on the right is the usual bulk contribution. The second and third terms depend on L and diverge logarithmically in $d = 4$. These kinds of uv-divergent finite-size terms are absent at the one-loop level.

After renormalizing Eq. (44) one can show that all the $\epsilon-$ poles cancel exactly, including the $L-$ dependent divergences noted above. The computation is straightforward, but tedious, and will not be reproduced here. Suffice it to say that the $L-$ dependent divergences introduced by the third and fourth terms of Eq. (44) are cancelled by those of the second and third terms on the right of Eq. (45).

After multiplication of Eq. (44) by L^d, one sees that the independent variables on the right hand side can be taken as $\kappa L, L^2 t, L^{d-2}M^2$, so that these will be the natural variables. The additive renormalizations have from the outset been taken as the same as for the bulk calculation, and after multiplication by L^d the contribution of the additive renormalizations to the singular part of the total free energy will involve only the combination $L^2 t$, leaving the choice of variables unmodified.

The renormalization group equations are solved by introducing the running coupling constants as discussed above. Combinations like $L^2 t$ become $L^2\rho^2 t(\rho)$ which generate the final scaling combinations. Hence, as described in the text, the singular part of the free energy may be expressed in terms of the scaling variables x, y, z introduced in Eq. (41).

References

[1]K. Binder, ed., Monte Carlo Methods in Statistical Physics (Springer, N. Y., 1979), and references therein.

[2]K. K. Mon and D. Jasnow, Phys. Rev. A **30**, 670 (1984).

[3]M. E. Fisher and M. N. Barber, Phys. Rev. Letters **28**, 1516 (1972).

[4]S. Singh and R. K. Pathria, Phys. Rev. A **31**, 1816 (1985).

[5]E. Brézin, J. Phys. (Paris) **43**, 15 (1982).

[6]M. Suzuki, Prog. Theor. Phys. (Japan) **58**, 1142 (1977).

[7]J. Rudnick, H. Guo, and D. Jasnow, J. Stat. Phys. **41**, 353 (1985).

[8]H. Guo and D. Jasnow, Phys. Rev. B **35**, 1846 (1987).

[9]H. Guo and D. Jasnow, Phys. Rev. B **39**, 753 (1989).

[10]H. Guo, *Scaling in Condensed Matter Physics*, PhD thesis, University of Pittsburgh, August, 1987.

[11]E. Brézin, J. C. LeGuillou, and J. Zinn-Justin, in *Phase Transitions and Critical Phenomena, Vol.6.*, edited by C. Domb and M. Green (Academic, N.Y., 1976).

[12]L. Schwartz, *Mathematics for the Physical Sciences* (Addison-Wesley, Reading, Mass., 1966).

[13]D. J. Amit, *Field Theory, the Renormalization Group and Critical Phenomena* (McGraw-Hill, New York, 1978).

[14]V. Privman and M. E. Fisher, Phys. Rev. B **30**, 322 (1984).

[15]D. Stauffer, M. Ferer, and M. Wortis, Phys. Rev. Letters **29**, 345 (1972).

[16]M. Ferer, M. A. Moore, and M. Wortis, Phys. Rev. B **8**, 5205 (1973).

[17]P. Gerber, J. Math. Phys. A **8**, 5205 (1973).

[18]C. Bervillier, Phys. Rev. B **14**, 4964 (1976).

[19]S. Singh and R. K. Pathria, Phys. Rev. B **31**, 4483 (1985).

[20]K. K. Mon, Phys. Rev. Letters **54**, 2671 (1985).

[21]K. K. Mon and D. Jasnow, J. Stat. Phys. **41**, 273 (1985).

[22]D. Jasnow, Rept. Prog. Phys. **47**, 1059 (1984).

[23]K. K. Mon and M. P. Nightingale, Phys. Rev. B **31**, 6137 (1985).

[24]F. Wegner, Phys. Rev. B **5**, 4529 (1972).

[25]H. Guo, 1987, unpublished.

[26]H. W. Diehl, in *Phase Transitions and Critical Phenomena, Vol.10.*, edited by C. Domb and J. L. Lebowitz (Academic, N.Y., 1986).

[27]D. Jasnow, in *Phase Transitions and Critical Phenomena, Vol.10.*, edited by C. Domb and J. L. Lebowitz (Academic, N.Y., 1986).

[28]S. Singh and R. K. Pathria, Phys. Rev. B **32**, 4618 (1985).

[29]J. Rudnick, H. Guo, and D. Jasnow, 1987, unpublished.

[30]K. K. Mon and D. Jasnow, Phys. Rev. Lett. **59**, 2983 (1987).

[31]J. Vinals and D. Jasnow, Mod. Phys. Lett. B **2**, 527 (1988).

[32]J. Viñals and D. Jasnow, Phys. Rev. B **37**, 9582 (1988).

[33]J. Cardy, ed., *Finite-Size Scaling* (North-Holland, N.Y.,1988).

[34]G. A. B. Jr., B. G. Nickel, M. S. Green, and P. I. Meiron, Phys. Rev. Letters **36**, 1351 (1976).

[35]I. D. Lawrie, J. Phys. A **9**, 961 (1976).

III. FULLY FINITE MEAN SPHERICAL MODELS

Joseph Rudnick

TABLE of CONTENTS

FULLY FINITE MEAN SPHERICAL MODELS

Joseph Rudnick

Department of Physics, UCLA
405 Hilgard Ave.
Los Angeles, California 90024

Introduction

The spherical model of magnetism is almost unique in statistical mechanics in that it combines exact solvability and direct relevance to real, physically well-motivated models [1,2]. While the global spherical constraint introduces effective interactions that span the physical dimensions of the system, and thus render it unsuitable as a model of systems with short range interactions, the spherical model provides a remarkably close parallel to the O(n) models (with $2 \leq n \leq \infty$) which describe classical, fixed moment ferromagnets, among other physical systems. This relevance is understandable, in part, given the correspondence established by H.E. Stanley between the spherical model and the $n=\infty$ limit of the O(n) model [3]. Among the properties of the O(n) models that are emulated by the spherical model are critical behavior in the immediate vicinity of the ferromagnetic phase transition. In fact, it is possible to analyze the spherical model via the Renormalization Group, and to illustrate the effects on the existence and properties of the ferromagnetic phase transition of such important parameters as spatial dimensionality [4].

Thus, while there does not appear to be any real physical system that is properly represented by the spherical model one has excellent reasons to turn to it for important qualitative signposts to the behavior of an important class of systems.

This review contains a description of some of the properties of the fully finite spherical model that have been established in recent calculations. The standard caveat applies to these results: given the long-range nature of the spherical constraint the model does not apply directly to any real physical system. However, as noted above, the relevance of the fully finite spherical model to real fully finite systems can be considerable. Furthermore, within the context of the spherical model, the results to be reported here are <u>exact.</u>

Given the limited scope of this review, there are important topics that have been ignored. For instance, there is no discussion of dimensional crossover in the case of a spherical model that is finite in a subset of its spatial dimensionalities. For an excellent treatment of this subject, see Barber and Fisher [5] and Singh and Pathria [6].

<u>Mean Spherical Model</u>

The actual model to be discussed here is the <u>mean</u> spherical model. A brief description of the model is as follows.

One starts with a set of gaussian-weight spins, σ_i, with ferromagnetic coupling strengths J_{ij}, so that the Boltzmann factor associated with the interaction between two spins, say σ_i and σ_j, is $e^{\beta J_{ij}\sigma_i\sigma_j}$., where $\beta = 1/kT$. We will assume that the gaussian weight is entirely taken into account by the Boltzmann factors. We now introduce a global constraint on the expectation value of the spin amplitudes. This constraint is

$$\sum_{i=1}^{N} <\sigma_i^2> = N \quad , \qquad (1)$$

Where N is the total number of spins in the system. The expectation value <...> is the standard one, i.e.

$$<\sigma_i^2> = \frac{\sum' \sigma_i^2 e^{-\beta H}}{\sum' e^{-\beta H}} \quad , \qquad (2)$$

where $e^{-\beta H}$ is the standard Boltzmann factor. The summation $\sum'$ in Eq. (2) is the sum over spin states, in this case a multiple integral over the σ_i's (i.e. $\sum' e^{-\beta H} \rightarrow \int e^{-\beta H} d\sigma_1 ... d\sigma_N$). The global constraint in Eq.(1) is enforced by introducing a Lagrange multiplier, λ,into the Hamiltonian H:

$$H \to H + \lambda \sum_{i=1}^{N} \sigma_i^2 \ .$$

$$= - \sum_{ij} J_{ij}\sigma_i\sigma_j \ -h \sum_{i=1}^{N} \sigma_i \ + \lambda \sum_{i=1}^{N} \sigma_i^2 \quad , \qquad (3)$$

where h is an effective magnetic field,and the multiplier λ is adjusted to guarantee the satisfaction of Eq.(1).

If the spins are on a d-dimensional hypercubic lattice and the interaction is between nearest neighbors only (i.e. J_{ij} = J if i and j are the indices of nearest neighbor sites, and J_{ij} =0 otherwise) the Hamiltonian can be diagonalized by fourier transforming the σ_i's. Eq. (1), takes the form

$$\beta = \left(\frac{1}{L^d} \sum_{n_1 \ldots n_d = 1}^{L} [\alpha + J \sum_{i=1}^{d} (1 - \cos \frac{2\pi n_i}{L})]^{-1} \right) + \frac{h^2}{\alpha^2} \quad , \qquad (4)$$

where

$$\alpha = \lambda - Jd \quad , \qquad (5)$$

and L is the length of a side of the hypercubic lattice.Eq. (5) defines the function $\alpha(t)$ implicitly. Given this function the thermodynamics and static correlations are readily obtained. For example, one has for the susceptibility, $\chi(t)$,

$$\chi(t, H{=}0) = \frac{1}{\alpha(T)} \quad . \qquad (6)$$

For the correlation length, $\xi(t)$, it is useful to utilize a form that is appropriate to a finite system, and that takes into account the periodic boundary conditions that apply here. The expression that will be utilized is [7]

$$\xi(t,H{=}0,L) = \frac{1}{\sqrt{\alpha(T) + \frac{2\pi^2}{L^2}}} \quad , \qquad (7)$$

For the spatial fourier transform of the correlation function, $G(\vec{k})$, one has

$$G(\vec{k}) = \beta^{-1}\left(\alpha(T)+J\sum_{i=1}^{d}(1-\cos k_i)\right)^{-1} . \tag{8}$$

In the infinite system one recovers the standard predictions for the spherical model. When the dimension d satisfies $2 < d \leq 4$ one has the correlation length exponent $\nu = 1/(d-2)$. The "magnetic" exponent, x, is given by $x=(d+2)/2$. All other exponents follow from hyperscaling, i.e. $\alpha = 2-d\nu = -2/(d-2)$, and so on [8].

Finite size scaling theory provides predictions concerning the modification of the critical behavior of key thermodynamic functions when the system is not infinite [9]. Given a quantity, such as the zero field isothermal susceptibility, $\chi(T,h,L)$ one has, when the reduced temperature, $t= (T-T_C)/T_C$, and the magnetic field, h, are small and a characteristic dimension, L, of the finite system is large,

$$\chi(T,h,L) = L^{\gamma/\nu}X(L^{1/\nu}\tilde{t},hL^{x/\nu}) + L^{\gamma/\nu+y}\,Y(L^{1/\nu}\tilde{t},hL^{x/\nu}) \tag{9}$$

where γ is the temperature critical exponent of χ when $L=\infty$ and the <u>shifted</u> reduced temperature $\tilde{t}$is given by $\tilde{t}$

$$\tilde{t}= \frac{T-T_C}{T_C} + \varepsilon(L) \quad . \tag{10}$$

The shift $\varepsilon(L)$ goes to zero when $L\rightarrow\infty$. According to the postulates of finite size scaling, one has $\varepsilon(L) \propto L^{-1/\nu}$. Eq. (9) also displays the effect of the leading order correction to scaling, contained in the second term on the right hand side, $L^{\gamma/\nu+y}\,Y(L^{1/\nu}\tilde{t},hL^{x/\nu})$.

In the thermodynamic limit $(N\rightarrow\infty)$ the sum over the n_i's in Eq. (4) can be replaced by a multiple integral in the standard way. However, in order to study finite size effects it is necessary to keep it in the form of a summation. The technical challenges that must be met in the study of the finite mean spherical model are all associated with the evaluation of the sum over the n_i's.

A variety of techniques exist for the evaluation of the kind of lattice sum contained in the spherical constraint equation (4) [10-12]. One particularly powerful approach is the Ewald summation technique [12], generally utilized to perform sums over real crystal lattices, as in the calculation of the Madelung constant or the evaluation of the electrostatic energy of an ionic crystal. This technique utilizes the Poisson sum formula and other tricks to cast the lattice sum in the form of an integral which contains the $N\to\infty$ limit and "correction" terms, some of which require numerical evaluation, but all of which can be fairly rapidly calculated to very high numerical accuracy. Examples of the application of the Ewald technique to finite size critical behavior can be found in Refs. [7] and [13].

Applying the Ewald technique to Eq. (4) yields the following form for the spherical constraint equation

$$\beta = \int_0^{cL^2}\left(I_0(Jx)\right)^d dx + \frac{1}{L^{d-2}} f(\alpha L^2) + \frac{g(\alpha L^2)}{L^d} \quad , \quad (11)$$

where $I_0(Jx)$ is a zero order modified Bessel function and the correction term $g(\alpha L^2)/L^d$ is due to the replacement of $\cos k_i$ by $1-k_i^2/2$ in the expression on the right hand side of Eq. (8). This correction term will henceforth be ignored, as it is always small compared to the second term on the right hand side of (11). The expression $f(\alpha L^2)$ in Eq.(11) is given by

$$f(\alpha L^2) = \frac{1}{\alpha L^2} e^{-\alpha L^2 c} + c \int_1^{\infty} e^{-\alpha L^2 cx}\left([S(2\pi^2 Jcx)]^d - 1\right) dx$$

$$+ \frac{1}{2^d}\left(\frac{1}{2\pi J}\right)^{d/2} c^{(2-d)/2} \int_1^{\infty} e^{-\alpha L^2 c/x}\, x^{(d-4)/2} \left([S(\frac{x}{2Jc})]^d - 1\right) dx \, .$$

(12)

The function S(y) in Eq. (12) is given by

$$S(y) = \sum_{n=-L/2}^{L/2} e^{-n^2 y} \qquad . \tag{13}$$

If c is chosen to be equal to $(2\pi J)^{-1}$ in Eq. (12) the summations involved in calculating the function S are always rapidly converging, and accurate numerical results are readily obtained. Because of the exponential convergence of the integrands in Eq. (12) for <u>all</u> real, positive values of α, $f(\alpha L^2)$ is a well behaved function of its argument with no singularities in any regime of interest to us here, except for the simple pole in the first term on the right hand side of Eq. (12).

After a bit of processing, mostly in the evaluation of the integral over $I_0(Jx)$, one obtains the formula for the spherical constraint

$$L^{d-2}\left\{ -\frac{1}{T}[t-\varepsilon(L)]\right\} = W(\alpha L^2) + \frac{C\alpha L^2}{L^{4-d}} + O\left(\frac{(\alpha L^2)^2}{L^{6-d}}\right) . \tag{14}$$

The term $\frac{C\alpha L^2}{L^{4-d}}$ in Eq. (14) represents the leading order correction to scaling terms, associated with the leading irrelevant variable. This variable is "dangerous" when $d \geqq 4$ and so cannot be ignored if that range of dimensions is to be considered. The final term on the right hand side of Eq. (14) represents the <u>next</u> to leading order correction and can, by contrast, be ignored. The finite size shift term, $\varepsilon(L)$, is given by

$$\varepsilon(L) = K_\varepsilon T L^{2-d} \qquad , \tag{15}$$

with K_ε a constant. This form for $\varepsilon(L)$ is consistent with finite size scaling (see below Eq. (8)).

Below, we will distinguish between a number of different dimensional regimes. Before discussing the general application of Eq. (14), and its relevance to finite size scaling, as it is manifested in the mean spherical model, the equation will be presented in a form that allows us to investigate <u>explicitly</u> the behavior of the spherical constraint equation, both in the finite size scaling region and in the low temperature phase. This form of Eq. (14) is as follows

$$L^{d-2}\left[\beta - \left(\frac{1}{2\pi}\right)^{d/2}\frac{2}{d-2} - \beta_\infty(d)\right] = -\frac{2}{d-2}\left(\frac{1}{2\pi}\right) + \beta_1(d) + \frac{1}{\alpha L^2} + A_1(d)\,\alpha L^2$$

$$-\frac{2\alpha L^2}{4-d}\left(\frac{1}{2\pi}\right)^2 + \frac{1}{2\pi}\sum_{n=2}^{\infty}\frac{2(-\alpha L^2/2\pi)^n}{n!(2n+2-d)}$$

$$+\left\{f(\alpha L^2) - \beta_1(d) - \frac{1}{\alpha L^2} - A_1(d)\,\alpha L^2\right\}$$

$$+L^{d-4}(\alpha L^2)\left[\left(\frac{1}{2\pi}\right)^{d/2}\frac{2}{4-d} + B_1(d)\right]. \tag{16}$$

The dimensionality-dependent constants, $\beta_1(d)$ and $A_1(d)$ are the zeroth and first order terms in the Laurent expansion of $f(\alpha L^2)$. As a result, the expansion in αL^2 of the expression in curly brackets on the right hand side of Eq. (16) contains as its lowest order component the term $(\alpha L^2)^2$. The terms contributing to the spherical constraint equation that behave in a singular manner at the upper and lower critical dimensions, d=2 and 4, have been explicitly displayed. They will play a crucial role in the critical behavior of the model at those special dimensions.

When the parameter combination αL^2 is small one can neglect all but the lowest order terms in Eq. (16). As αL^2 is not large in that region, the approximation that results from discarding second order and higher terms is reasonably good throughout the region in which finite size effects dominate the behavior of the model. Eq. (16) becomes

$$L^{d-2}\left[\beta - \left(\frac{1}{2\pi}\right)^{d/2}\frac{2}{d-2} - \beta_\infty(d)\right] = -\frac{2}{d-2}\left(\frac{1}{2\pi}\right) + \beta_1(d) + \frac{1}{\alpha L^2} + A_1(d)\,\alpha L^2$$

$$-\frac{2\alpha L^2}{4-d}\left(\frac{1}{2\pi}\right)^2$$

$$+L^{d-4}(\alpha L^2)\left[\left(\frac{1}{2\pi}\right)^{d/2}\frac{2}{4-d} + B_1(d)\right] . \tag{17}$$

2<d<4

This is the range of dimensions that is most commonly studied. The infinite mean spherical model undergoes a phase transition form a paramagnetic to a ferromagnetic state, and the phase transition is characterized by non-mean-field critical exponents. By contrast, the finite mean spherical model has no phase transition at non-zero temperature. What this last statement means, strictly speaking, is that there is no singularity in the thermodynamic functions of this model at any $T \neq 0$. In a more, model-specific, sense, the statement means that there is no ferromagnetic order in the finite mean spherical model when the temperature is non-zero and the ordering field, h, is.

To obtain explicit results for the behavior of $\alpha(T)$, and, hence, of the various thermodynamic functions of the finite mean spherical model in the regime $T \lesssim T_c$, we rewrite Eq. (17) as follows:

$$L^{d-2}[\beta - \beta_c] = \frac{1}{\alpha L^2} - y - \alpha L^2[c + uL^{d-4}] \quad , \tag{18}$$

where the expressions for the quantities β_c, y, c and u are easily obtained by inspection of Eq. (17). The above equation is a simple quadratic one for αL^2. The solution of interest is

$$\alpha L^2 = \frac{1}{2[c+uL^{d-4}]}\left\{-y-L^{d-2}(\beta-\beta_c) + \sqrt{[y+L^{d-2}(\beta-\beta_c)]^2+4[c+uL^{d-4}]}\right\}$$

$$= \frac{1}{2[c+uL^{d-4}]}\left\{L^{d-2}(\beta-\beta_c(L)) + \sqrt{[L^{d-2}(\beta-\beta_c(L))]^2+4[c+uL^{d-4}]}\right\} . \tag{19}$$

In Eq. (19),

$$\beta_c(L) = \beta_c - yL^{-(d-2)} \quad . \tag{20}$$

The size dependence of the shift in the nominal dependence of the critical temperature is as expected (see Eq. (15)). Eq. (19) has the proper scaling form:

$$\alpha L^2 = A(L^{1/\nu}\tilde{t},\ uL^y) = A_0(L^{1/\nu}\tilde{t}) + uL^y A_1(L^{1/\nu}\tilde{t}) + ... \tag{21}$$

where the quantity u is the dominant correction-to-scaling variable. Note that the power series expansion in u indicated above is explicitly allowed by Eq. (19). In the range $2 < d < 4$, the irrelevant variable u is <u>not</u> dangerous.

Given Eqs. (19)-(21) and (6) one has immediately, for the isothermal susceptibility, $\chi(T,L)$,

$$\chi(T,L) = L^2 X(L^{1/\nu}\tilde{t}) + L^{2-(4-d)} Y(L^{1/\nu}\tilde{t}) \quad , \tag{22}$$

which from the results for the critical exponents ν and γ written above Eq. (9), yields the scaling result displayed in that equation, in the limit $h=0$.

It is interesting that the low temperature behavior ($T<T_c$, or $\beta > \beta_c$) of the mean spherical model is properly predicted by Eq. (19). In the limit of very large β, one has

$$\alpha L^2 \rightarrow \frac{L^{2-d}}{\beta}$$

so that, as $T \rightarrow 0$, $\alpha \rightarrow kTL^{-d}$, and $\chi \rightarrow CL^d$. This is the low temperature behavior of the finite mean spherical model, and, indeed, of a wide class of finite non-itinerant spin systems. At low temperatures such systems behave as if all spins were locked together into one "giant" moment with a net spin proportional to the volume of the system. One recovers as the limiting low temperature behavior that of a single, giant, spin [14].

When β is fixed, however, and as $L\rightarrow\infty$, Eq. (7) yields a correlation length that goes as the expected power of L, i.e. L^2. Inspection yields a correction term with a relative amplitude proportional to $L^{-(4-d)}$. The fact that the correlation length "sticks" at a very large value in the low temperature phase is a consequence of the gaplessness of the infinite spherical model when $T<T_c$. In the O(n) models, this gaplessness is the result of Goldstone Bosons induced by the fact that the ferromagnetic phase is a phase of broken continuous symmetry [15].

d=2

This is an important borderline dimensionality for the spherical model. At and below this dimension the infinite spherical model with short range interactions does not support a phase transition at non-zero temperature. Indeed, d=2 is special for all O(n) models with $n \geq 2$ [15-17]. One way to see how, and with what signature, the finite temperature phase transition disappears as the dimensionality of the mean spherical model approaches two from above is to note the following. First, the critical temperature in the infinite system goes to zero as $d\rightarrow 2^+$. This is reflected in the divergence, as 1/(d-2), of the second term in brackets on the right hand side of Eq. (17). Secondly, from the relationship $\nu=1/(d-2)$, it is clear that certain critical exponents will also diverge as $d\rightarrow 2^+$. In particular, the exponent describing the disappearance of α as the critical temperature is approached becomes infinite.

It is important to keep in mind that there is, indeed, a phase transition in the two dimensional mean spherical model. This transition, however, occurs at T=0 <u>exactly</u>. The infinite value for the exponent ν in two dimensions is reflected in the fact that $\alpha(T)$ behaves as follows

$$\alpha(T) \propto e^{-2\pi\beta} \quad . \tag{23}$$

That is, $\alpha(T)$, and hence $\xi_\infty(T)$, the correlation length of the infinite system, exhibits an essential singularity at the critical point.

Referring to Eq. (17) we see that there are two terms, each of which diverges as $d \to 2$. However, their individual singularities cancel, and, in the limit d=2, one has for the approximate spherical constraint equation:

$$\beta - b_c = -B\alpha L^2 + \frac{1}{\alpha L^2} + \frac{1}{\pi}\ln(L) \quad , \tag{24}$$

where we have ignored the leading order correction to scaling. As in the case of Eq. (18) the quantities b_c and B an be readily obtained in terms of quantities appearing in Eq. (17). Solving Eq.(24) for αL^2 :

$$\alpha L^2 = \frac{1}{2B}\left\{\sqrt{[\beta - b_c - \frac{1}{\pi}\ln(L)]^2 + 4B} - [\beta - b_c - \frac{1}{\pi}\ln(L)\right\} \quad . \tag{25}$$

When $(\beta, L) >> 1$, we have

$$\alpha L^2 \to \beta - \frac{1}{\pi}\ln(L) \quad , \tag{26}$$

Which means that

$$\chi(T,L) \propto \frac{1}{\alpha} = \frac{L^2}{\beta - \frac{1}{\pi}\ln(L)} \quad . \tag{27}$$

The explicit finite size scaling form of $\chi(T,L)$ is not expressed manifestly in Eq. (27), or in Eq. (25). This form is, in fact not correctly anticipated in Eq. (9). A more general set of relations, encompassing Eq. (9), are as follows [18]

$$\chi(T,L) = \chi_\infty(T)\ Z(\xi_\infty/L) \quad , \tag{28}$$

$$\xi(T,L)/L = Y(\xi_\infty/L) \quad . \tag{29}$$

To see how these forms arise in the case at hand we utilize the following identity:

$$\beta - \frac{1}{\pi}\ln(L) = \frac{1}{2\pi}\ln\left(e^{2\pi\beta}/L^2\right)$$

$$= \frac{1}{2\pi} \ln\left[\left(\xi_\infty(T)/L\right)^2\right] \quad . \qquad (30)$$

Then, from Eq.(25),

$$\alpha L^2$$

$$= \frac{1}{2B}\left\{\sqrt{\left[\frac{1}{2\pi}\ln\left[\left(\xi_\infty(T)/L\right)^2\right]-b_c\right]^2+4B} - \left[\frac{1}{2\pi}\ln\left[\left(\xi_\infty(T)/L\right)^2\right]-b_c\right\}$$

$$\equiv A(\xi_\infty/L) \, . \qquad (31)$$

As $L \to \infty$,

$$\xi(T,L) = \frac{L}{\sqrt{\alpha}} = \left(A(\xi_\infty/L)\right)^{-1/2} \quad , \qquad (32)$$

and

$$\chi(T,L) = \frac{L^2}{\alpha L^2} = \frac{L^2}{A(\xi_\infty/L)} = \xi_\infty^{\ 2}\frac{\left(L/\xi_\infty\right)^2}{A(\xi_\infty/L)}$$

$$= X_\infty Z(\xi_\infty/L) \quad . \qquad (33)$$

The proper scaling form is recovered.

d<2

Here, again, there is a zero temperature phase transition in both the finite and the infinite system. In the infinite system the spherical constraint yields

$$\beta = a\alpha^{(d-2)/2} + \beta' + b\alpha + \ldots \quad , \qquad (34)$$

so, as $\beta \to \infty$ $(T \to 0)$,

$$\alpha \propto T^{2/(2-d)} \quad . \qquad (35)$$

The appropriate version of Eq. (17) for the finite system is

$$L^{d-2}(\beta-\beta_0) = -B\alpha L^2 + \frac{1}{\alpha L^2} + C \quad , \tag{36}$$

where, as above, the parameters β_0, B and C can be expressed in terms of quantities appearing in Eq. (17). Once again, the term going like $L^{-(d-4)}\alpha L^2$ has been suppressed. Solving Eq. (36) for αL^2:

$$\alpha L^2 = \frac{1}{2B}\left\{ C-L^{d-2}(\beta+\beta_0)+\sqrt{[C-L^{d-2}(\beta-\beta_0)]^2+4B} \right\} . \tag{37}$$

Note that when β and L are large this solution for αL^2 approaches the following form

$$\alpha L^2 = \mathrm{B}(kTL^{2-d}) = \mathrm{B}(kTL^{1/\nu}) = \mathrm{E}(\xi_\infty/L) \quad . \tag{38}$$

The function $\mathrm{B}(x)$ has a finite limit at infinite argument, while, as $x\to 0$, $\mathrm{B}(x)\to x$. Thus, for a finite system one recovers $\alpha \propto L^{-2}$ for fixed temperature as $L\to\infty$, while, as $T\to 0$ for fixed L we have $\alpha\to kTL^{-d}$. The scaling form embodied in Eq. (37) interpolates between the expected finite size results ($\xi^2 \propto L^2$) and the giant moment behavior that is achieved at T=0.

d=4

At this, the upper critical dimensionality, the critical exponents take on their mean field values. In particular, the correlation length exponent ν is equal to 2 in d=4. As is the case with other spin systems, there are logarithmic corrections to the power laws [19]. These corrections are associated with the crossover from renormalized to mean field behavior that occur at this borderline dimensionality. The mathematical cause of the logarithms is the degeneracy that occurs as the exponent of the leading order correction to scaling variable goes through zero.

In the infinite system the spherical constraint takes the form

$$\beta - \beta_c = \frac{1}{2\pi^2}\alpha \ln \alpha + \ldots\ldots \quad , \tag{39}$$

which yields, when $T>T_c$,

$$\alpha \propto \frac{T-T_c}{\ln(T-T_c)} \quad , \tag{40}$$

which, via Eqs. (6) and (7) (with $L=\infty$) implies

$$\xi \propto \sqrt{\frac{\ln(T-T_c)}{T-T_c}} \quad , \tag{41}$$

and

$$\chi \propto \frac{\ln(T-T_c)}{T-T_c} \quad . \tag{42}$$

The critical exponents ν and γ are equal to 1/2 and 1. respectively, but the dependence on the reduced temperature of the quantities χ and ξ is no longer pure power law. There are multiplicative logarithmic corrections.

The approximate spherical constraint equation for the finite four dimensional system is as follows:

$$L^2\,[\beta - \beta_c + \varepsilon L^{-2}] = \frac{1}{\alpha L^2} - \left(\frac{1}{2\pi}\right)^2 \ln(L)\ \alpha L^2 + C \quad . \tag{43}$$

The solution for αL^2 is

$$\alpha L^2 = \frac{2\pi^2}{\ln(L^2)} \left\{ L^2[\tilde{\beta}_c(L)-\beta] + \sqrt{\left(L^2[\tilde{\beta}_c(L)-\beta] + \frac{4}{(2\pi)^2}\ln\left(L^2\right)\right)} \right\} .$$
(44)

In Eq. (44) the quantity $\tilde{\beta}(L)$ is the critical value of β, with the appropriate finite size shift. Note that at the nominal critical temperature, $\beta=\beta_c$, the quantity αL^2 is proportional to $\left(\ln\left(L^2\right)\right)^{-1/2}$.

There is thus a logarithmic correction to the low temperature behavior of the finite system correlation length. Once again, in the limit $\beta\rightarrow\infty$, we have $\alpha\rightarrow L^{-4}$, uncorrected by logarithms. This is just the expected giant moment behavior.

d>4

There are three major differences between the thermodynamics of the mean spherical model above four dimensions and its corresponding behavior when 2<d<4. First, the critical exponents of the mean spherical model are all mean-field-like above four dimensions [19]. In particular ν=2 and γ=1. Second, the susceptibility of this model is finite along the h=0 coexistence line for temperatures below T_c. The consequences of the broken continuous symmetry (in the language appropriate to O(n) models) are not as significant as they are when d<4 . Higher derivatives of the thermodynamic potentials (e.g. $\frac{\partial\chi}{\partial h}$) are singular, however. Third, the leading order correction-to-scaling variable is now dangerous, in that one is no longer allowed to expand certain thermodynamic functions as Taylor series like expansions in that variable. This last feature of the model provides an explanation for the violation of hyperscaling above four dimensions, as is also the case for O(n) models [20].

The appropriate version of the spherical constraint equation for the finite mean spherical model, as approximated by Eq. (17) is, when d>4,

$$L^2[\beta -\tilde{\beta}_c(L)] = -A\alpha L^2 + L^{-(d-4)}\frac{u}{\alpha L^2} \qquad , \qquad (45)$$

where $\tilde{\beta}_c(L) = \beta_c+\varepsilon L^{-(d-2)}$, and the quantities β_c, ε, A and u in Eq. (45) are, as in all previous cases, obtainable from Eq. (17) by inspection. A remarkable feature of Eq. (45) is that the finite size shift in β_c is considerably smaller than one expects on the basis of naive finite size scaling, e.g. $\tilde{\beta}_c(L) = \beta_c+\varepsilon L^{-\nu}$, with ν =1/2 This can be explained in terms of the dangerous irrelevant variable [21]. The quantity u is, as it turns out, equal to 1. This quantity will be treated as if it were an adjustable parameter, because it is, in fact, the coefficient of the dangerous irrelevant operator in the mean spherical model Hamiltonian. It turns out that one can formulate a variant of the mean spherical model in which the coefficient of $1/\alpha L^2$ in the equation corresponding to Eq. (45) is, indeed, adjustable, and, in fact, can be taken to the (unacceptable) limit u=0 [7].

The solution of Eq. (45) for αL^2 is

$$\alpha L^2 = \frac{1}{2A}\left\{-L^2[\beta-\tilde{\beta}_c(L)]+\sqrt{\left(L^2[\beta-\tilde{\beta}_c(L)]\right)^2+4AuL^{-(d-4)}}\right\}$$

$$= f(\tilde{t}L^{1/\nu}, uL^{-(d-4)}) \qquad (46)$$

Thus,

$$\chi = \frac{L^2}{\alpha L^2} = L^{1/\nu}\left(f(\tilde{t}L^{1/\nu}, uL^{-(d-4)})\right)^{-1} \qquad (47)$$

and both $\alpha(T,L)L^2$ and $\chi(T,L)L^2$ have the expected form. However, when $\tilde{t}<0$ $(\beta > \tilde{\beta}_c)$ One can verify by inspection of Eqs. (45) and (46) that the susceptibility cannot be expanded as a Taylor series in the combination $uL^{-(d-4)}$. In fact, when $\tilde{t}<0$ and $uL^{-(d-4)}$ is small one has

$$\chi(T,L) = L^2\left[\frac{L^2[\beta-\tilde{\beta}_c(L)]+\sqrt{\left(L^2[\beta-\tilde{\beta}_c(L)]\right)^2+4AuL^{-(d-4)}}}{2uL^{-(d-4)}}\right]$$

$$\rightarrow \frac{L^4[\beta-\tilde{\beta}_c(L)]}{uL^{-(d-4)}} \qquad (48)$$

The leading order contribution to $\chi(T,L)$ diverges as $u\rightarrow 0$, In fact, the dependence on u is such that $\chi(T,L) \propto L^d$, and the giant moment behavior that characterizes the magnetic response of finite O(n) spin systems as $\beta\rightarrow 0$ is recovered. The term proportional to u is essential (and dangerous) because it stabilizes the low temperature phase. In this respect it plays the same role as the fourth order term in the standard Ginzburg-Landau-Wilson spin model, which is irrelevant above four dimensions, but which <u>cannot</u> be set equal to zero if the low temperature phase is to be non-pathological, in the absence, of course, of higher order stabilizing terms.

Susceptibility when $T<T_c$. Crossover from infinite system to finite system behavior

The previous discussion contained allusions to the "expected" finite size dependence of the susceptibility of the mean spherical model in the limit h=0. However, it is well known that in the infinite spherical model (i.e. in the thermodynamic limit) the behavior of the low temperature isothermal susceptibility is fundamentally different. When 2<d<4 one has, if $T<T_c$, $\chi(h,T) \propto h^{-(d-4)/2}$ as $h\to 0$, while when d>4, $\chi(h,T)$ has a finite limit when h=0. At d=4 exactly, $\chi(h,T)$ diverges as $|\ln h|$ when $h\to 0$ [2].

On the other hand, the fact that $\chi = \frac{\partial M}{\partial h}$ is never singular in a finite system follows from the analytic, nonsingular dependence af all thermodynamic functions in such a system on the variables h and T. This regular behavior is a straightforward consequence of the well-behaved nature of the Boltzmann factors that are summed over in the grand canonical ensemble.

By the same token, one expects fundamentally different behavior of the magnetic order, $M = \frac{1}{N}\sum_{i=1}^{N} s_i$, as $h\to 0$ in the finite system when $T<T_c$, than applies in the thermodynamic limit. In the latter case $M\to M_0(T)$, a non-zero remanent value, in the ferromagnetic phase ($T<T_c$), as $h\to 0$. This spontaneous symmetry breaking is realized only in the thermodynamic limit. The symmetry with respect to a sign change in h that is a property of the underlying Hamiltonian of the mean spherical model is respected as long as the system is finite. What this means is that the free energy of the finite mean spherical model contains only even powers of h so that, for small h the free energy, F(T,h,L) has the form

$$F(T,h,L) = F_0(T,L) + F_2(T,L)\, h^2 + O(h^4) \quad , \qquad (49)$$

and

$$M(T,h,L) = -\frac{\partial F}{\partial h} \propto h + O(h^3) \qquad . \qquad (50)$$

Thus $\lim_{h\to 0} M(T,h,L) = 0$.

The crossover that occurs between the finite and thermodynamic behavior outlined above is consistent with scaling. To be specific, in fewer than four dimensions the critical contribution to the free energy of the mean spherical model, F(T,h,L), has the following scaling form

$$F(T,h,L) = f(tL^{d-2}, hL^{(d+2)/2}) \qquad . \qquad (51)$$

This means that

$$M(T,h,L) = -\left(\frac{\partial F}{\partial h}\right)L^{-d} = L^{(2-d)/2}m(tL^{d-2}, hL^{(d+2)/2}) \; , \qquad (52)$$

and

$$\chi(T,h,L) = \frac{\partial m}{\partial h} = L^2 X(tL^{d-2}, hL^{(d+2)/2}) \qquad . \qquad (53)$$

Eq. (53) is just a restatement of Eq. (9) without the correction-to-scaling contribution. The scaling forms embodies in Eqs. (51)-(53) are immediately derivable from the following expression for the mean spherical model free energy

$$L^{-d}F(T,h,L) = \left(-\frac{h^2}{\alpha} + L^{-d}\ln\alpha + L^{-d} \sum_{n1...nd=1}^{L} \ln[\alpha + J\sum_{i=1}^{d}(1-\cos\frac{2\pi n_i}{L})] - \lambda \right) \qquad (54)$$

from which one generates the spherical constraint equation, Eq. (4), by extremizing with respect to λ. Recall that the quantity α, as defined in Eq. (5), depends linearly on λ.

Because of the extremum condition, one has immediately

$$-\frac{1}{L^d}\frac{\partial F}{\partial h} = M(t,h,L) = \frac{hL^2}{\alpha(T,h,L)L^2} \quad . \tag{55}$$

With the addition of an applied magnetic field the approximate constraint equation is amended as follows

$$L^{d-2}[\beta - \beta_c] = \frac{h^2 L^{d+2}}{(\alpha L^2)^2} + \frac{1}{\alpha L^2} - y - \alpha L^2[c + uL^{d-4}] \quad , \tag{56}$$

where the quantities β_c, y, c and u are the same as in Eq. (18). It is manifestly obvious that the solution of this new equation for the combination αL^2 will fall into the scaling form

$$\alpha(T,h,L)\ L^2 = A(tL^{d-2}, hL^{(d+2)/2}) \quad . \tag{57}$$

In the thermodynamic limit, the leading contributions to the constraint equation yield the following infinite-system-limit equation

$$-t = \frac{h^2}{\alpha^2} - C\alpha^{(d-2)/2} \quad . \tag{58}$$

Multiplying this equation by L^{d-2} we obtain a scaling form appropriate to the solution for $\alpha(T,h,L)$ displayed in Eq. (57). Solving Eq. (58) for α we obtain

$$\alpha = \frac{h}{\sqrt{-t+C\alpha^{(d-2)/2}}} \approx \frac{h}{\sqrt{-t+C\left(\frac{h}{\sqrt{-t}}\right)^{(d-2)/2}}} \tag{59}$$

so that, when t<0

$$M = \frac{h}{\alpha} \approx |t|^{1/2} + h^{(d-4)/2}\ |t|^{d/4} \quad . \tag{60}$$

We see immediately from Eq. (60) and the relation $\chi = \frac{\partial M}{\partial h}$ that (1) there is a spontaneous magnetization when h=0, and (2) the susceptibility, $\chi(T,h,L{=}\infty)$ diverges as $h \to 0$. Both Eq. (59) and Eq. (60) are consistent with the scaling forms , Eqs. (52) and (53), for M(T,h,L) and χ(T,h,L), respectively.

When L is finite we return to Eq. (56). This version of the constraint equation is readily transformed to a cubic equation for αL^2. As the reduced temperature t passes through zero the solution for αL^2 that applies to the case at hand interpolates between the limiting results $\alpha L^2 \propto L^{d-2}t$ when t is positive and $\alpha L^2 \propto \frac{1}{L^{d-2}|t|} + h^2 L^{d+2}$ when t is negative. In the negative t regime the finite system displays the expected nonsingular behavior.

When d>4 the approximate constraint equation for the finite mean spherical model in the presence of an ordering field takes the form

$$L^2[\beta - \tilde{\beta}_c(L)] = -A\alpha L^2 + L^{-(d-4)}\frac{u}{\alpha L^2} + L^{-(d-4)}u\frac{(hL^{(d+2)/2})^2}{(\alpha L^2)^2} \quad . \qquad (61)$$

Once again, the parameter u has been inserted by hand. As previously, this was done to elucidate the L-dependence of the terms in the constraint equation that have scaling properties associated with multiplication by the leading order correction-to-scaling variable. Clearly, the general solution will be of the form $\alpha L^2 = A(\tilde{t}L^2, hL^{(d+2)/2}, uL^{-(d-4)})$. The actual solution of Eq. (61) for $\alpha(T,h,L)L^2$ interpolates between $L^2\tilde{t} = L^{1/\nu}\tilde{t}$ when $\tilde{t} > 0$ and

$$L^2\alpha(T<T_c,h,L,u) \approx \frac{uL^{-(d-4)}}{L^2[\beta - \tilde{\beta}_c(L)]} \quad . \qquad (62)$$

Thus,

$$M(T,h,L,u) = \frac{L^{(d+2)/2}}{uL^{-(d-4)}}\, m(\tilde{t}L^2, hL^{(d+2)/2}, uL^{-(d-4)}) \quad . \qquad (63)$$

The L, t, and h dependence of M is as expected. Once again, the giant moment behavior is recovered at low temperatures.

On the other hand, when $L \to \infty$ the spherical constraint takes the form

$$-t = u\,\frac{h^2}{\alpha^2} - Cu\alpha + O(\alpha^{(d-2)/2}) \quad . \qquad (64)$$

When h is small and t is less than zero, this yields for $M=h/\alpha$

$$\begin{aligned} M &= |t|^{(d-2)/4} f(u|t|^{(d-4)/4}) \\ &= |t|^{(d-2)/\nu} f(u|t|^{(d-4)/\nu}) \\ &= \lim_{L\to\infty} L^{(d-2)/2} m(uL^{-(d-4)}, tL^2, hL^{(d+2)/2}) \; . \end{aligned} \qquad (65)$$

Once again, the scaling form is recovered, with u as a dangerous irrelevant variable. For additional discussion of the magnetism and susceptibility of the mean spherical model see Singh and Pathria. [22].

Long-range interactions

Systems in which the interactions between spins extend to distant pairs of sites can exhibit critical behavior that is substantially different than is manifested in systems with short range interactions. One of the most commonly studied type of long range interaction has power law decay as the separation between the spins becomes large. If one has a two-spin interaction like $J(\vec{r}_i - \vec{r}_j)s_i s_j$, $\vec{r}_i$ and $\vec{r}_j$ being the locations of the two spins, and

$$J(\vec{R}) \propto |\vec{R}|^{-\sigma-d} \quad , \qquad (66)$$

where d is the physical dimensionality of the system and σ is an additional exponent, then, when $\sigma<2$ the critical behavior can be substantially modified [23].

The property of $J(\vec{R})$ that leads to a change in the critical properties is the absence of a finite second spatial moment of the interaction, which precludes a simple gradient expansion of $J(\vec{R})$. In particular, if $J(\vec{R})$ behaves as in Eq. (66), then its spatial fourier transform, $\mathcal{J}(\vec{k})$ has the following form

$$\mathcal{J}(\vec{k}) = \mathcal{J}_0 + K|\vec{k}|^{\sigma} \quad . \tag{67}$$

The thermodynamic functions of the model are then obtained exactly as they are for the mean spherical model with short range interactions.

As an important aside it must be noted that the proper form of $\mathcal{J}(\vec{k})$ must take into account the boundary conditions of the system. In the case of periodic boundary conditions one appropriate form for $\mathcal{J}(\vec{k})$ is

$$\mathcal{J}(\vec{k}) = \mathcal{J}_0 + \mathcal{J}_1\left[\sum_{i=1}^{d}(1-\cos k_i a)\right]^{\sigma/2} \quad . \tag{68}$$

Two important features of this $\mathcal{J}(\vec{k})$ are (1) there is a cubic anisotropy in $\mathcal{J}(\vec{k})$ (i.e. it is not solely dependent on the amplitude of $\vec{k}$, and (2) $J(\vec{R})$ now depends on the size of the system [24]. This seems unavoidable if one hopes to combine power law decay of the interactions with proper respect for periodic boundary conditions. The infinite range of the interaction causes it to effectively "sample" the boundary. the effect is not, however, large.

The mathematical difficulties that are confronted in the calculation of the partition function of this version of the finite mean spherical model are all in the evaluation of the lattice sums over the k_i's in the spherical constraint equation. Of the various methods of accomplishing these summations two in particular have been proposed [25-27]. One due to Brankov and Tonchev, and to J. Shapiro [26,27] utilizes the following result

$$\frac{1}{r+|w|^{\sigma}} = \int_0^{\infty}\frac{Q(r,z)}{z^2+w^2}\,dz \quad , \tag{69}$$

where

$$Q(r,z) = \frac{1}{\pi}\frac{\sin\left(\frac{\pi\sigma}{2}\right)z^{\sigma/2}}{\left[r^2+2r\cos\left(\frac{\pi\sigma}{2}\right)z^{\sigma/2}+z^{\sigma}\right]} . \tag{70}$$

This equality can be established via contour integration.

With the use of Eqs. (69) and (70) the construction of the spherical constraint equation entails a single integral over the lattice sum evaluated in the study of the short range mean spherical model. The constraint equation takes the form

$$L^{d-\sigma}[\beta - \tilde{\beta}_c(L)] = f(\alpha L^{\sigma}, h^{(d+\sigma)/2}, uL^{|d-2\sigma|}) . \tag{71}$$

The correction to scaling variable u has been inserted by hand, as previously.

One may now distinguish between various dimensionality regimes. As what follows closely parallels the discussion in corresponding sections earlier in this article, the discussion will be in an abbreviated form.

$2\sigma > d > \sigma$

Here the exponents are as follows. The correlation length exponent ν is given by $\nu = 1/(d-\sigma)$. The "magnetic" exponent, x, is given by $x=(d+\sigma)/2$. All other exponents follow from hyperscaling, i.e. $\alpha = 2-d\nu$, and so on. At lowest order in αL^{σ} the spherical constraint equation is of the form (when h=0),

$$L^{d-\sigma}[\beta - \tilde{\beta}_c(L)] = \frac{1}{\alpha L^{\sigma}} + \alpha L^{\sigma}(B+CuL^{\sigma-2d}) + \ldots \tag{72}$$

The solution for αL^{σ} interpolates between

$$\alpha L^{\sigma} \approx L^{d-\sigma}\tilde{t}(L) , \tag{73}$$

when $\tilde{t} > 0$ and

$$\alpha L^{\sigma} \approx \frac{L^{\sigma - d}}{kT} \qquad , \qquad (74)$$

as T→0. As in the case of the finite spherical model with short range interactions, there is no phase transition at non-zero temperature. The shift in the nominal critical temperature due to the finite size of the system is proportional to $L^{-(d-\sigma)} = L^{1/\nu}$. As can be inferred from Eqs. (73) and (6) the low temperature behavior of the susceptibility is giant-moment-like.

d = σ

Here, at the lower critical dimensionality, the transition temperature of the infinite system is suppressed down to T=0. a generalized from of finite size scaling is recovered, that takes into account the fact that the correlation length exhibits an essential singularity at the zero temperature phase transition. To be specific,

$$\alpha(T,L)L^{\sigma} = A(\xi_{\infty}/L) \qquad , \qquad (75)$$

where

$$\xi_{\infty} \propto e^{C/T} \qquad . \qquad (76)$$

and

$$\xi(T,L) = \frac{1}{\alpha(T,L)} = L^{\sigma}\Big(A(\xi_{\infty}/L)\Big)^{-1} \qquad . \qquad (77)$$

The specific form of $A(\xi_{\infty}/L)$ yields, at low temperature,

$$\chi(T,L) \propto L^{\sigma} \ln(\xi_{\infty}/L) \rightarrow \frac{L^{\sigma}}{T} = \frac{L^{d}}{T} \qquad , \qquad (78)$$

the latter limit being achieved as T → 0. Giant moment behavior is once again regained.

d < σ

Equation (72) applies here. There is a zero temperature phase transition in both the finite and the infinite system. In the finite system α is of the following form as low temperatures

$$\alpha = T^{\sigma/(\sigma - d)} \quad , \tag{79}$$

yielding a susceptibility χ that diverges as $T^{-\sigma/(\sigma - d)}$. Utilizing the following asymptotic form of Eq. (72):

$$L^{d-\sigma} \beta = \frac{1}{L^{\sigma}\alpha} \quad , \tag{80}$$

we have

$$\alpha = \frac{kT}{L^{d}} \quad , \tag{81}$$

and

$$\chi \propto \frac{L^{d}}{T} \quad . \tag{82}$$

d = 2σ

At this, the upper critical dimensionality, one encounters mean field behavior modified by logarithmic corrections. The quantity $\alpha(T,L)$ behaves as follows in and below the critical region

$$\alpha(T,L)L^{\sigma} = \frac{1}{2A\ln(L)}\left\{L^{2}[\tilde{\beta}_{c}(L)-\beta] + \sqrt{\left(L^{2}[\tilde{\beta}_{c}(L)-\beta] + 4A\ \ln(L)\right)}\right\} \tag{83}$$

2σ < d

Here, as in the corresponding dimensionality range of the nearest neighbor spherical model, there is a dangerous irrelevant variable that affects the scaling behavior. The approximate constraint equation is

$$L^{\sigma}[\beta - \tilde{\beta}_{c}(L)] = -A\alpha L^{\sigma} + \frac{uL^{-(d-2\sigma)}}{\alpha L^{\sigma}} \quad , \tag{84}$$

with u the coefficient of the dangerous irrelevant variable. Solving foe αL^{σ},

$$\alpha L^{\sigma} = \frac{1}{2A}\left\{ -L^{\sigma}[\beta - \tilde{\beta}_c(L)] + \sqrt{\left(L^{\sigma}[\beta - \tilde{\beta}_c(L)] \right)^2 + 4AuL^{-(d-2\sigma)}} \right\}$$

$$= f(\tilde{t}L^{1/\nu}, uL^{-(d-4)}) \quad . \tag{85}$$

When $\tilde{t} < 0$ the quantity u cannot be neglected. This is is obvious, as in the short range case, when we consider the susceptibility, $\chi(T,L)=1/\alpha(T,L)$, which diverges as 1/u in the limit u=0. In the absence of this dangerous variable the the behavior is the type of mean field behavior that applies to systems with long range interactions ($\nu = 1/\sigma$, $\gamma = 1$, and so on).

Privman and Fisher have considered in detail finite size crossover at the coexistence line ($T<T_c$) in systems with long range interactions [24]. Their discussion parallels and extends considerably the portion of this review that concerned similar issues for the short range spherical model.

The Privman-Fisher hypothesis

Privman and Fisher have proposed that a nontrivial extension of two scale factor universality holds for fully finite systems [28]. In its original formulation, by Stauffer, Ferer and Wortis[29], the hypothesis of two scale factor universality asserts that the singular part of the free energy in a volume ξ^d (i.e. the volume circumscribed by a correlation length) will be a universal quantity. This hypothesis has been tested against experiment and has been shown to be consistent with the results of Renormalization Group calculations [30]. In addition it has also been borne out in the context of the infinite mean spherical model [31].

If one accepts that the effective correlation length in a finite system will be, essentially, a characteristic linear dimension of the system at the nominal critical temperature, then the conjecture of Privman and Fisher is, at least, plausible. What they have hypothesized is that the total singular free energy in a finite system will be a universal constant - that is, the same number for every system in a given universality class - at the bulk critical temperature. The singular part of the free energy is the continuation to a finite system of that contribution to the free energy that is singular in the bulk.

Singh and Pathria have tested the Privman-Fisher hypothesis within the context of the mean spherical model [32]. Using techniques that they had developed to study the fully finite ideal Bose gas, and, in particular, Bose condensation in that system, they were able to verify that the hypothesis does, indeed, hold. Furthermore, they have written down an explicit expression for the universal number, involving numerically tractable infinite sums over modified Bessel functions.

That the Privman-Fisher hypothesis holds when there are long range interactions of the type discussed in this review has been established by Brankov [26], who developed extensions of the methods of Singh and Pathria and utilized them to calculate the universal constant in the range $\sigma < d < 2\sigma$.

Antiperiodic boundary conditions

In all preceding discussions periodic boundary conditions have been assumed. Such boundary conditions are especially useful (if highly unphysical) in the theoretical study of finite systems. Probably the most useful consequence of periodic boundary conditions is that they remove all surfaces from the system. It is then commonly the case that the approach to bulk behavior is exponentially rapid in as the size of the system is increased, if $T>T_c$. More quantitatively, if Δ is the difference between some intrinsic thermodynamic quantity in a finite system of characteristic linear dimension, L, and its bulk ($L\to\infty$) limit, then

$$\Delta \propto e^{-L/C} \tag{86}$$

The constant C is generally proportional to the correlation length. This kind of convergence does <u>not</u> occur when there are power law interactions.

Singh and Pathria have studied the mean spherical model with <u>antiperiodic</u> boundary conditions [33]. They have found that, while the convergence to bulk behavior is generally exponential when $T>T_c$, the isothermal susceptibility approaches the bulk limit as a power law, exhibiting surface-type corrections. Singh, Pathria and Fisher have shown how this behavior arises from variations in the local susceptibility, induced by the antiperiodic boundary conditions. They also find that below T_c such boundary conditions yield new exponents for the local susceptibility [34].

Acknowledgements

The author gratefully acknowledges useful interactions with many colleagues, especially Professors Vladimir Privman and George Gaspari, and Dr. Jonathan Shapiro. Much of this review concerns work performed by Dr. Shapiro, whose contributions the the education of the author on the subject have been substantial.

References

1. T. H. Berlin and M. Kac, Phys. Rev. 86, 821 (1952).

2. For an overview of the model and additional references, see G. S. Joyce in Phase Transitions and Critical Phenomena, Vol. 2, C. Domb and M. S. Green, eds. (Academic Press, New York, 1972).

3. H. E. Stanley, Phys. Rev. 176, 718 (1968).

4. S.-k. Ma Rev. Mod. Phys. 45, 589 (1973).

5. M. Barber and M. E. Fisher, Ann. Phys. 77, 1 (1973).

6. S. Singh and R. K. Pathria, Phys. Rev. B32, 4618 (1985). See also E. Brézin, Ann. de Physique **43**:15 (1982).

7. J. Shapiro and J. Rudnick, J. Stat. Phys. 43, 51 (1986).

8. M. E. Fisher in Critical Phenomena, Lecture Notes in Physics, 186 (Springer Verlag, Berlin, 1983) p. 1

9. M. Fisher in Critical Phenomena, Proc. 51st Enrico Fermi Summer School, Varenna, M. S. Green, ed. (Academic Press, New York, 1972).

10. E. Brézin, J. Physique, **43**; 15 (1982).

11. S. Singh and R. K. Pathria, Phys Rev. B31, 4483 (1985); Phys Rev. B32, 4618, (1985).

12. P. P. Ewald, Ann. Phys. 64, 253 (1921).

13. J. Rudnick, H. Guo and D. Jasnow, J. Stat. Phys. 41, 353, (1985).

14. L. Néel, C. R. Acad. Sci. 228, 664 (1949); Ann. Geophys. 5, 99 (1949). See also section 5 in the contribution by V. Privman in this volume.

15. N. N. Bogoliubov, Phys. Abhandl. S. U. 6, 113 (1962).

16. N. Mermin and H. Wagner, Phys. Rev. Lett. 17, 1133 (1966).

17. P. C. Hohenberg, Phys. Rev. 158, 385 (1967)

18. E. Brézin, Ann N. Y. Acad. Sci. 410, 339 (1983). See also J. M. Luck, Phys. Rev. B31, 3069 (1985).

19. K. Wilson and J. Kogut, Phys. Rep. 12, 75 (1974).

20. M. E. Fisher in Renormalization Group in Critical Phenomena and Field Theory, J. D. Gunton and M. S. Green, eds. (Temple University, Philadelphia) p. 65.

21. J. Rudnick, G. Gaspari and V. Privman, Phys. Rev. B32, 7594 (1985).

22. S. Singh and R. K. Pathria, Phys. Rev. B37, 7806 (1988).

23. M. E. Fisher, S.-k. Ma and B. G. Nickel, Phys. Rev. Lett. 29, 917 (1972).

24. M. E. Fisher and V. Privman, Commun. Math. Phys. 103, 527 (1986).

25. J. G. Brankov, preprint

26. J. G. Brankov and N. S. Tonchev, J, Stat. Phys. 52, 143, (1988).

27. J. Shapiro, Ph.D. thesis, UCLA (1988).

28. V. Privman and M. E. Fisher, Phys. rev. B30, 322 (1984).

29. D. Stauffer, M. Ferer and M. Wortis, Phys Rev. Lett. 29, 345 (1972). See also M. Ferer, M. Moore and M. Wortis, Phys. Rev. B8, 5205 (1973), and M. Ferer, D. Stauffer and M. Wortis Proc. AIP Conference on Magnetism and Magnetic Materials, Vol. 10 (American Institute of Physics, New York, 1973), p. 836.

30. A. Aharony, Phys Rev. B9, 2107, (1974).

31. P. R. Gerber, J. Phys. A: Math, Gen. 8, 67 (1975).

32. See the first of refs. 11.

33. S. Singh and R. K. Pathria, Phys. Rev. B31, 4483 (1985); Phys Rev. B32, 4618 (1985).

34. S. Sing, R. K. Pathria and M. E. Fisher Phys. Rev. B33, 6415 (1986).

IV. SOME RECENT PROGRESS IN THE PHENOMENOLOGICAL THEORY OF FINITE SIZE SCALING AND APPLICATION TO MONTE CARLO STUDIES OF CRITICAL PHENOMENA

Kurt Binder

TABLE of CONTENTS

SOME RECENT PROGRESS IN THE PHENOMENOLOGICAL THEORY OF FINITE SIZE SCALING AND APPLICATION TO MONTE CARLO STUDIES OF CRITICAL PHENOMENA

Kurt Binder

Institut für Physik
Johannes Gutenberg–Universität Mainz
Postfach 3980, 6500 Mainz
Federal Republic of Germany

1. INTRODUCTION

In this brief review, some recent developments concerning the understanding of finite size effects on second order phase transitions will be reported.

Statistical mechanics tells us that a singular behavior of the free energy, a phase transition, can occur in thermodynamic limit only, $N \to \infty$: in a finite system, the free energy is regular, there is simply no phase transition whatsoever! [1–14].

Finite size scaling theory [3,6,10,11,14] now describes how it happens that phase transitions occur after the limit $N \to \infty$ is taken, although there are none for any finite N, even if N is very large. What happens is that the transition in the finite system is rounded and shifted [2–6] over some region: this region smoothly shrinks to zero as $N \to \infty$.

For instance, let us consider a phase transition occurring as a function of temperature T at a critical temperature $T_c(\infty)$ in the infinite system, and we deal with a finite d–dimensional box with all its linear dimensions equal to L.

The transition then is "smeared out" ("rounded") over a temperature region ΔT (Fig. 1)

$$\Delta T_{\text{rounding}} \sim L^{-\Theta}. \tag{1}$$

The center of this region occurs at $T_c(L)$, i.e. at a temperature that may be shifted away from $T_c(\infty)$

$$T_c(L) - T_c(\infty) \sim L^{-\lambda}. \tag{2}$$

A quantity, which in the thermodynamic limit shows a critical divergence, such as the susceptibility χ' at a transition from a paramagnetic to a ferromagnetic state, m being the magnetization

$$\chi' = (L^d/k_B T)(\langle m^2\rangle_T - \langle|m|\rangle_T^2) \sim |T - T_c(\infty)|^{-\gamma}, \tag{3}$$

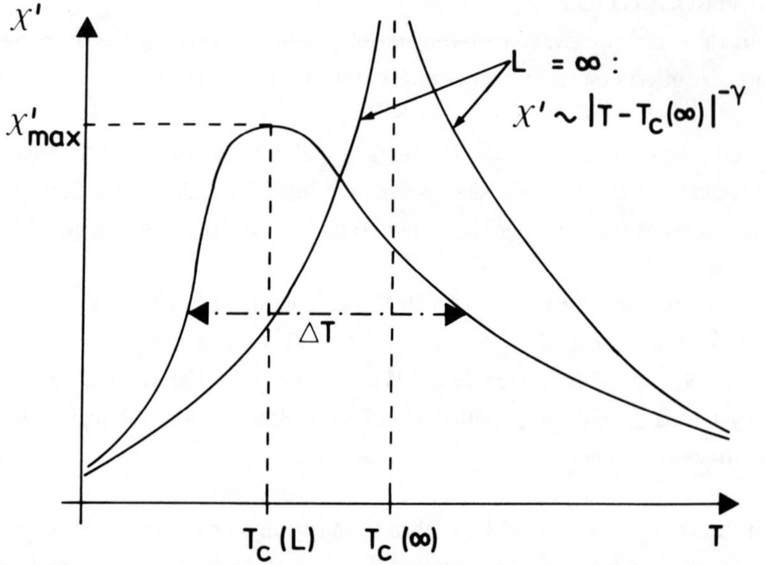

Fig. 1 Qualitative sketch of the susceptibility χ' of a system undergoing a phase transition from a ferromagnetic phase to the paramagnetic phase, as the temperature T is raised. Both the behavior in the thermodynamic limit and of the finite system are shown.

in the finite system reaches only a maximum of finite height χ_{max},

$$\chi'_{max} \sim \chi(L,T{=}T_c) \sim L^{\gamma_c}. \tag{4}$$

Similarly, the order parameter $<|m|>_{T,L}$ does not vanish at T_c, but rather we have

$$<|m|>_{T_c,L} \sim L^{-\beta_c}. \tag{5}$$

By these equations, various critical exponents have been introduced dealing with the size dependence in the critical region: the rounding exponent Θ, the shift exponent λ, and exponents describing the size dependence at T_c for various critical quantities, such as β_c for the magnetization, γ_c for the susceptibility, etc. First of all, we want to understand these exponents.

These questions not only are of interest as a fundamental question of statistical physics: they are also relevant for the understanding of bulk critical phenomena. An experimental system always is homogeneous over a region of finite extent only. This limitation is particularly relevant for the study of quasi–two–dimensional phase transitions in adsorbed monolayers at surfaces, where the substrate surface is ideal typically over an area of (100 Å)^2 only [15].

Even more important are these questions for the approach of "computer simulation", i.e. Monte Carlo [5,8–10,12–14,16–20] and molecular dynamics [19] computer experiments. These simulations always deal with rather small systems, and at phase transitions then finite size effects are very pronounced. Understanding finite size effects is absolutely crucial for the useful application of computer simulation [20].

Now a very simple answer to those questions has been given by Fisher's finite size scaling theory [3,6,11]. One agrees that near a critical point there is only one important characteristic length, the correlation length ξ of order parameter fluctuations: then it is plausible that finite size effects are controlled by a comparison of lengths, the linear dimension L and this correlation length ξ.

Since ξ diverges as {$\hat{\xi}_\pm$ are "critical amplitudes" for $T \gtrless T_c$}

$$\xi = \hat{\xi}_\pm |1-T/T_c|^{-\nu}, \ T \to T_c, \qquad (6)$$

the condition that rounding and shifting of the transition sets in when $L \approx \xi$ immediately yields

$$|\Delta T_{\mathrm{rounding}}/T_c| \approx L \to \Delta T_{\mathrm{rounding}} \sim L^{-1/\nu}. \qquad (7)$$

From this argument and an analogous consideration for the shift we hence find, comparing to the definitions Eqs. (1) and (2) that

$$\Theta = 1/\nu,\ \lambda = 1/\nu. \tag{8}$$

Expressing the singularities of χ {Eq. (3)} and of $<|m|>_{T,L\to\infty} \equiv M_{sp}$, the spontaneous magnetization

$$M_{sp} = \hat{B}(1-T/T_c)^{\beta},\ T\to T_c \tag{9}$$

in terms of ξ,

$$<|m|> \sim \xi^{-\beta/\nu},\ \chi \sim \xi^{\gamma/\nu} \tag{10}$$

the idea that the growth of ξ near T_c is limited by L immediately implies that

$$\chi'_{max} \sim L^{\gamma/\nu},\ <|m|>_{T_c} \sim L^{-\beta/\nu}. \tag{11}$$

Comparing Eqs. (4),(5) and (11) we conclude

$$\beta_c = \beta/\nu,\ \gamma_c = \gamma/\nu. \tag{12}$$

Of course, this simple principle that "L scales with ξ" is not the whole story. In this review, first the application of this rather standard form of finite size scaling to computer simulations will be elaborated (Sec. 2), emphasizing the finite size scaling properties of the order parameter probability distribution $P_L(m)$ and the cumulant intersection method for the location of T_c [10,12–14]. Then problems will be considered where the situation is fundamentally more complicated:

(i) Geometries involving several linear dimensions, e.g. $L_\perp x L_{||}$ in two dimensions [2,11,21–26] or quasi–one–dimensional systems with d–1 linear dimensions $L_\perp$ and one linear dimension $L_{||} >> L_\perp$. We want to understand how quantities such as χ depend on both linear dimensions $L_{||}$, $L_\perp$ (Sec. 3).

(ii) Anisotropic critical phenomena: at uniaxial Lifshitz points [27–32], for instance, we have the behavior that the correlation function decays with correlation lengths $\xi_{||}$, $\xi_{\perp}$ in different lattice directions:

$$\xi_{||} \sim |1-T/T_c|^{-\nu_{||}}, \; \xi_{\perp} \sim |1-T/T_c|^{-\nu_{\perp}}. \tag{13}$$

The generalization of finite size scaling to such anisotropic situations (see also [33–40]) will be discussed in Sec. 4.

(iii) Systems above their marginal dimension d* (Sec. 5). Remember that the statement that there is only one relevant length near a critical point for isotropic systems as well is only true for d<d*, the dimension where Landau theory starts to be valid [41]. For d>d*, the hyperscaling relation between the critical exponents

$$d\nu = \gamma + 2\beta, \tag{14}$$

which ensures that there is only one relevant length, the correlation length ξ, fails. E.g., for the Ising model we have

$$d^* = 4 \tag{15}$$

and with the Landau theory exponents [41]

$$\gamma = 1, \; \beta = \frac{1}{2}, \; \nu = \frac{1}{2} \tag{16}$$

the hyperscaling relation, Eq. (14), fails in d = 5 dimensions. In the simple geometry where all linear dimensions are equal and periodic boundary conditions are applied, the correlation length ξ simply gets replaced by another length, the thermodynamic length ℓ_T [42,43] which will be defined in Sec. 5. For the geometry with $L_{||} \neq L_{\perp}$, however, the situation is much more complicated [22,25, 44–46].

(iv) Finally, we can consider the combination of the complications (ii) + (iii), anisotropic critical phenomena above their marginal dimension d*. While for

Lifshitz points this is a rather academic problem, it is probably relevant for nonequilibrium phase transitions in driven systems! A simple example is the Ising lattice gas problem where the charged particles hopping from lattice site to lattice site [47] experience an electric field [48–53] (Fig. 2). This system shows a symmetry breaking into domains of low density and high density, with domain walls running parallel to the direction of the electric field. For this system it is predicted [54,55] that correlation lengths in the directions parallel and perpendicular to the field diverge with different exponents, and the (generalized) hyperscaling relation does not hold.

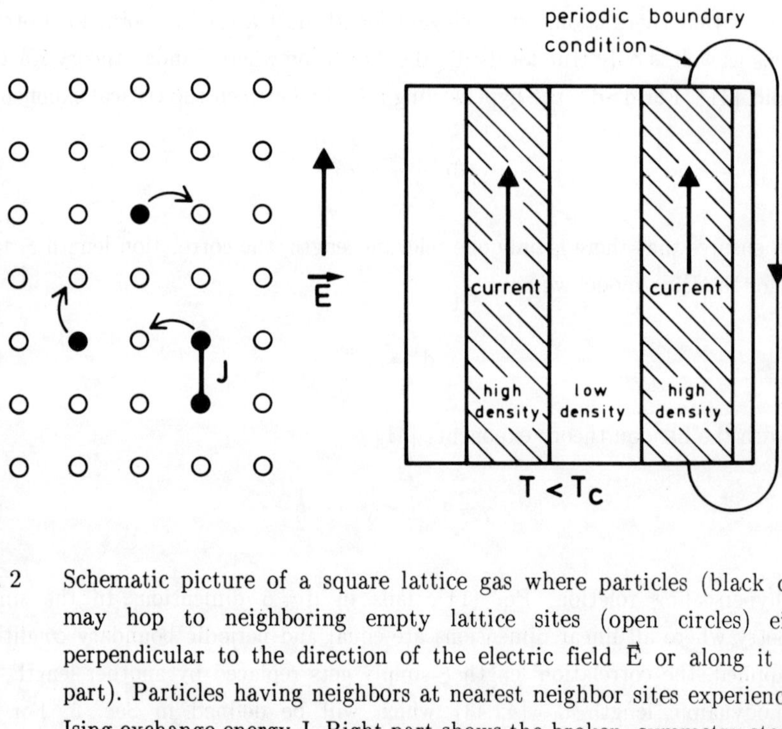

Fig. 2 Schematic picture of a square lattice gas where particles (black dots) may hop to neighboring empty lattice sites (open circles) either perpendicular to the direction of the electric field $\vec{E}$ or along it (left part). Particles having neighbors at nearest neighbor sites experience an Ising exchange energy J. Right part shows the broken–symmetry state at $T<T_c$, where domain walls separating a high density from a low density phase run through the system along the direction of the current (and the field, respectively).

2. THE ORDER PARAMETER PROBABILITY DISTRIBUTION FUNCTION

The statement that in a finite system a phase transition cannot occur, is equivalent to saying that a spontaneously broken symmetry cannot occur in finite systems. Thus, in the absence of symmetry breaking fields the thermal average of the magnetization is identically zero at all nonzero temperatures,

$$\langle m \rangle_{T,L} \equiv 0. \tag{17}$$

Of course, there is no contradiction between Eq. (17) and Eq. (5) because of the absolute value of the order parameter used in Eq. (5): the vanishing of $\langle m \rangle_{T,L}$ is just due to the fact, that both signs of the magnetization are equally probable. This is true both above and below T_c: in both cases the probability distribtution $P_L(m)$ of the magnetization m in the finite cube of volume L^d is symmetric around m=0, Fig. 3, although $P_L(m)$ does change its <u>shape</u> appreciably when T is raised from T less than T_c to $T>T_c$: $P_L(m)$ for an Ising–type model has a double peak structure, the peaks being centered around the two values ($\pm M_{sp}$) of the spontaneous magnetization. In fact, near these peaks $P_L(m)$ can be written approximately as the sum of two gaussians,

$$P_L(m) \sim \exp\left[-\frac{(m-M_{sp})^2 L^d}{2k_BT\chi'}\right] + \exp\left[-\frac{(m+M_{sp})^2 L^d}{2k_BT\chi'}\right], \tag{18}$$

here χ' is the susceptibility below T_c, and agrees with the quantity defined in Eq. (3) for $L\to\infty$.

As T is raised through T_c, the two peaks described by Eq. (18) move towards each other, since $M_{sp}\to 0$; very near T_c, where ξ reaches the size L, the peaks merge into a nontrivial critical distribution function, which will be discussed below, while far above T_c (where again $L>>\xi$) a single gaussian centered around zero magnetization results:

$$P_L(m) \sim \exp\left[-\frac{m^2 L^d}{2 k_B T \chi}\right], \; T>T_c, \; L>>\xi . \tag{19}$$

The susceptibility χ above T_c entering here satisfies the standard thermodynamic fluctuation relation

$$\chi = (L^d/k_B T)(<m^2>_T - <m>_T^2). \tag{20}$$

Since for finite systems in the absence of symmetry–breaking fields we can use Eq. (17), Eq. (20) is equivalent to

$$\chi = (L^d/k_B T)<m^2>_T . \tag{21}$$

If the thermodynamic limit is taken before one lets the symmetry–breaking field go to zero, the susceptibility defined in Eq. (20) reduces automatically to χ' defined in Eq. (3) for $T<T_c$ and to χ defined by Eq. (21) above T_c. However, the susceptibility χ in Eq. (21) is a monotonically increasing function as T is lowered through T_c – it then can be written essentially as $\chi \approx (L^d/k_B T)M_{sp}^2$, for $T<T_c$ and $L>>\xi$. On the other hand, if χ' {Eq. (3)} is used above T_c, it yields $\chi' \approx \chi(1-2/\pi)$ for $L>>\xi$. Thus χ' diverges with the same exponent γ as χ does, but its amplitude is reduced by a factor $1-2/\pi$ [10,20]. Thus, in computer simulations one must use different expressions for the susceptibility above T_c {namely χ, Eq. (21)} and below T_c {namely χ', Eq. (3)}, if one wants to estimate the critical behavior of the susceptibility including prefactors correctly. This problem was often ignored in early Monte Carlo work, where often Eq. (20) was used both above and below T_c. Due to metastability effects, one finds at low temperatures often $<m>_T = M_{sp}$ or $<m>_T = -M_{sp}$, depending on the initial condition, instead of Eq. (17). Due to this nonergodic behavior, which occurs because the "observation time" of the simulation is shorter than the "ergodic time" needed for the system to make transitions from M_{sp} to $-M_{sp}$ or vice versa, Eq. (20) then gives essentially the same result as Eq. (3). However, the peak seen in χ defined from Eq. (20) occurs for the temperature where the "ergodic time" matches the "observation time", since there occurs a transition from the behavior described by Eq. (21) to the behavior described by Eq. (3). Neither the peak height nor the peak temperature of the susceptibility defined in this way are of interest to us here: these quantities depend both on the observation time and the

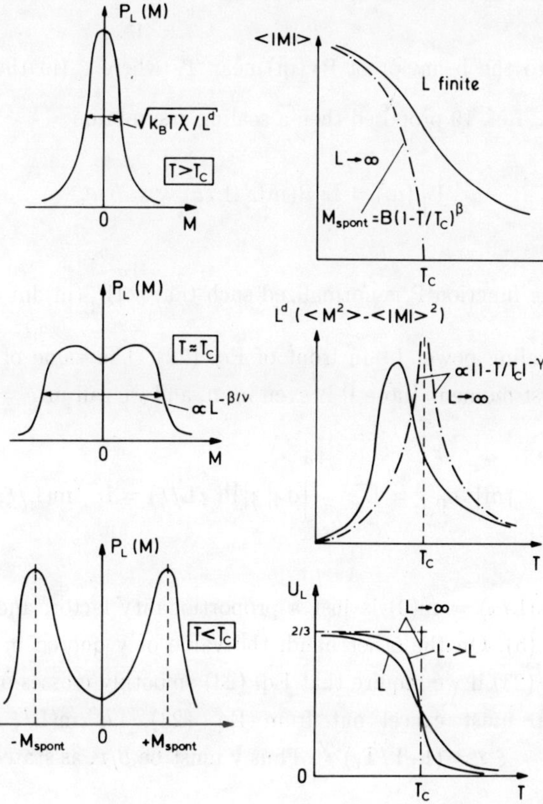

Fig. 3 Schematic evolution of the order parameter probability distribution $P_L(m)$ from $T<T_c$ to $T>T_c$ (left part) and corresponding temperature dependence of the mean order parameter $<|m|>$, the order parameter susceptibility $k_BT\chi' = L^d(<m^2>-<|m|>^2)$, and the reduced fourth order cumulant, $U_L = 1-<m^4>_L/(3<m^2>^2)$. Dash–dotted curves indicate the singular variation which results in the thermodynamic limit, $L\to\infty$.

details of the dynamics produced by the Monte Carlo algorithm for the considered model, and moreover for large observation time do not occur in the critical region but deeply in the regime of the ordered phase.

We now turn to the behavior of $P_L(m)$ near T_c where ξ (in the infinite system would exceed L: Ref. 10 proposed then a scaling assumption

$$P_L(m) = L^y \tilde{P}(mL^y, L/\xi), \; y = \beta/\nu. \tag{22}$$

Here the scaling function $\tilde{P}$ is normalized such that $\int_{-\infty}^{+\infty} P_L(m)dm = 1$ – which also required the scaling power L^y in front of Eq. (22). The value of y is fixed if we consider the first moment {note $\tilde{P}$ is even in m, and we call $mL^y = z$}

$$<|m|>_{T,L} = L^{-y} \int_{-\infty}^{+\infty} dz |z| \tilde{P}(z, L/\xi) = L^{-y} \tilde{m}(L/\xi). \tag{23}$$

For $T = T_c$ $\tilde{m}(L/\infty) = \tilde{m}(0)$ is just a proportionality factor, and thus Eqs. (22), (23) yield Eq. (5). On the other hand, the value of y quoted in Eq. (22) is also fixed from Eq. (23) if we require that Eq. (23) smoothly crosses over to Eq. (9) as $L/\xi \to \infty$: then L must cancel out from Eq. (23), i.e. $\tilde{m}(L/\xi \to \infty) \to (L/\xi)^y$ and $<|m|>_{T,L\to\infty} \sim \xi^{-y} = (1-T/T_c)^{\nu y}$. Thus y must be β/ν, as stated in Eq. (12).

Of course, a similar reasoning yields $<m^2> = L^{-2y}\tilde{\chi}(L/\xi)$, and hence Eq. (21) yields

$$\chi = (L^{d-2\beta/\nu}/k_B T)\tilde{\chi}(L/\xi). \tag{24}$$

Now, provided the hyperscaling relation Eq. (14) holds, we thus indeed find $\chi_{T=T_c} \sim L^{\gamma/\nu}$ (and similarly $\chi'_{max} \sim L^{\gamma/\nu}$), as stated in Eq. (12).

For locating the critical temperature in simulations, it turns out to be most convenient to use the normalized fourth–order cumulant of the distribution,

$$U_L = 1 - \langle m^4 \rangle_L / (3 \langle m^2 \rangle_L^2). \tag{25}$$

While far above T_c where the distribution is a single gaussian, $U_L=0$, we have $U_L \approx 2/3$ for T far below T_c where Eq. (18) holds {since then in Eq. (25) we may use $\langle m^2 \rangle \approx M_{sp}^2$, $\langle m^4 \rangle \approx M_{sp}^4$}. However, at T_c itself U_L tends to a nontrivial "fixed point value" U^*, since Eq. (22) implies that U_L scales as follows

$$U_L = \tilde{U}(L/\xi) \to U_L(T=T_c) = \tilde{U}(0) = U^*. \tag{26}$$

If we thus plot U_L vs. T for different choices of L, these curves should have a unique intersection point U^*, and the temperature where this occurs is T_c – in this way an estimate of T_c can be obtained, which is not biased by any assumptions about critical exponents. Originally, this method was successfully demonstrated for Ising models [10,12,13], but it has also proven useful for the study of two–dimensional XY models with cubic anisotropy [56], clock models [57], models for polymer mixtures [58] (Fig. 4), etc.

From the fact that U_L must be a regular function of temperature, since for finite size systems all properties are regular functions of temperature, we conclude that near T_c {$U^*_{(1)}$ is another constant}

$$U_L \approx U^* + U^*_{(1)} L^{1/\nu}(1-T/T_c) + \ldots \tag{27}$$

which yields for the slope of the curve U_{bL} vs. U_L at U^*

$$\partial(U_{bL})/\partial U_L \big|_{U^*} = b^{1/\nu}. \tag{28}$$

Similarly, the exponent β/ν can be obtained from the ratio of second moments,

$$W_\beta(T=T_c) = -\ln\left[<m^2>_{bL,T_c}/<m^2>_{L,T_c}\right]/\ln b = 2\beta/\nu. \tag{29}$$

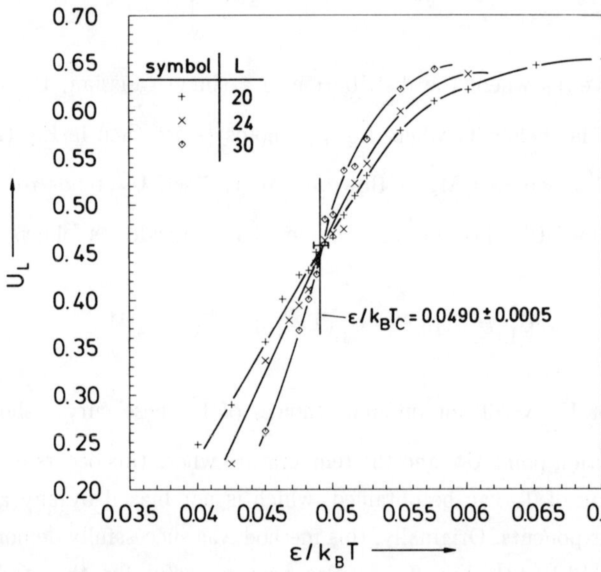

Fig. 4 Cumulant intersection plot for Monte Carlo simulations of the Flory–Huggins lattice model of a polymer mixture: polymers A,B are represented by self–avoiding walks of N=32 steps on a simple cubic lattice. If two neighboring lattice sites are taken by monomers of the same kind, an energy ϵ is won. Monomer volume fractions are $\phi_A = \phi_B = 0.2$, a fraction ϕ_V of sites is left vacant. Within the accuracy of the Monte Carlo data, a unique intersection of the cumulants occurs for $\epsilon/k_BT_c \cong 0.049$. From Sariban and Binder [58].

A particular merit of this technique is that corrections to finite size scaling can be taken into account systematically [10,12–14,59]. Remember that Eqs. (1,2,4,5,11, 22–24,26–29) are asymptotic laws, which hold in the scaling limit $L\to\infty$, $T\to T_c$, L/ξ finite, while for fixed finite L correction terms are expected: For instance, the second moment can be written as [10]

$$<m^2>_{L,T_c} = L^{-2\beta/\nu}\tilde{\chi}(0)\left[1+\chi_c L^{-x_c}+...\right], \tag{30}$$

where x_c is the leading correction exponent and χ_c the associate amplitude factor. Using Eq. (30) instead of the leading term $L^{-2\beta/\nu}\tilde{\chi}(0)$ alone, which has been used so far, Eq. (29) gets replaced by (scale factor $b>1$)

$$W_\beta(T=T_c) \cong 2\beta/\nu - \frac{\chi_c L^{-x_c}}{\ln b}(1-b^{-x_c}). \tag{31}$$

Thus the recipe is to evaluate W_β for many pairs of sizes (L,bL), and to extrapolate the resulting estimates for $2\beta/\nu$ versus $(\ln b)^{-1}$, treating the smaller block size L as a parameter (Fig. 5). The implication of Eq. (31) is that the variation with $(\ln b)^{-1}$ should be linear; all the different curves for the various choices of L should intersect at $(\ln b)^{-1}=0$ in the same value, which then is the final estimate for $2\beta/\nu$. This type of "phenomenological renormalization" of Monte Carlo results yields exponent estimates nearly competitive with the best renormalization group [60], high temperature series [61,62] and Monte Carlo renormalization group [63] estimates. In fact, early series estimates [61] were quite outside of the range of the Monte Carlo estimates, as Fig. 5 shows.

The linear dimensions shown in Fig. 5 are rather small, and perhaps the success of this work should be considered somewhat fortuitous. With the Santa Barbara Ising special purpose computer [64] lattices up to 128^3 were studied, and the cumulant intersection method has yielded a very accurate estimate for T_c, namely $J/k_BT_c = 0.221650 \pm 5$. Unfortunately, this work has not yielded a meaningful analysis of critical exponents: some finite size scaling assumptions used were erroneous, as pointed out in Ref. 42, and in addition there was a spurious irregular size–dependence of some Monte Carlo data, which most probably is due to the use of pseudorandom numbers of bad quality [65]. Thus, a really accurate finite–size scaling analysis of the three–dimensional Ising model on up–to–date computers (Fig. 5 presents data taken at an IBM 3033 scalar computer) remains to be given!

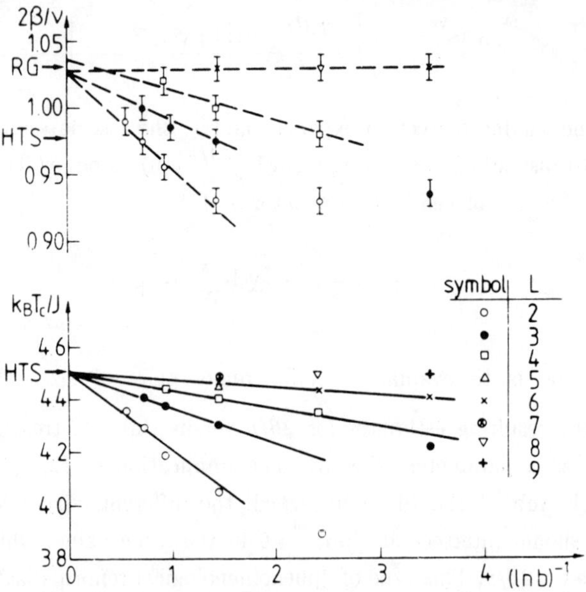

Fig. 5 Estimates for $2\beta/\nu$ for three–dimensional subsystem blocks of 24^3 Ising simple cubic lattice (upper part) and estimates for k_BT_c for fully finite blocks with free boundary conditions (lower part). Estimates from field–theoretic renormalization [60] (denoted by an arrow marked RG) and from high temperature series [61] denoted by an arrow marked HTS) are shown for comparison. From Ref. 10.

3. SYSTEM SHAPE EFFECTS IN FINITE SIZE SCALING

In this section, a very simple extension of finite–size scaling theory will be treated: namely the case of two linear dimensions $L_\perp$, $L_\|$ in the case of the two–dimensional Ising system. Hence, we ask the question: How do the relations for the LxL system, $\chi'_{max} \sim \chi(T_c) \sim L^{\gamma/\nu}$, $<|m|>_{T_c,L} \sim L^{-\beta/\nu}$ generalize?

One important feature to realize is the fact that this generalization depends on the type of boundary condition. Here we briefly follow arguments described in more detail in Ref. 25, and consider first a $L_{\|} x L_{\perp}$ subsystem of an infinite system: we write the susceptibility χ as a sum over correlations (Fig. 6)

$$k_B T \chi = \frac{1}{L_{\|} L_{\perp}} \sum_{x_1=1}^{L_{\|}} \sum_{y_1=1}^{L_{\perp}} \sum_{x_2=1}^{L_{\|}} \sum_{y_2=1}^{L_{\perp}} \langle S(x_1,y_1) S(x_2,y_2) \rangle_T . \tag{32}$$

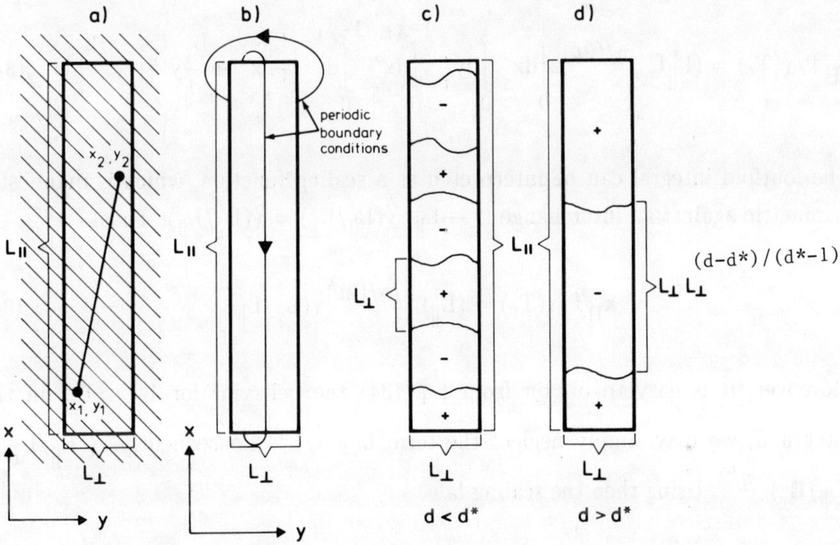

Fig. 6 Geometry of the $L_{\|} x L_{\perp}$ subsystem (a) and a fully periodic $L_{\|} x L_{\perp}$ system (b). Slightly below the critical temperature, case (a) is uniformly ordered while case (b) is nonuniformly ordered, with domain walls an average distance $L_{\perp}$ apart (c). Case (d) refers to a system with d>d* (Sec. 5) slightly below T_c: there the order for $L_{\|} >> L_{\perp}$ is also nonuniform, but the average distance between walls L_o is much larger than $L_{\perp}$.

Using the scaling form for the correlation function [66]

$$<S(0,0)S(x,y)>_T \approx (x^2+y^2)^{-\eta/2} f\left[\frac{x}{\xi},\frac{y}{\xi}\right],\ T\to T_c, \tag{33}$$

where f is some scaling function which needs not to be specified here, we can for $L_\perp >> 1$, $L_{||} >> 1$ convert the sums into integrals near T_c. After rescaling the integration variables one finds

$$k_B T_c \chi(T_c) = (L_{||}L_\perp)^{\gamma/(2\nu)} 2\int_0^1 dx_1' \int_0^1 dy_1' \int_0^{1-x_1'} dx' \int_0^{1-y_1'} dy' \left[\frac{L_{||}}{L_\perp}x'^2 + \frac{L_\perp}{L_{||}}y'^2\right]^{-\eta/2}. \tag{34}$$

The fourfold integral can be interpreted as a scaling function, which is manifestly symmetric against an interchange $L_\perp \leftrightarrow L_{||}$, $\tilde{\chi}(L_{||}/L_\perp) = \tilde{\chi}(L_\perp/L_{||})$. Thus

$$k_B T_c \chi(T_c) = (L_{||}L_\perp)^{\gamma/(2\nu)} \tilde{\chi}(L_\perp/L_{||}). \tag{35}$$

Moreover, it is easy to obtain from Eq. (34) the behavior for $L_{||} >> L_\perp$: in the integrand, we may simply neglect the term $L_\perp y'^2/L_{||}$. Consequently, $\tilde{\chi}(L_{||}/L_\perp) \sim (L_{||}/L_\perp)^{-\eta/2}$. Using then the scaling law

$$\gamma/\nu = 2-\eta, \tag{36}$$

the result can be written as follows,

$$\chi(T_c) \sim (L_{||}L_\perp)^{1-\eta/2}(L_{||}/L_\perp)^{-\eta/2} = L_\perp L_{||}^{1-\eta} = L_\perp L_{||}^{\gamma/\nu-1}. \tag{37}$$

Next we discuss the absolute value of the magnetization $<|m|>_{T_c}$ in the $L_{||} x L_\perp$ subsystem block. We start from the argument that the probability distribution of the order parameter in a finite system with $L_{||} \neq L_\perp$ at T_c scales in a way analogous to Eq. (22),

$$P_{L_{\|},L_{\perp}}(m) = L_{\|}^{\beta/\nu}\tilde{P}(L_{\|}^{\beta/\nu}m, L_{\|}/L_{\perp}). \tag{38}$$

Therefore

$$<|m|>_{T_c} \sim \sqrt{<m^2>_{T_c}} = \sqrt{\frac{k_B T_c \chi(T_c)}{L_{\|}L_{\perp}}} = \left[(L_{\|}L_{\perp})^{\frac{\gamma}{2\nu}-1}\right]^{1/2}\left[\tilde{\chi}\left(\frac{L_{\perp}}{L_{\|}}\right)\right]^{1/2}. \tag{39}$$

Due to the hyperscaling relation {Eq. (14)} the exponent $\gamma/2\nu)-1 = -\beta/\nu$, and thus we find

$$<|m|>_{T_c} \sim (L_{\|}L_{\perp})^{-\beta/2\nu}\,\tilde{m}(L_{\|}/L_{\perp}) \underset{L_{\|}>>L_{\perp}}{\sim} L_{\|}^{-\beta/\nu}. \tag{40}$$

For $L_{\|}>>L_{\perp}$ the order parameter hence becomes independent of $L_{\perp}$.

The behavior, however, is very different if one considers a fully finite system (either with periodic boundary conditions {Fig. 6b} or with free boundary conditions or with mixed free/periodic boundary conditions [26]). For simplicity, we mainly consider the fully periodic geometry here. Again we proceed integrating the correlation function

$$k_B T_c \chi(T_c) \approx \int_0^{L_{\|}} dx \int_0^{L_{\perp}} dy <S(0,0)S(x,y)>_{T_c,L_{\|},L_{\perp}}, \tag{41}$$

but now the answer is different! The reason is that the correlation function is no longer simply given by a power law decay, but it is modified itself by a finite size scaling function $\tilde{F}$

$$<S(0,0)S(x,y)>_{T_c} \approx (x^2+y^2)^{-\eta/2}\,\tilde{F}\left[\frac{x}{L_{\perp}},\frac{y}{L_{\perp}},\frac{L_{\|}}{L_{\perp}}\right]. \tag{42}$$

For $L_{\|}>>L_{\perp}$ and $L_{\perp}<<x<<L_{\|}$ there appears an exponential decay in the

correlation function,

$$<S(0,0)S(x,y)>_{T_c} \sim L_\perp^{-\eta} \exp\{-\text{const}\ (x/L_\perp)\},\ L_\perp << x << L_\parallel. \tag{43}$$

The physical interpretation of this behavior is the following: what matters in the finite size scaling of a system which is finite in all its dimensions is its smallest linear dimension, $L_\perp$. So, when we approach criticality, the correlation length first grows uniformly until $\xi \approx L_\perp$; then the system feels its finite size already very strongly, and therefore ξ stays of the order of $L_\perp$ throughout the critical region. Thus we must have an exponential decay in x–direction, with a correlation length proportional to $L_\perp$. At and slightly below T_c the system order hence is nonuniform, see Fig. 6c.

Although the scaling structure which results from this ansatz is the same as above,

$$\chi(T_c) = (L_\parallel L_\perp)^{\gamma/(2\nu)} \tilde{\chi}(L_\perp/L_\parallel), \tag{44}$$

the scaling function now is very different: for $L_\parallel >> L_\perp$ we basically have

$$k_B T_c \chi(T_c) \approx L_\perp^{1-\eta} \int_0^\infty dx\ \exp\left\{-\text{const.}(x/L_\perp)\right\} \sim L_\perp^{2-\eta} = L_\perp^{\gamma/\nu}. \tag{45}$$

Note that in the integral the upper limit $L_\parallel$ could be replaced by infinity; thus for large $L_\parallel$ $\chi(T_c)$ does not depend on $L_\parallel$, unlike the subsystem case {Eq. (37)}.

Applying again the argument relating the order parameter at T_c to the susceptibility,

$$<|m|>_{T_c} \sim \sqrt{<m^2>_{T_c}} = \sqrt{k_B T_c \chi(T_c)/(L_\parallel L_\perp)} \underset{L_\parallel >> L_\perp}{\sim} L_\perp^{-\beta/\nu} (L_\perp/L_\parallel)^{1/2}. \tag{46}$$

Again the result differs from the corresponding result for the subsystem geometry {Eq. (40)}.

Next we turn to discuss Monte Carlo tests of these ideas. Jian–Sheng Wang and the author [25] have simulated subsystems of rectangular $L_{||} x L_{\perp}$ geometry with $4 \leq L_{||}, L_{\perp} \leq 128$. The size of the total system was 1024 x 1024. Ideally, one would like to simulate subsystems of an infinite system: this is impossible, of course! Already the system of size 1024 x 1024 is the largest system that ever was equilibrated right at the critical point [25]. This calculation was possible only by applying the novel Swendsen–Wang cluster algorithm [67], which largely removes the problem of "critical slowing down" [16–18]. While in this way, only 5 hours CPU time on the Cyber 205 of the John von Neumann Center at Princeton were needed, fast multispin coding algorithms [68] also available on this machine but relying on standard Monte Carlo methods would have taken at least 500 hours CPU time: Critical slowing down implies a correlation time [16–18] $\tau \approx L_{max}^{z}$, and since $z \approx 2$ and $L_{max}=1024$, the time needed to pass that subsequent configurations are uncorrelated is of the order of 10^6 Monte Carlo steps (MCS)/spin for standard Monte Carlo methods.

This example shows that in order to make progress with computer simulations today one needs both new ideas about their analysis and new ideas about better algorithms.

Thus with the present calculation we were able to verify both the predicted scaling property and to compare with the scaling function (Fig. 7) which for the susceptibility can be calculated exactly from the critical correlations [69]. This scaling function crosses over from one straight line

$$\chi(T_c)L_{||}^{-\gamma/\nu} \sim (L_{\perp}/L_{||})^{\gamma/2\nu} \quad \text{for } L_{\perp} \geq L_{||}. \tag{47}$$

to another straight line

$$\chi(T_c)L_{||}^{-\gamma/\nu} \sim L_{\perp}/L_{||} \quad \text{for } L_{\perp} << L_{||}. \tag{48}$$

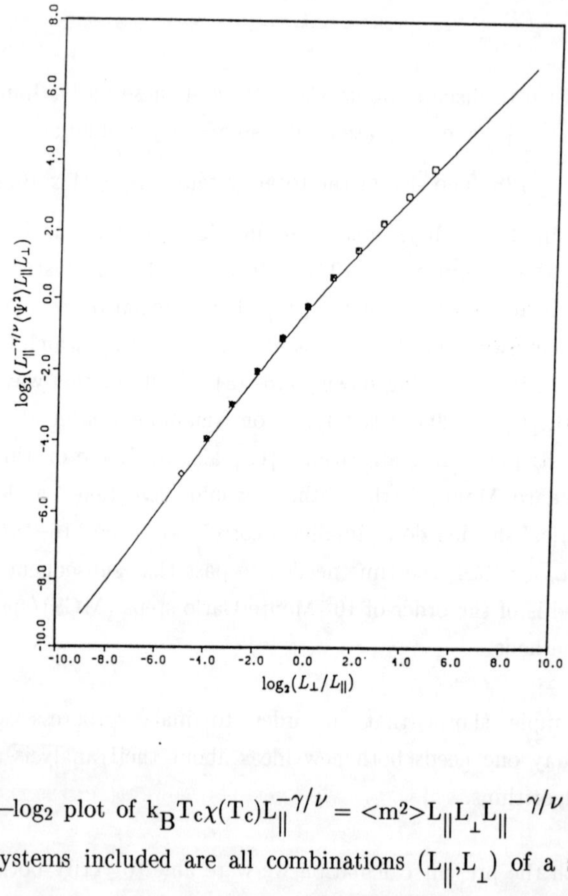

Fig. 7 Log$_2$–log$_2$ plot of $k_B T_c \chi(T_c) L_{\parallel}^{-\gamma/\nu} = \langle m^2 \rangle L_{\parallel} L_{\perp} L_{\parallel}^{-\gamma/\nu}$ versus $L_{\perp}/L_{\parallel}$. Subsystems included are all combinations $(L_{\parallel}, L_{\perp})$ of a system of size L_{max} with L_{max}=1024, with linear dimensions $L_{\parallel}, L_{\perp}$ being 4,8,16,32,64 and 128, respectively. The smooth curve is a calculation of the scaling function based on the critical correlations. From Binder and Wang [25].

For the order parameter (Fig. 8) the scaling function drawn here is that for $\sqrt{\langle m^2 \rangle_{T_c}}$ rather than that for $\langle |m| \rangle_{T_c}$, and hence gives only the qualitative trend; but again the data scale well.

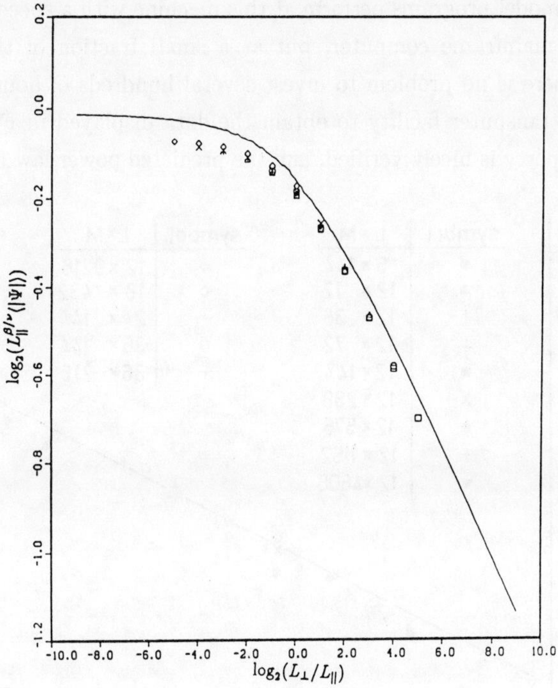

Fig. 8 $\log_2$–$\log_2$ plot of $L_{\|}^{\beta/\nu}<|m|>$ versus $L_\perp/L_{\|}$, using the same subsystems and geometry as in Fig. 7. The curve is the scaling function for $L_{\|}^{\beta/\nu}\sqrt{<m^2>}$, which can be related to the scaling function of Fig. 7. From Binder and Wang [25].

For fully finite systems rather than subsystems of a large system, calculations need to be done for many different geometries $L_\perp x L_{\|}$ at T_c. Now critical slowing down is not such a problem to warrant the use of the Swendsen–Wang [67] cluster algorithm – but still a large effort in computing time is needed for taking accurate data at T_c for so many geometries. For this purpose, it has been found advantageous to use [26] the multi–transputer–facility at the University of Mainz

[70,71]. Ising model programs perform at this machine with a speed comparable to an IBM 3081 mainframe computer, but at a small fraction of the costs of the latter. Thus there is no problem to invest several hundreds of hours of CPU time on this multi–transputer facility to obtain the data displayed in Figs 9, 10: again the scaling property is nicely verified, and the predicted power law for the

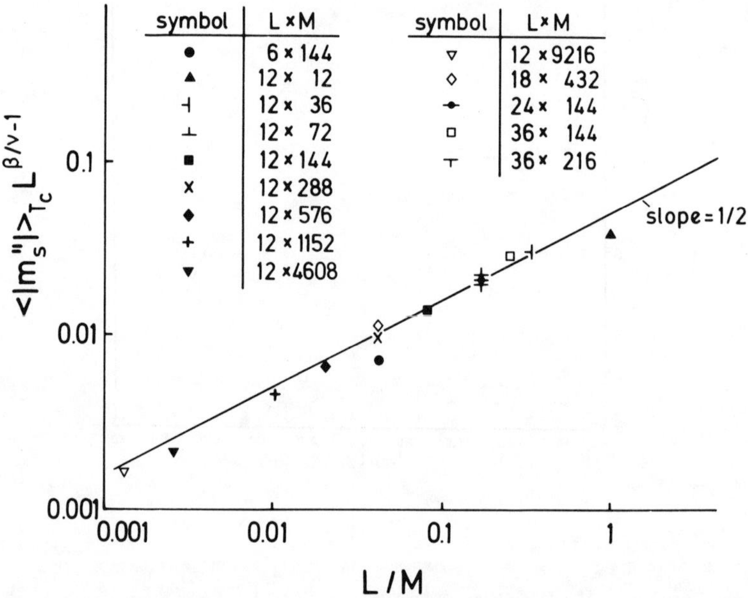

Fig. 9 Log–log plot of the scaled magnetization at Tc, $<|m|>_{T_c} L^{\beta/\nu}$, versus the ratio L/M, for LxM ferromagnetic nearest neighbor Ising square lattices, with two free boundaries of length M and otherwise periodic boundary conditions. Individual choices of LxM are indicated by different symbols. Straight line shows the predicted power law, Eq. (46). From Albano et al. [26].

magnetization $\{<|m|>_{T_c} L_\perp^{\beta/\nu} \sim (L_\perp/L_{||})^{1/2}$, Eq. (40)$\}$ is confirmed, just as the fact that $\chi(T_c)$ is independent of $L_{||}$ for $L_{||} >> L_\perp$ (Fig. 10).

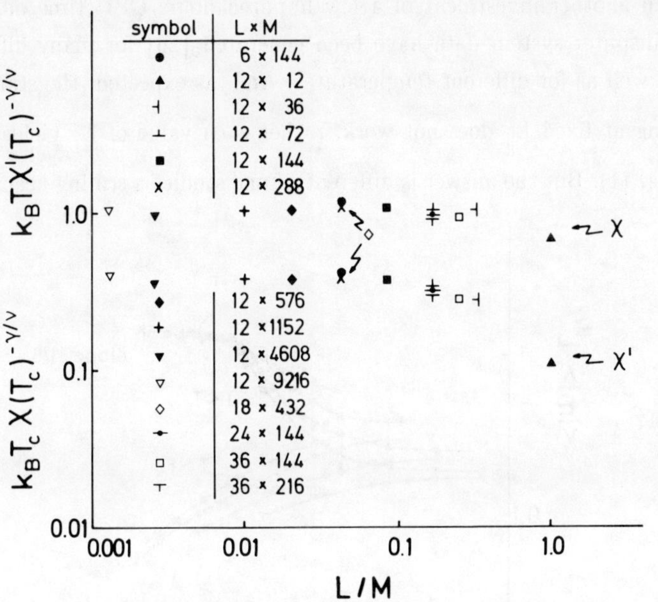

Fig. 10 Log–log plot of the scaled susceptibilities $k_B T_c \chi(T_c) L^{-\gamma/\nu}$ (upper part) and $k_B T_c \chi'(T_c) L^{-\gamma/\nu}$ (lower part) versus L/M, for the same model as Fig. 9. Individual choices of LxM are indicated by different symbols. From Albano et al. [26].

Of course, it is not difficult to extend this shape–dependent finite–size scaling to temperatures off T_c – one just has to add a second scaling variable ($t=T/T_c-1$),

$$<|m|>_T = L_\perp^{-\beta/\nu} \tilde{m}(L_\perp t^\nu, L_\perp/L_{||}), \tag{49}$$

$$\chi(T) = L_\perp^{\gamma/\nu} \tilde{\chi}(L_\perp t^\nu, L_\perp/L_{||}). \tag{50}$$

Thus at fixed length $L_{||}$ the standard [8,9,20] finite size scaling data collapsing plot $\{<|m|>_T L_\perp^{\beta/\nu}$ versus $L_\perp |t|^\nu\}$ should not work, due to the presence of the second variable!

With another investment of a few hundred hours CPU time on the Mainz multi–transputer system data have been generated [26] for many different sizes $L_{\|}xL_{\perp}$ as well as for different temperatures. And, as expected, the standard finite size scaling at fixed $L_{\|}$ does not work, rather each value of $L_{\perp}$ yields a separate curve (Fig. 11). But the answer is different, if one studies a scaling keeping the

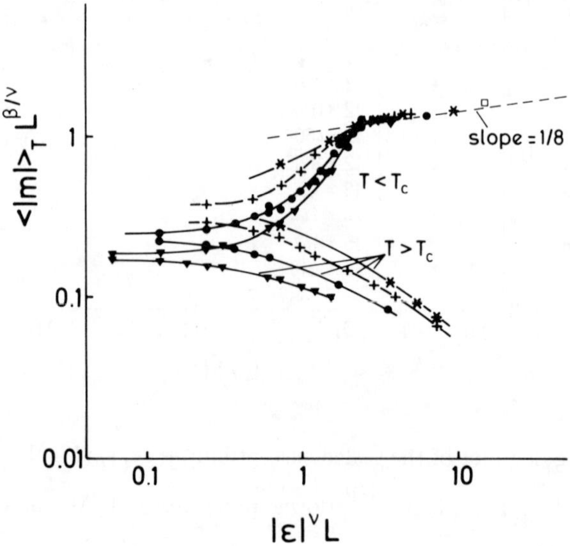

Fig. 11 Finite–size scaling plot of the scaled magnetization, $<|m|>_T L^{\beta/\nu}$, versus scaled temperature, $|t|^y L$, for LxM lattices, with two free boundaries of length M=144, and periodic boundary conditions in the other direction. Data included are for L=6 (triangles), 12 (dots), 24 (crosses), 36 (stars) and 144 (squares). Curves are only guides to the eye. From Albano et al. [26].

"aspect ratio" $L_{\perp}/L_{\|}$ fixed: now finite size scaling in the usual form works (Fig. 12), as expected {Eq. (49)}. Thus we conclude, corroborating Refs. 21,24,25, and 26, that shape effects on finite size scaling are well understood for systems with isotropic critical behavior. In particular, for quasi–one dimensional systems a characteristic feature is that at T_c and slightly below T_c the system exhibits nonuniform structure (Fig. 6c), domains form with an average distance of order L at T_c between domain walls. The crossover from a multidomain state to a monodomain state is responsible for the steep rise in the $<|m|>_T$ vs. t plots {Figs. 11,12}. Of course, it is also possible to identify directly the domain patterns in snapshot pictures of the spin configurations of the model (Fig. 13).

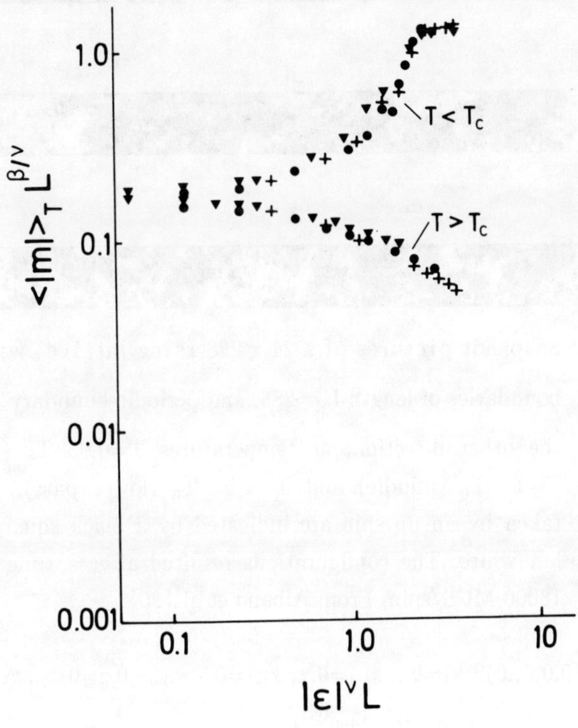

Fig. 12 Finite size scaling plot of the scaled magnetization $<|m|>_T L^{\beta/\nu}$ versus scaled temperature $|t|^{\nu}L$, for the same model as in Figs. 9–11, keeping now the ratio L/M=1/6 fixed. Dots show data for L x M = 12 x 72; crosses for 24 x 144; and triangles for 36 x 216, respectively. From Albano et al. [26].

4. FINITE SIZE SCALING FOR ANISOTROPIC CRITICAL BEHAVIOR

We assume correlation functions to scale as follows

$$<S(0,\ldots,0)S(x_1=0,\ldots,x_{d-1}=0,z)> \cong z^{-(d-2+\eta_{\parallel}^{RS})}\tilde{G}(z/\xi_{\parallel}), \tag{51}$$

a)

b)

c)

Fig. 13: Snapshot pictures of a 24 x 288 Ising lattice, with two free boundaries of length $L_{\parallel}=288$, and periodic boundary conditions in the other direction, at temperatures T=0.95 T_c (upper part), T=1.0 T_c (middle) and T=1.05 T_c (lower part). Lattice sites taken by an up spin are indicated by a black square, others are left white. The configurations resulted after a time evolution of 18000 MCS/spin. From Albano et al. [26].

$$<S(0,0,\ldots,0)S(x_1=0,\ldots x_{j-1}=0,x_j,x_{j+1}=0,\ldots x_{d-1}=0,z=0)>_T \cong$$
$$x_j^{-(d-2+\eta_{\perp}^{RS})}\tilde{G}_{\perp}(x_j/\xi_{\perp}), \tag{52}$$

with

$$\xi_{\parallel} \sim t^{-\nu_{\parallel}}, \; \xi_{\perp} \sim t^{-\nu_{\perp}}, \tag{53}$$

and also decay exponents $\eta_{\parallel}^{RS}$, $\nu_{\perp}^{RS}$ different from each other for the decay of correlations at T_c in parallel and perpendicular directions have been introduced.

In this case the fluctuation relation for the susceptibility is

$$k_B T\chi' = L_{\|} L_{\perp}^{d-1} (<m^2>_T - <|m|>_T^2). \tag{54}$$

We assume that the generalization of finite–size scaling to this situation is simply [25,35]

$$<|m|>_T = L_{\|}^{-\beta/\nu_{\|}} f_1\left[\frac{\xi_{\|}}{L_{\|}}, \frac{\xi_{\perp}}{L_{\perp}}\right], \tag{55}$$

$$k_B T\chi' = L_{\|}^{\gamma/\nu_{\|}} f_2\left[\frac{\xi_{\|}}{L_{\|}}, \frac{\xi_{\perp}}{L_{\perp}}\right], \tag{56}$$

with f_1, f_2 suitable scaling functions.

First of all it will be shown [25] that this description makes sense only if a generalized hyperscaling relation [27–29] holds, namely

$$(d-1)\nu_{\perp} + \nu_{\|} = \gamma + 2\beta. \tag{57}$$

In order to show this, we only have to require that $<|m|>_T^2$ and $<m^2>_T$ scale in the same way, as we did above. So

$$<m^2>_T = L_{\|}^{-\beta/\nu_{\|}} f_1'(\xi_{\|}/L_{\|}, \xi_{\perp}/L_{\perp}), \tag{58}$$

with f_1' another scaling function. Now from Eqs. (55), (58), and (54) an alternative expression for the susceptibility χ' is constructed,

$$k_B T\chi' = L_{\perp}^{d-1} L_{\|}^{1-\beta/\nu_{\|}} \{f_1'(\xi_{\|}/L_{\|}, \xi_{\perp}/L_{\perp}) - f_1^2(\xi_{\|}/L_{\|}, \xi_{\perp}/L_{\perp})\}. \tag{59}$$

Next the curly bracket is rewritten in a manner which defines another scaling function f_2,

$$\{...\} = \left[\frac{L_\perp}{\xi_\perp} / \left(\frac{L_{||}}{\xi_\perp}\right)^{\nu_\perp/\nu_{||}}\right]^{1-d} f_2\left[\frac{\xi_{||}}{L_{||}},\frac{\xi_\perp}{L_\perp}\right] = L_\perp^{-(d-1)} L_{||}^{(d-1)(\nu_\perp/\nu_{||})} f_2. \tag{60}$$

Note that the factor is the square bracket in Eq. (60), split off from the curly bracket, is constructed such as to yield a factor $L_\perp^{-(d-1)}$, in order to cancel the corresponding factor $L_\perp^{d-1}$ in Eq. (59) when Eqs. (59),(60) are combined:

$$k_B T\chi' = L_{||}^{1-2\beta/\nu_{||}+(d-1)\nu_\perp/\nu_{||}} f_2\left[\frac{\xi_{||}}{L_{||}},\frac{\xi_\perp}{L_\perp}\right]. \tag{61}$$

Now we see that Eqs. (56),(61) are consistent with each other if the function f_2 defined in Eq. (60) reduces to the f_2 in Eq. (56), and if the powers of the $L_{||}$–prefactors are equal: $\gamma/\nu_{||} = 1-2\beta/\nu_{||}+(d-1)\nu_\perp/\nu_{||}$. But the latter condition is nothing else than the generalized hyperscaling relation, Eq. (57)!

Let us now assume, that this is true, and we use Eqs. (55),(56) to consider the behavior right at T_c. These two arguments of f_1,f_2 would be infinite – thus we first have to rewrite $f_1(\xi_{||}/L_{||},\xi_\perp/L_\perp)$ as follows

$$f_1(\xi_{||}/L_{||},\xi_\perp/L_\perp) = f_1''\left\{\xi_{||}/L_{||},\left(\frac{\xi_{||}}{L_{||}}\right)^{-\nu_\perp/\nu_{||}}(\xi_\perp/L_\perp)\right\} = f_1''\left[\frac{\xi_{||}}{L_{||}},\frac{L_{||}^{\nu_\perp/\nu_{||}}}{L_\perp}\right], \tag{62}$$

before letting $T\to T_c$. Thus one immediately predicts [25]

$$k_B T_c\chi(T_c) = L_{||}^{\gamma/\nu_{||}} \tilde{\chi}\left[L_\perp/L_{||}^{\nu_\perp/\nu_{||}}\right]. \tag{63}$$

For a fully finite geometry, this reduces to the following limiting behavior [25]

$$k_BT_c\chi(T_c) \sim L_{\|}^{\gamma/\nu_{\|}}, \quad L_{\|}<<L_{\perp}^{\nu_{\|}/\nu_{\perp}}, \tag{64a}$$

and

$$k_BT_c\chi(T_c) \sim L_{\perp}^{\gamma/\nu_{\perp}}, \quad L_{\|}>>L_{\perp}^{\nu_{\|}/\nu_{\perp}}. \tag{64b}$$

This reasoning can be extended to the order parameter, and to the subsystem geometry, etc.; we refer the reader to Ref. [25] for details. But the main point of this section is: a simple finite size scaling now no longer requires a constant shape ratio $L_{\perp}/L_{\|}$, as in the previous section, but rather the ratio $L_{\perp}/L_{\|}^{\nu_{\perp}/\nu_{\|}}$ needs to be kept constant! Since in the cases of interest the ratio $\nu_{\perp}/\nu_{\|}$ is not known beforehand, finite size scaling is difficult to apply. For Lifshitz points [27–32], $\nu_{\|}\neq\nu_{\perp}$ is expected but there the additional difficulty arises to accurately locate the Lifshitz–point in the parameter space of the model [30–32]. Nontrivial finite size effects have been noted [32] but a quantitatively meaningful finite size scaling analysis for a Lifshitz point is still lacking. Perhaps models such as directed percolation [33,34] are more suitable for a test of these ideas.

One example for anisotropic critical behavior in two dimensions is the exactly solved Kasteleyn model [72], for which Bhattacharjee and Nagle [73] analysed finite size scaling of the specific heat C. It scales as follows

$$C = L_{\|}^{\alpha/\nu_{\|}}\,\tilde{C}\left[L_{\|}^{1/\nu_{\|}}, L_{\perp}^{1/\nu_{\perp}}/L_{\|}^{1/\nu_{\|}}\right], \tag{65}$$

where both the exponents are known exactly (α=1/2, $\nu_{\perp}$=1, $\nu_{\perp}$=1/2) [73], satisfying the anisotropic hyperscaling relation $\nu_{\|}+\nu_{\perp}=2-\alpha$, and also the scaling function $\tilde{C}$ has been calculated [73]. It has also been shown that for small $L_{\perp}$ pronounced corrections to finite size scaling occur [73].

This model is of great physical interest, because it is isomorphic to domain–wall models of px1 commensurate–incommensurate phase transitions [74,75] in adsorbed layers at surfaces.

5. FINITE SIZE SCALING IN ISOTROPIC SYSTEMS WITHOUT HYPERSCALING

After the above remarks about the anisotropic case with hyperscaling here the opposite case is considered, isotropic systems without hyperscaling. Since we have in mind systems above their marginal dimension d*, this case can be understood by suitable generalizations of Landau theory! This will be described next.

We wish to calculate the partition function with a Ginzburg–Landau type Hamiltonian

$$Z = \int D\psi(\vec{r}) \exp[-\mathcal{H}_{\text{eff}}\{\psi\}] \tag{66}$$

with $[t = T/T_c - 1]$ is defined without the absolute value here

$$\mathcal{H}_{\text{eff}}\{\psi\} = \int d\vec{r} \left\{ \frac{1}{2}(\nabla\psi)^2 + \frac{1}{2} t\psi^2 + \frac{u_o}{4!}\psi^4 \right\}, \tag{67}$$

where u_o is a constant. Units of the order parameter field ψ and of length are here defined such that constants in front of the $(\nabla\psi)^2$, ψ^2 terms are absorbed. Following Zinn–Justin and Brézin [45] ψ is decomposed in terms of an average value $\overline{\psi}$ and Fourier compoments $\delta\psi_{\vec{q},q_{\|}}$

$$\psi(\vec{x}_{\perp}, z) = \overline{\psi} + \sum_{\vec{q}_{\perp} \neq 0} \sum_{q_{\|} \neq 0} \exp(i\vec{q}_{\perp} \cdot \vec{x}_{\perp} + iq_{\|} z)\, \delta\psi_{\vec{q}_{\perp}, q_{\|}}\,, \tag{68}$$

where $\vec{q}_{\perp}$ is "quantized" in units of $2\pi/L_{\perp}$ and $q_{\|}$ is quantized in units of $2\pi/L_{\|}$. For $d > d^*$, the fluctuations $\delta\psi_{\vec{q},q_{\|}}$ can be integrated out perturbatively, if $L_{\|}$, $L_{\perp}$

are of comparable size. This yields

$$Z = \int d\overline{\psi} \exp[-\mathcal{H}(\overline{\psi})], \tag{69}$$

$$\mathcal{H}(\overline{\psi}) = L_{\|} L_{\perp}^{d-1} \left[\frac{1}{2} t \overline{\psi}^2 + \frac{u_o}{4!} \overline{\psi}^4\right]. \tag{70}$$

Thus the probability $P_{L_{\|},L_{\perp}}(\overline{\psi})$ to find an order parameter $\overline{\psi}$ in a system with linear dimensions $L_{\|},L_{\perp}$ simply is [42]

$$P_{L_{\|},L_{\perp}}(\overline{\psi}) \sim \exp\left[-\left[\frac{1}{2} t \overline{\psi}^2 + \frac{u_o}{4!} \overline{\psi}^4\right] L_{\|} L_{\perp}^{d-1}\right]. \tag{71}$$

Introducing the spontaneous order parameter ψ_{sp} and rescaling the order parameter field by ψ_{sp},

$$\psi_{sp} = (-6t/u_o)^{1/2}, \ \tilde{\psi} = \overline{\psi}/\psi_{sp}, \tag{72}$$

the probability distribution can be rewritten as

$$P_{L_{\|},L_{\perp}}(\overline{\psi}) \sim \exp\left[-\frac{1}{4}(\tilde{\psi}^2-1)^2 \left[\frac{L_{\|}}{\ell_T}\right] \left[\frac{L_{\perp}}{\ell_T}\right]^{d-1}\right] \tag{73}$$

where the <u>thermodynamic length ℓ_T</u> is defined as [42,43]

$$\ell_T = (u_o/3)^{1/d} |t|^{-2/d}. \tag{74}$$

This length for d>d* has a temperature dependence manifestly different from the correlation length ξ. E.g., for the five–dimensional Ising model we have

$$\ell_T \sim |t|^{-2/5}, \ \xi \sim |t|^{-1/2}. \tag{75}$$

Taking the first and second moment of Eq. (73) one finds

$$<|m|>_T = \left[L_{\|}L_{\perp}^{d-1}\right]^{-1/4} \tilde{m}(L_{\|}/\ell_T, L_{\perp}/\ell_T), \tag{76}$$

$$k_B T_c \chi(T) = \left[L_{\|}L_{\perp}^{d-1}\right]^{1/2} \tilde{\chi}(L_{\|}/\ell_T, L_{\perp}/\ell_T). \tag{77}$$

The "renormalized coupling constant" $g_L = -3U_L$ {Eq. (25)} becomes

$$g_L = \tilde{g}(L_{\|}/\ell_T, L_{\perp}/\ell_T). \tag{78}$$

Moreover, from Eq. (73) one can work out the scaling functions $\tilde{m}$, $\tilde{\chi}$ and $\tilde{g}$ explicitly [45]. Numerical results for the five-dimensional Ising system with $L_{\|}=L_{\perp}=L$ (Fig. 14) are nicely consistent with this description (Fig. 15). So when $L_{\|}$ and $L_{\perp}$ are comparable, there exists for $d>d^*$ a very simple generalization of finite size scaling: just the correlation length ξ must be replaced by the thermal length ℓ_T. Of course, from Eq. (18) it is clear that <u>for all d</u> the probability distribution $P_L(m)$ for $L>>\xi$ and $T<T_c$ can be written in the form of Eq. (73), with the thermal length ℓ_T in general being defined as

$$\ell_T = [k_B T_c \chi'(T)/M_{sp}^2(T)]^{1/d} \sim (-t)^{-(2\beta+\gamma)/d}. \tag{79}$$

But due to the hyperscaling relation, Eq. (14), for $d<d^*$ we simply have $\ell_T \sim \xi$ and thus there is no contradiction then with the treatment of Sec. 2.

The situation for $d>d^*$ gets more complicated, however, when we consider shapes with $L_{\|}>>L_{\perp}$: while it is still o.k. to integrate out all transverse fluctuations, inhomogeneous ordering in the longitudinal direction must be allowed for! Thus we introduce a Fourier transform in the transverse direction only,

$$\psi(\vec{x}_{\perp},z) = \psi(z) + \sum_{\vec{q}_{\perp}\neq 0} \exp(i\vec{q}_{\perp}\cdot\vec{x}_{\perp})\delta\psi_{\vec{q}}(z). \tag{80}$$

Integrating out the $\delta\psi_{\vec{q}_\perp}(z)$ then yields a functional integral for a one–dimensional Hamiltonian,

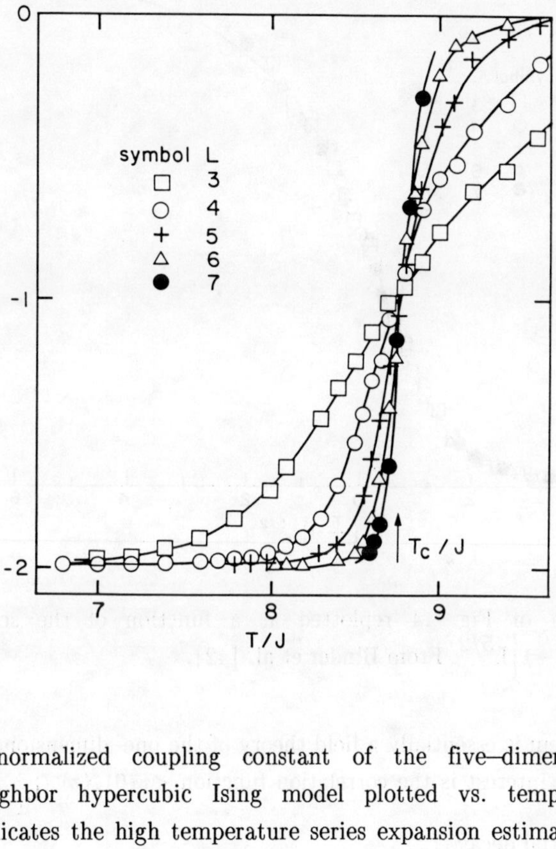

Fig. 14 Renormalized coupling constant of the five–dimensional nearest neighbor hypercubic Ising model plotted vs. temperature; arrow indicates the high temperature series expansion estimate of $T_c(k_B \equiv 1)$ due to Ref. [76]. Note that all curves $g_L(T)$ intersect at T_c in a unique point, consistent with Eq. (78), but contrary to an erroneous scaling ansatz made in Ref. [64]. From Binder et al. [42].

$$Z = \int \mathscr{D}\psi(z) \exp[-\mathscr{H}_{\text{eff}}\{\psi\}], \qquad (81)$$

$$\mathscr{H}_{\text{eff}}(\{\psi\}) = L_\perp^{d-1} \int_0^{L_\|} dz \left\{ \frac{1}{2}\left[\frac{d\psi}{dz}\right]^2 + \frac{1}{2} t \psi^2 + \frac{u_o}{4!}\psi^4 \right\}. \tag{82}$$

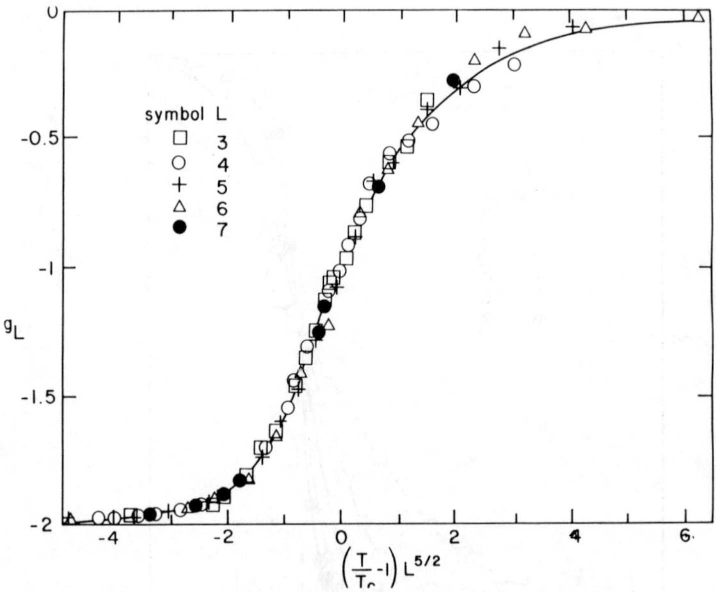

Fig. 15 Data of Fig. 14 replotted as a function of the scaling variable $\left[\frac{T}{T_c} - 1\right] L^{5/2}$. From Binder et al. [42].

Since the problem is essentially a field theory of the one–dimensional Ising model, the quantity of interest is the correlation function $<\psi(0)\psi(z)>_{L_\perp,T}$, which has a simple exponential decay

$$<\psi(0)\psi(z)>_{L_\perp,T} \approx A(L_\perp,T)\exp[-z/\xi_\|(L_\perp,T)],\ L_\perp << z << L_\|. \tag{83}$$

From the correlation we can again obtain susceptibility and order parameter,

$$k_B T\chi = L_\perp^{d-1} \int_0^{L_{\|}} \langle\psi(0)\psi(z)\rangle_{L_\perp,T} dz \underset{L_{\|}\to\infty}{\approx} A(L_\perp,T)L_\perp^{d-1}\xi_{\|}(L_\perp,T), \tag{84}$$

$$\langle|\psi|\rangle \sim \sqrt{\langle\psi^2\rangle} = [k_B T\chi/L_{\|}L_\perp^{d-1}]^{1/2} \approx [A(L_\perp,T)\xi_{\|}(L_\perp,T)/L_{\|}]^{1/2}. \tag{85}$$

We now estimate both $\xi_{\|}(L_\perp,T)$ and $A(L_\perp,T)$ from the simple dimensional analysis [25] arguments: the scaling behavior of the theory can thus be found without the need to precisely evaluate the functional integral, Eq. (82). Our reasoning is as follows: For $t \geq 0$ the minimum of $\mathcal{H}_{eff}(\psi)$ is given by $\mathcal{H}_{eff}(\psi_{min})=0$ with $\psi_{min}(z)=0$. Now only such paths $\psi(z)$ will contribute to the path integral significantly for which $\mathcal{H}_{eff}(\psi)$ is of order of unity – paths $\psi(z)$ for which $\mathcal{H}_{eff}(\psi)>>1$ have negligible statistical weight.

We consider "typical" paths of the trial function form [25]

$$\psi(z) = \psi_o \exp(-z/\xi_{\|}) \tag{86}$$

consistent with the expected correlation function form, Eq. (83). Here ψ_o and $\xi_{\|}$ have to be chosen such that $\mathcal{H}_{eff}(\psi)\approx 1$ for such a typical path. The three terms in the integrand contribute to $\mathcal{H}_{eff}$ the terms I, II, III (again we let $L_{\|}\to\infty$)

$$\mathcal{H}_{eff}(\{\psi\}) = L_\perp^{d-1} \int_0^\infty dz \left\{\frac{1}{2}\left[\frac{d\psi}{dz}\right]^2 + \frac{1}{2}t\psi^2 + \frac{u_o}{4!}\psi^4\right\} = I+II+III, \tag{87}$$

with

$$I \sim L_\perp^{d-1}\psi_o^2/\xi_{\|},\ II \sim L_\perp^{d-1}t\psi_o^2\xi_{\|},\ III \sim L_\perp^{d-1}u_o\psi_o^4\xi_{\|}. \tag{88}$$

For $t>0$ terms I and II dominate. Hence we put, requiring that both these terms are of order unity

$$L_\perp^{d-1}\psi_o^2/\xi_{||} = L_\perp^{d-1} t\psi_o^2 \xi_{||} = 1, \tag{89}$$

which yields

$$\xi_{||} \sim t^{-1/2},\ \psi_o \sim t^{-1/2}/L_\perp^{d-1},\ \chi \sim t^{-1}. \tag{90}$$

For the last equation, we invoked $\psi_o^2 \sim A(L_\perp,T)$ and used Eq. (84). As expected, in Eq. (90) we just have recovered the standard results of Landau theory.

But a new situation emerges for $T>T_c$, where II in Eq. (87) vanishes and I is balanced by III in Eq. (87). Again requiring that these terms are of order unity,

$$L_\perp^{d-1}\psi_o^2/\xi_{||} = L_\perp^{d-1} u_o \psi_o^4 \xi_{||} = 1, \tag{91}$$

we obtain, calling $\xi_{||}(t{=}0) = \xi_o$ [25]

$$\xi_o \sim u_o^{-1/2}\, L_\perp^{(d-1)/3}, \tag{92}$$

$$\psi_o^2 \sim u_o^{-1/3}\, L_\perp^{-2(d-1)/3}, \tag{93}$$

and invoking once more $\psi_o^2 \sim A(L_\perp,T)$ and Eq. (84) we get

$$k_B T_c \chi(T_c) \sim L_\perp^{2(d-1)/3},\ L_{||}\to\infty, \tag{94}$$

$$<|\psi|>_{T_c} \sim L_\perp^{-(d-1)/6}\, L_{||}^{-1/2},\ L_{||}\to\infty. \tag{95}$$

What Eq. (92) says is that the length $L_{||}$ scales with ξ_o: in fact, the generalization of Eq. (94) is [25]

$$k_B T_c \chi(T_c) = L_\perp^{d-2} L_\perp^{(d-4)/6} \tilde{\chi}\left[\frac{L_\perp}{L_\parallel} L_\perp^{(d-4)/3}\right]. \qquad (96)$$

It is instructive to compare this expression with the result valid in the regime where hyperscaling holds, namely {cf. Eq. (35)} $k_B T_c \chi(T_c) = L_\perp^{\gamma/\nu} \tilde{\chi}(L_\perp/L_\parallel)$. It is seen that this expression agrees with Eq. (96) for d=d*=4, where γ=1, ν=1/2, γ/ν=d−2=2. The enhancement factors in Eq. (96) which are powers of $L_\perp^{(d-4)/3}$ then vanish. Thus while for d≤d* $L_\parallel$ scales with a correlation length $\xi_o \sim L_\perp$ – and thus the scaling function has in its argument the "aspect ratio" $L_\perp/L_\parallel$ – this is no longer true for d>d*: $L_\parallel$ scales with the length $\xi_o \sim L_\perp(L_\perp^{(d-4)/3})$ larger than $L_\perp$. This also implies that for d>d* at $T \lesssim T_c$ the typical distance between domain walls in a system with $L_\parallel >> L_\perp$ is not of order of $L_\perp$, but of the order of this larger length ξ_o (Fig. 6d). It is also gratifying to note that putting $L_\parallel \approx \xi_o$ in Eq. (77) for t=0, $k_B T_c \chi(T_c) \sim \left[L_\parallel L_\perp^{d-1}\right]^{1/2} \sim \left[L_\perp^{4(d-1)/3}\right]^{1/2}$ one recovers Eq. (94): the expression for the susceptibility appropriate for a finite system that orders uniformly {Eq. (77)} for $L_\parallel \approx \xi_o$ gives the same order of magnitude as the result appropriate for nonuniform order, Eq. (94), suggesting again a smooth crossover between these two behaviors.

At finite non–zero temperature t the length $\xi_\parallel$ in the parallel direction crosses over from ξ_o to the standard correlation length [45],

$$\xi_\parallel = t^{-1/2} \tilde{\xi}\left[\frac{L_\perp}{t^{-1/2}} L_\perp^{(d-4)/3}\right]. \qquad (97)$$

So there are three characteristic lengths that we have identified as playing a rôle in finite size scaling for d>d*: the thermal length ℓ_T, the correlation length ξ, and the above length $\xi_\parallel$ which crosses over into ξ_o {Eq. (92)} as $T \to T_c$.

6. ANISOTROPIC SYSTEMS WITHOUT HYPERSCALING

The approach [25] described here can easily be generalized to other models with different marginal dimensionalities. As a first example, let us consider isotropic and uniaxial Lifshitz points [27]. The effective Hamiltonians are, for an isotropic Lifshitz point

$$\mathcal{H}_{\text{eff}}(\psi) = \int d\vec{r}\left\{\frac{1}{2}(\nabla^2\psi)^2 + \frac{1}{2}t\psi^2 + \frac{u_o}{4!}\psi^4\right\}, \tag{98}$$

while for an uniaxial Lifshitzpoint

$$\mathcal{H}_{\text{eff}}(\psi) = \int d\vec{r}\left\{\frac{1}{2}\sum_{j=1}^{d-1}(\partial\psi/\partial x_j)^2 + \frac{1}{2}\left[\frac{\partial^2\psi}{\partial z^2}\right]^2 + \frac{1}{2}t\psi^2 + \frac{u_o}{4!}\psi^4\right\}. \tag{99}$$

Note that within Landau theory in the disordered phase the wavevector–dependent susceptibility $\chi(q)$ for these systems becomes

$$\chi^{-1}(q) \sim t+q^4 = t(1+q^4\xi^4), \tag{100}$$

for the isotropic Lifshitz point, while for the uniaxial Lifshitz point

$$\chi^{-1}(q) \sim t+q_\perp^2+q_\parallel^4 = t(1+q_\perp^2\xi_\perp^2+q_\parallel^4\xi_\parallel^4). \tag{101}$$

From Eq. (100) one can read off the (mean–field) exponents

$$\gamma = 1,\ \nu = 1/4, \tag{102}$$

which together with β=1/2 yield d*=8 in the hyperscaling relation $d^*\nu = \gamma+2\beta = 2$, for the isotropic Lifshitz point. Similarly, Eq. (101) yields

$$\gamma = 1,\ \nu_\perp = \frac{1}{2},\ \nu_\parallel = \frac{1}{4}, \tag{103}$$

and together with β=1/2 the generalized hyperscaling relation $(d^*-1)\nu_\perp+\nu_\parallel = \gamma+2\beta = 2$, this yields d*=4.5 for the uniaxial Lifshitz point [27].

For the isotropic Lifshitz point, the treatment of Sec. 5 presented in Eqs. (68–79) still holds when $L_{||}$ and $L_{\perp}$ are of the same order. But for $L_{||} >> L_{\perp}$ we now have [25]

$$\mathcal{H}_{\mathrm{eff}}(\psi) = L_{\perp}^{d-1} \int_0^{L_{||}} dz \left\{ \frac{1}{2} \left[\frac{d^2\psi}{dz^2} \right]^2 + \frac{1}{2} t\psi^2 + \frac{u_o}{4!} \psi^4 \right\} = \mathrm{I+II+III}, \tag{104}$$

where we again estimate the order of magnitude of the integrals I,II,III using Eq. (86) as a trial function, $\psi = \psi_o \exp(-z/\xi_{||})$. This yields

$$\mathrm{I} \sim L_{\perp}^{d-1} \psi_o^2/\xi_{||}^3, \ \mathrm{II} \sim L_{\perp}^{d-1} t \psi_o^2/\xi_{||}, \ \mathrm{III} \sim L_{\perp}^{d-1} u_o \psi_o^4/\xi_{||}. \tag{105}$$

As in Sec. 5, for $t>0$ and $L_{\perp} \to \infty$ we may ignore III and again requiring

$$L_{\perp}^{d-1} \psi_o^2/\xi_{||}^3 = L_{\perp}^{d-1} t \psi_o^2 \xi_{||} = 1, \tag{106}$$

we conclude

$$\xi_{||} \sim t^{-1/4}, \ \psi_o^2 \sim t^{-3/4}/L_{\perp}^{d-1}, \ \chi \sim \psi_o^2 \xi_{||} L_{\perp}^{d-1} \sim t, \tag{107}$$

in agreement with Eq. (102). But for $t=0$ we match I and III

$$L_{\perp}^{d-1} \psi_o^2/\xi_{||}^3 = L_{\perp}^{d-1} u_o \psi_o^4/\xi_{||} = 1 \tag{108}$$

to conclude [25]

$$\xi_o \sim u_o^{-1/7} L_{\perp}^{(d-1)/7} = u_o^{-1/7} L_{\perp} L_{\perp}^{(d-8)/7} \underset{L_{\perp} \to \infty}{>>} L_{\perp}. \tag{109}$$

The crossover scaling expression analoguous to Eq. (97) is

$$\xi_{\|}(L_{\perp},T) = t^{-1/4}\tilde{\xi}\left[\frac{L_{\perp}}{t^{-1/4}}\,L_{\perp}^{(d-8)/7}\right]. \tag{110}$$

Now for the uniaxial Lifshitz point with inhomogeneous ordering in z–direction the calculation is exactly the same as presented in Eqs. (104)–(110): the distinction only is that these results now should hold for $d>d^* = 4.5$, while for isotropic Lifshitz points they should hold only at correspondingly higher dimensionalities ($d>8$). It is useful to bring out the appropriate marginal dimension in the notation clearly. This can simply be done rewriting ξ_o in Eq. (109) as follows

$$\xi_o \sim L_{\perp}^{(d-1)/7} = L_{\perp}^{1/2}\,L_{\perp}^{(d-4.5)/7}. \tag{111}$$

So one obtains at T_c [25]

$$k_B T_c \chi(T_c) = L_{\perp}^{d/2-1/4}\,L_{\perp}^{(d-4.5)/14}\,\tilde{\chi}\left[\frac{L_{\perp}^{1/2}}{L_{\|}}\,L_{\perp}^{(d-4.5)/7}\right],\ d>d^*=4.5. \tag{112}$$

Once more it is instructive to compare this result with the corresponding result in the regime $d<d^*$ where hyperscaling is valid, Eq. (63)

$$k_B T_c \chi(T_c) = L_{\perp}^{\gamma/\nu_{\perp}}\,\tilde{\chi}\left[L_{\perp}^{\nu_{\|}/\nu_{\perp}}/L_{\|}\right]. \tag{113}$$

For $d=d^*=4.5$ where Eq. (103) holds, if logarithmic corrections are neglected, both Eqs. (112) and (113) reduce to $k_B T_c \chi(T_c) = L_{\perp}^2 \tilde{\chi}(L_{\perp}^{1/2}/L_{\|})$, and thus there is a smooth crossover as a function of d, as expected.

Again one finds that Eq. (112) crosses over for $L_{\|} \approx \xi_o = L_{\perp}^{1/2}\,L_{\perp}^{(d-4.5)/7}$ to the susceptibility in the homogeneous case, Eq. (77) $\{k_B T_c \chi(T_c) \sim (L_{\|} L_{\perp}^{d-1})^{1/2} = L_{\perp}^{d/2-1/4}\,L_{\perp}^{(d-4.5)/14}\}$, as it should. Thus for a uniaxial Lifshitz system homogeneous ordering occurs only for systems with $L_{\|} < \xi_o << L_{\perp}$: systems with $L_{\|} = L_{\perp}$ are not simple, they exhibit nonuniform order at T_c and slightly below it.

As last point of this section, we briefly return to the lattice gas model under the influence of an electric field [48–53], which has been mentioned in the introduction (Fig. 2). Field–theoretic renormalization group work [54,55] suggested that for 2<d<5 the following critical exponents are exact:

$$\beta=\tfrac{1}{2},\ \gamma=1,\ \nu_\perp=\tfrac{1}{2},\ \nu_{||}=1+(5-d)/6. \tag{114}$$

While the exponents $\beta,\gamma,\nu_\perp$ are just those of standard Landau theory, $\nu_{||}$ is nontrivial. If Eq. (114) is true, then this model belongs in the present section: the generalized hyperscaling relation {Eq. (57)} $(d-1)\nu_\perp+\nu_{||}=\gamma+2\beta$ with Eq. (114) is satisfied only for d*=2!. But note that here for d>d* no simple meanfield theory can hold, due to the nontrivial value of $\nu_{||}$. In order to understand finite size effects in this model, the author and Wang [25] speculatively suggested to work with a "generalized" Ginzburg–Landau–theory [77] where one again modifies the gradient energy term:

$$\mathcal{H}_{\rm eff}\{\psi\} = L_\perp^{d-1}\int_0^{L_{||}} dz\left\{\frac{1}{2}\left[\frac{d^\kappa\psi}{dz^\kappa}\right]^2+\frac{1}{2}t\psi^4+\frac{u_o}{4!}\psi^4\right\} = I+II+III, \tag{115}$$

with $\kappa=[2(1+\frac{5-d}{6})]^{-1}$. Again the method of estimating the terms in the path integral for $\psi=\psi_o\exp(-z/\xi_{||})$ can be applied: the three terms of the integral are estimated as

$$I\sim L_\perp^{d-1}\psi_o^2\xi_{||}^{1-2\kappa},\ II\sim L_\perp^{d-1}t\psi_o^2\xi_{||},\ III\sim L_\perp^{d-1}u_o\psi_o^4\xi_{||}. \tag{116}$$

For t>0 again we put

$$L_\perp^{d-1}\psi_o^2\xi^{1-2\kappa} = L_\perp^{d-1}t\psi_o^2\xi = 1, \tag{117}$$

to conclude $\xi\sim t^{-1/(2\kappa)} = t^{-(1+(5-d)/6)} = t^{-\nu_{||}}$, as expected. For t=0 the condition

$$L_\perp^{d-1}\psi_o^2\xi_o^{1-2\kappa} = L_\perp^{d-1}u_o\psi_o^4\xi_o = 1 \tag{118}$$

now yields

$$\xi_o \sim L_\perp^3 L_\perp^{-(d-2)\frac{d-7}{d+1}}. \tag{119}$$

Thus one predicts that the characteristic length for this model in d=2 is $\xi_o \sim L_\perp^3 = L_\perp^{\nu_\perp/\nu_\|}$, while for d=3 it is $\xi_o \sim L_\perp^4$ ($L_\perp^{\nu_\|/\nu_\perp}$ in d=3 would be much smaller, namely $L_\perp^{8/3}$).

Of course, it is very doubtful whether the above generalized Ginzburg–Landau approach works for such driven systems far from equilibrium. Monte Carlo work motivated by the ideas described above [53] did not yield any significant conclusions whether Eq. (119) is correct – but it should be noted that earlier Monte Carlo work [48–52] did not corrobate the exponent predictions [54,55] of Eq. (114) either! So more work is needed to clarify the status of finite size scaling in this driven Kawasaki spin–exchange model.

7. CONCLUSIONS

This review has the aim to demonstrate that the theory of finite size scaling still poses challenging problems to our theoretical understanding of critical phenomena. While shape effects in Ising systems in d=2 are probably well understood, more work needs probably done for higher dimensionalities, where also the treatment of thin–film geometry deserves discussion, which was deliberately left out here, although it has been considered already in some early work [78,79]. Also this review focused entirely on scalar order parameters – XY, Heisenberg models, etc. have not been discussed explicitly. But even in the framework of the Ising type models, we have only described phenomenological ideas [25,42] and Monte Carlo work [25,26,42,43] – neither exact results based on conformal invariance [24] or on ϵ=4–d renormalization group expansions, nor results obtained in conjunction with the transfer matrix approach {see e.g. [11,80–82]} have been given any consideration. Thus, this review is essentially centered around work in which the author was involved, rather than giving an unbiased overview over this very rich field. And, as a final disclaimer, even our phenomenological scaling theories are somewhat incomplete: we have completely ignored that at $d=d^*$ in the various models logarithmic correction factors $(\ln L)^p$, where p is some exponent, should occur modifying the power laws. Such factors have been identified in a few cases [46,59].

Also with respect to numerical results, there is still need of more work concerning shape effects for $d>2$, and as yet there exist no complete studies of size effects in systems with anisotropic critical behavior. And particularly limited is our understanding of size effects in driven systems far from equilibrium, both from the analytical and from the numerical side.

ACKNOWLEDGEMENTS:

Large parts of the work that was described here originated in a fruitful and pleasant collaboration with Jian–Sheng Wang (Ref. [25]). The author is particularly grateful to him for this cooperation.

Thanks are also due to E.V. Albano, D.W. Heermann and W. Paul for their stimulating collaboration that led to the work of [26] which was mentioned in Sec. 3, to V. Privman, M. Nauenberg and A.P. Young for their stimulating collaboration that led to the work of Ref. [42] described in Sec. 5, and to A. Sariban for her fruitful collaboration on applying finite size scaling ideas to polymer mixtures (Fig. 4 [58]).

Finally, the author's interest in finite size scaling and his understanding of the subject has also profited much from an intensive collaboration with D.P. Landau and K.K. Mon on related subjects. Also support from the Materialwissen–schaftliches Forschungszentrum (MWFZ) of the University of Mainz is acknowledged.

REFERENCES

1. T.L. Hill, Statistical Mechanics of Small Systems (Benjamin, New York 1956), Vols I and II
2. A.E. Ferdinand and M.E. Fisher, Phys. Rev. 185, 832 (1969)
3. M.E. Fisher, in Critical Phenomena (M.S. Green, ed.) p.1 (Academic Press, New York and London 1971)
4. Y. Imry and D.J. Bergman, Phys. Rev. A3, 1416 (1971)
5. K. Binder, Physica 62, 508 (1972)
6. M.E. Fisher and M.N. Barber, Phys. Rev. Lett. 28, 1516 (1972)
7. P.G. Watson, in Phase Transitions and Critical Phenomena, Vol. II (C. Domb and M.S. Green, eds) p. 101 (Academic Press, New York and London 1972)
8. D.P. Landau, Phys. Rev. B13, 2997 (1976); B14, 255 (1976)
9. K. Binder and D.P. Landau, Phys. Rev. B20, 1941 (1980)
10. K. Binder, Z. Physik B43, 119 (1981)
11. M.N. Barber, in Phase Transitions and Critical Phenomena, Vol. VIII (C. Domb and J.L. Lebowitz, eds) p. 146 (Academic Press, New York and London 1983)
12. K.K. Kaski, K. Binder and J.D. Gunton, Phys. Rev. B29, 3996 (1984)
13. D.P. Landau and K. Binder, Phys. Rev. B31, 5946 (1985)
14. K. Binder, Ferroelectrics 73, 43 (1987)
15. R. Marx, Phys. Repts. 725, 1 (1985)
16. K. Binder (ed.), Monte Carlo Methods in Statistical Physics (Springer, Berlin–Heidelberg–New York, 1979)
17. K. Binder, in Phase Transitions and Critical Phenomena, Vol Vb (C. Domb and M.S. Green, eds) p. 1 (Academic Press, New York and London 1976)
18. K. Binder (ed.), Applications of the Monte Carlo Method in Statistical Physics (Springer, Berlin–Heidelberg–New York 1984)
19. D.W. Heermann, Computer Simulation Methods in Theoretical Physics (Springer, Berlin–Heidelberg–New York 1986)
20. K. Binder and D.W. Heermann, Monte Carlo Simulation in Statistical Physics: An Introduction (Springer, Berlin–Heidelberg–New York 1988)

21. P. Kleban and G. Akinci, Phys. Rev. B28, 1466 (1983); Phys. Rev. Lett. 51, 1058 (1988)
22. V. Privman and M.E. Fisher, J. Stat. Phys. 33, 385 (1983)
23. P. Kleban, G. Akinci, R. Hentschke, and R. Brownstein, J. Phys. A19, 437 (1986); Surface Science 166, 159 (1986)
24. J.L. Cardy, in Phase Transitions and Critical Phenomena, Vol. XI (C. Domb and J.L. Lebowitz, eds) p. 55 (Academic Press, New York and London 1987)
25. K. Binder and J.–S. Wang, J. Stat. Phys., in press (Vol. 55, No. 1/2, April 1989)
26. E.V. Albano, K. Binder, D.W. Heermann, and W. Paul, preprint
27. R.M. Hornreich, M. Luban and S. Shtrikman, Phys. Rev. Lett. 35, 1678 (1975)
28. R.M. Hornreich and A.D. Bruce, J. Phys. A11, 595 (1978)
29. G.S. Grest and J. Sak, Phys. Rev. B17, 3607 (1978)
30. W. Selke, Z. Physik B29, 133 (1978)
31. W. Selke, Z. Physik B43, 335 (1981)
32. K.K. Kaski and W. Selke, Phys. Rev. B31, 3128 (1985)
33. W. Kinzel, in Percolation Structures and Processes, Annals of the Israel Physical Society No. 5 (G. Deutscher, R. Zallen, and J. Adler, eds) p. 425 (Adam Hilger, Bristol 1983)
34. W. Kinzel and J.M. Yeomans, J. Phys. A14, L163 (1981)
35. A.M. Szpilka and V. Privman, Phys. Rev. B28, 6613 (1983)
36. V. Privman, J. Phys. A18, L63 (1985); G. Forgacs and V. Privman, J. Stat. Phys. 49, 1165 (1987)
37. W. Selke, D.A. Huse, and D.M. Kroll, J. Phys. A17, 3019 (1984)
38. R. Lipowsky and G. Gompper, Phys. Rev. B29, 5213 (1984)
39. M. Gelfand and R. Lipowsky, Phys. Rev. B36, 8725 (1987)
40. V. Privman and N.M. Svrakic, Phys. Rev. B37, 3713 (1988)
41. D.J. Amit, Field Theory, the Renormalization Group and Critical Phenomena (McGraw Hill, New York 1987)
42. K. Binder, M. Nauenberg, V. Privman, and A.P. Young, Phys. Rev. B31, 1498 (1985)
43. K. Binder, Z. Physik B61, 13 (1985)
44. E. Brézin, J. Phys. (Paris) 43, 15 (1982)

45. J. Zinn–Justin and E. Brézin, Nucl. Phys. B257 [FS14], 867 (1985)
46. J.M. Luck, Phys. Rev. B31, 3069 (1985)
47. K. Kawasaki, in Phase Transitions and Critical Phenomena, Vol. II (C. Domb and M.S. Green, eds) p. 443 (Academic Press, New York and London 1972)
48. S. Katz, J.L. Lebowitz and H. Spohn, Phys. Rev. B28, 1655 (1983); J. Stat. Phys. 34, 497 (1984)
49. J. Marro, J.L. Lebowitz, H. Spohn, and M.H. Kalos, J. Stat. Phys. 38, 725 (1985)
50. J.L. Vallés and J. Marro, J. Stat. Phys. 43, 411 (1986); 49, 89 (1987); J. Marro and J.L. Vallés, J. Stat. Phys. 49, 121 (1987)
51. J. Marro, J.L. Vallés, and J.M. Gonzáles–Miranda, Phys. Rev. B35, 3372 (1987)
52. M.Q. Zhang, J.–S. Wang, J.L. Lebowitz, and J.L. Vallés, J. Stat. Phys. (1988)
53. J.–S. Wang, J.L. Lebowitz, and K. Binder, preprint
54. J.K. Janssen and B. Schmittmann, Z. Physik B64, 503 (1986)
55. K.–T. Leung and J. Cardy, J. Stat. Phys. 44, 567 (1986)
56. D.P. Landau, J. Magn. Mag. Mat. 31–34, 11 (1983)
57. C.S.S. Murty and D.P. Landau, J. Appl. Phys. 55, 2429 (1984); M.S.S. Challa and D.P. Landau, Phys. Rev. B33, 437 (1986)
58. A. Sariban and K. Binder, Macromolecules 21, 711 (1988)
59. H. Bernreuther and M. Göckeler, Nucl. Physics B [F], (1988)
60. J.C. Le Guillou and J. Zinn–Justin, Phys. Rev. B21, 3976 (1980)
61. C. Domb, in Phase Transitions and Critical Phenomena, Vol. III (C. Domb and M.S. Green, eds) (Academic Press, New York and London 1974)
62. B.G. Nickel, Physica 106A 48 (1981)
63. G.S. Pawley, R.H. Swendson, D.J. Wallace, and K.G. Wilson, Phys. Rev. B29, 4030 (1984)
64. M.N. Barber, R.B. Pearson, D. Touissant, and J.L. Richardson, Phys. Rev. B32, 1720 (1985)
65. A. Hoogland, A. Compagner, and H.W. Blöte, Physica A132, 593 (1985)
66. E. Barouch, B.M. McCoy, and T.T. Wu, Phys. Rev. Lett. 31 1409 (1973)
67. R.H. Swendsen and J.–S. Wang, Phys. Rev. Lett. 58, 86 (1987)

68. S. Wansleben, Computer Phys. Commun. 43, 375 (1987)
69. T.T. Wu, Phys. Rev. 149, 380 (1966)
70. R.C. Desai and D.W. Heermann, Comp. Phys. Comm. (1988)
71. R.C. Desai, D.W. Heermann, and K. Binder, J. Stat. Phys. 53, 795 (1988)
72. P.W. Kasteleyn, J. Math. Phys. 4, 287 (1963)
73. S.M. Bhattacharjee and J.F. Nagle, Phys. Rev. A31, 3199 (1985); see also C.S.O. Yokoi, F. Nagle, and S.R. Salinas, J. Stat. Phys. 44, 729 (1986)
74. H.J. Schultz, Phys. Rev. B22, 5274 (1980)
75. D.A. Huse and M.E. Fisher, Phys. Rev. B29, 239 (1984)
76. A.J. Guttmann, J. Phys. A14, 233 (1981)
77. M. Luban, in Phase Transitions and Critical Phenomena, Vol. Vb, (C. Domb and M.S. Green, eds) p. 35 (Academic Press, New York and London 1976)
78. G.A.T. Allan, Phys. Rev. B1, 352 (1970)
79. K. Binder, Thin Solid Films 20, 367 (1974)
80. T.W. Burkhardt and B. Derrida, Phys. Rev. B32, 7273 (1985)
81. M.N. Barber, Phys. Rev. B27, 5879 (1983)
82. B. Derrida and H.J. Herrmann, J. Phys. (Paris) 44, 1365 (1983)

V. MONTE CARLO STUDIES OF FINITE SIZE EFFECTS AT FIRST AND SECOND ORDER PHASE TRANSITIONS

D. P. Landau

TABLE of CONTENTS

(continued)

MONTE CARLO STUDIES OF FINITE SIZE EFFECTS AT FIRST AND SECOND ORDER PHASE TRANSITIONS

D. P. Landau
Center for Simulational Physics
University of Georgia
Athens, GA 30602, U S A

1. INTRODUCTION

The connection between finite size scaling theory and Monte Carlo simulations of spin systems dates back more than a decade. On the one hand Monte Carlo simulations have been used to test and verify various aspects of finite size scaling theory, and on the other hand finite size scaling theory has been used to analyze and interpret simulations data. Perhaps the clearest example of the interplay between theory and simulations enters in the examination of systems undergoing phase transitions of unknown order. In such cases simulations will always produce smooth behavior regardless of the order of the transition in the thermodynamic limit. Singularities in the derivatives of the free energy can occur when the sample size, N, is infinite; thus, the discontinuous nature of a first–order transition is smeared out for finite N and one finds only smooth behavior in the thermodynamic quantities, similar to that at a continuous transition. Finite size effects at second–order transitions have been well–understood for a long time; the dependence on lattice size is determined by infinite lattice critical exponents [1–3]. Thus, for example, the manner in which the rounded specific heat peak for finite systems diverges as the thermodynamic limit is approached is governed by the universality class of the model. Our understanding of finite size effects at first–order transitions has developed more recently. Conclusions based on a variety of different arguments [4–10] reach a consensus: as the sample's linear dimension $L \to \infty$, the response functions diverge as L^d. (Here, as in the rest of this paper, the sample is a hypercube of spatial dimensionality d and side L, i.e., $N = L^d$. We shall restrict our discussion to isotropic systems. For a thorough description of the extension to anisotropic systems, see the article by K. Binder in this volume.)

These features for a first–order transition are illustrated in the following example where we preview some of the results of Section 3. Consider the Ising ferromagnet when a magnetic field, H, is varied at a temperature below the critical temperature. While the magnetization per particle, <s>, of the infinite system jumps from $+M_{sp}$ as $H \to 0^+$ and to $-M_{sp}$ as $H \to 0^-$, the behavior of the finite lattice magnetization $<s>_L$ vs. H is perfectly smooth for finite L. As shown schematically in Fig. 1, the transition is rounded over a region of field of the order of $|H| = k_BT/M_{sp}L^d$ where the susceptibility, $\chi_L = \partial <s>_L/\partial H$ is finite and of the order of $M_{sp}^2L^d/k_BT$. This behavior is in marked contrast to the δ–function singularity which appears for $L = \infty$ and is qualitatively similar to the

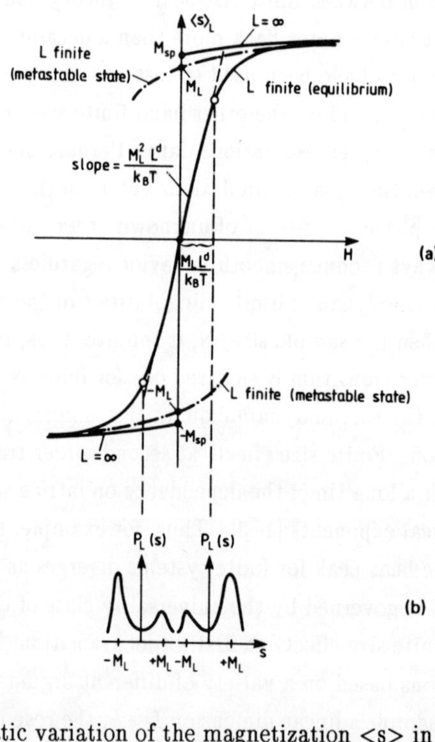

Fig. 1 (a) Schematic variation of the magnetization <s> in a finite L×L Ising ferromagnet and in an infinite Ising ferromagnet plotted vs. magnetic field (solid curves). The dash–dotted line shows the result of a Monte Carlo simulation with sufficiently short observation time that metastable branches occur. $\pm M_{sp}$ is the spontaneous magnetization of the infinite system, and $\pm M_L$ is the most probable value of the magnetization in the finite system at =0. (b) Schematic probability distribution $P_L(s)$ of the magnetization for two fields [open circles in (a)] where the magnetization is in between $\pm M_L$.

finite lattice behavior at a second order transition: in both cases there is a peak in χ_L which grows and sharpens as $L \to \infty$. (In general, the finite size of L also induces a shift in the location of the peak although this is not present at the field driven transition in the Ising model because of spin–inversion symmetry).

The objective of the present paper is to review the progress made in studying and understanding finite size effects using Monte Carlo simulations with particular emphasis placed on first–order phase changes. We present finite size scaling theory as utilized by the simulations and thermodynamic fluctuation theory to study the magnetization and energy distributions, deriving explicit predictions for the moments and comparing the theory with data from Monte Carlo simulations of well–known systems undergoing second–order transitions, in Section 2, and first–order transitions in Section 3. No attempt will be made to achieve completeness, and the focus of this article will be on the research carried out in the Center for Simulational Physics. In the final section of this article we shall summarize and draw final conclusions.

2. FINITE SIZE EFFECTS AT SECOND ORDER PHASE TRANSITIONS

2.1 Theory

Finite size scaling theory describes the bulk properties of a finite system of linear dimension L with periodic boundaries in terms of the relative size of the correlation length ξ and L. Near T_c, $\xi \sim |T-T_c|^{-\nu}$ and hence finite size effects appear when $|T-T_c|^{-\nu} \sim L$. This approach replaces the "singular" part of the free energy F(L,T) for a finite system at temperature T by a scaling form which depends upon a single variable $x=tL^{1/\nu}$:

$$F(L,T) = L^{(2+\alpha)/\nu} f(x) . \tag{1}$$

where $t = |1-T_c/T_c|$ is the reduced distance from the critical point. As a consequence of this scaling expression all of the bulk properties, which are merely appropriate derivatives of the free energy, can also be expressed in simple scaling forms:

$$m = L^{-\beta/\nu} \underline{X}(x) \tag{2a}$$

$$\chi = L^{\gamma/\nu}\,\underline{Y}\,(x) \tag{2b}$$

$$C\text{–}b = L^{\alpha/\nu}\,\underline{Z}\,(x) \tag{2c}$$

for the order parameter, susceptibility, and divergent part of the specific heat respectively. Thus at T_c , i.e. x=0, all bulk properties exhibit power law behavior as a function of L determined uniquely by the corresponding infinite lattice critical exponents. In the opposite limit, $x \to \infty$ and $T \neq T_c$, the L dependence must cancel out and the usual power law dependence on $t = (T-T_c)/T_c$ must emerge.

An "effective transition temperature" $T_c(L)$ can be defined, for example, as the position of the maximum of the specific heat, and this may be related to the critical temperature of the infinite lattice by

$$T_c(L) = T_c + aL^{-1/\nu}. \tag{3}$$

All of these expressions are valid only for sufficiently large L and for t in the asymptotic critical regime, but if these criteria are fulfilled, Eqns. 2a–2c must apply over the entire range of L and T. This scaling behavior means that in a log–log plot produced using Eqn. (2) all data must fall on a single universal curve and the large x behavior must be linear with a slope given by the approprite critical exponent. Although the scaling functions $\underline{X}$, $\underline{Y}$ and $\underline{Z}$ are not generally known, they can be determined graphically from the data, e.g., from Eqn. (2a) we find that $\underline{X}(x) = mL^{\beta/\nu}$.

An additional important finite size scaling relation applies to the behavior of the site–site correlation function $\Gamma(r,T)$ where a second scaling variable $y = r/L$ is needed

$$\Gamma = L^{-2\beta/\nu}\,G(x,y)\ . \tag{4}$$

This expression means that the scaling of the correlation function is described by a surface instead of a single curve as in Eqn. (2). If free boundaries are used instead, finite size scaling still holds but the asymptotic form includes a boundary term which also depends only upon infinite lattice critical exponents.

As long as hyperscaling is obeyed, the expressions presented above will be valid, and the implementation of finite size scaling for systems which violate hyperscaling will be discussed in the following chapter by Binder.

2.2 Simulation Results for the Ising Square Lattice Ferromagnet

These scaling forms (represented by Eqns. 1–4) can be readily tested using Monte Carlo data for simple Ising systems. In Fig. 2a we show data from Ref. 2 for the Ising square lattice using a range of different lattice sizes, and in Fig. 2b we replot the same data in scaling form. Not only do the data scale extremely well but in the large x regime where $L \to \infty$ they do indeed fall on a straight line with slope equal to β for $T < T_c$. The variation of the effective transition temperature as shown in Fig. 3 is well described by Eqn. (3) for both periodic and free boundaries.

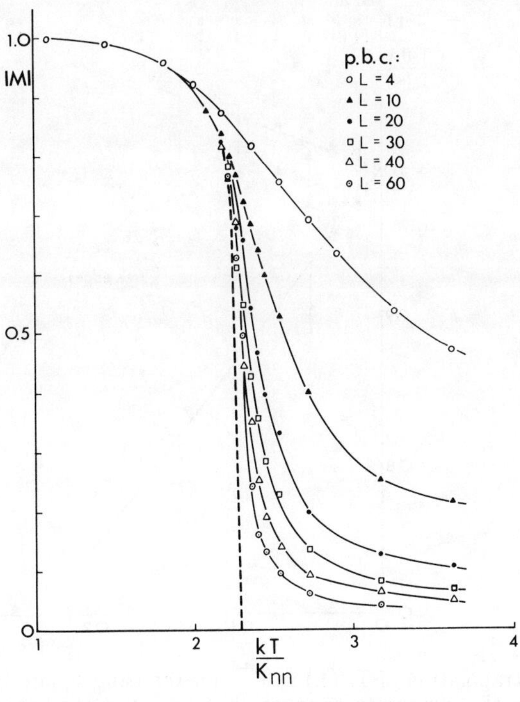

Fig. 2a Temperature variation of the order parameter for L×L Ising square lattices. The dashed lines represent the exact $L = \infty$ curve. (From Landau [2].)

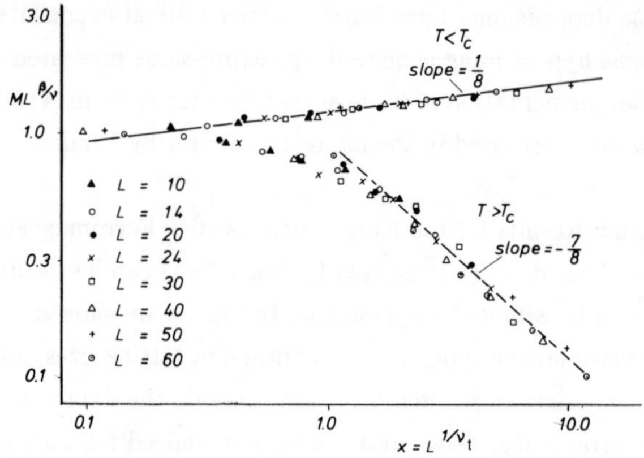

Fig. 2b Finite–size scaling plot for the order parameter for L×L Ising square lattices with p.b.c. For $T < T_c$ the solid line is $1.22\ x^{1/8}$; for $T > T_c$ the dashed line is $0.92\ x^{-7/8}$. (From Landau [2].)

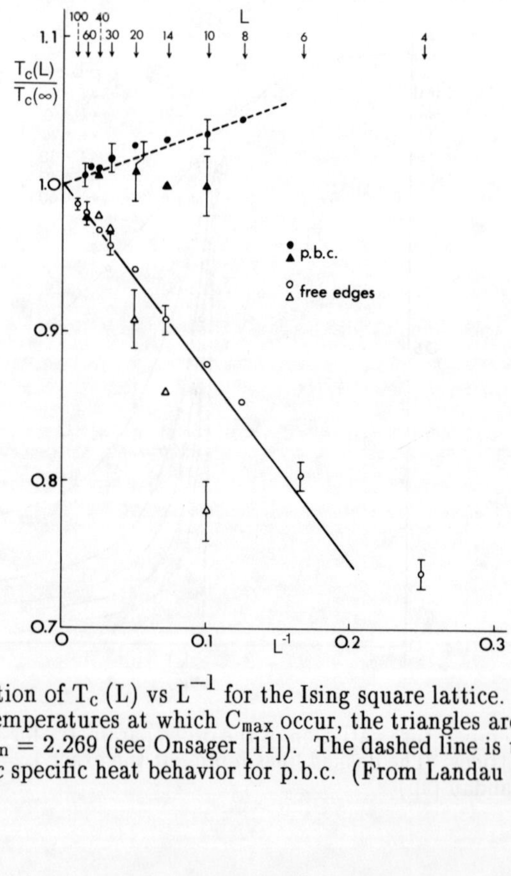

Fig. 3 Extrapolation of T_c (L) vs L^{-1} for the Ising square lattice. The circles give the temperatures at which C_{max} occur, the triangles are for χ_{max}. $kT_c(\infty)/J_{nn} = 2.269$ (see Onsager [11]). The dashed line is the *asymptotic* specific heat behavior for p.b.c. (From Landau [2].)

The scaling properties of the correlation function have been determined by making cuts through the surface described by Eqn. (4). In Fig. 4 we show scaling plots for fixed values of $y=r/L$. The result is a family of scaling curves which are described by Eqn. (4) with the known infinite lattice values of β and ν.

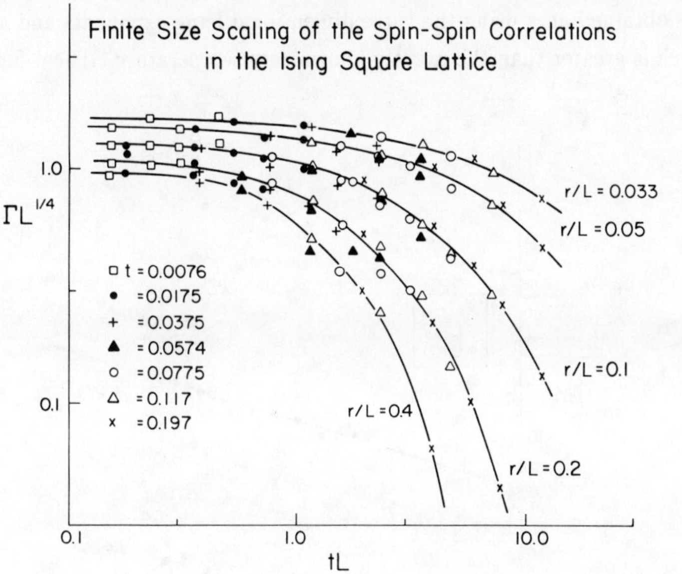

Fig. 4 Finite size scaling of the spin–spin correlations in L×L Ising square lattices with $t=(T-T_c)/T_c$. Data are for $10 \leq L \leq 60$. (From Landau [2].)

2.3 Simulations Results for the Body Centered Cubic Ising Antiferromagnet

In the previous section we have described finite size scaling behavior near a ferromagnetic critical point. The Ising model properties are well known, but finite size scaling can also be used to extract information about systems which are not exactly soluble in the thermodynamic limit. An example of such a system is the body centered cubic Ising antiferromagnet which is antiferromagnetic within a region of H and T enclosed by a second order phase boundary [12]. Mean field theory [13] suggests that this phase boundary decreases non–monotonically with increasing temperature and early Monte Carlo results seemed to verify this behavior. On the other hand, those models which were really well known, the square lattice Ising model [14] and the superexchange Ising model [15], showed monotonic decay and even series expansion results [16] for the bcc lattice could be reasonably extrapolated to zero temperature in a monotonic fashion. Finite size scaling also applies along the entire second order phase boundary, and we can use this property to locate it with quite high precision. In Fig. 5 we analyze the variation of the order parameter (the staggered magnetization) with applied magneic field and lattice size; proper scaling is obtained only using the three dimensional Ising exponents and a critical field which is greater than the exactly known zero temperature critical field.

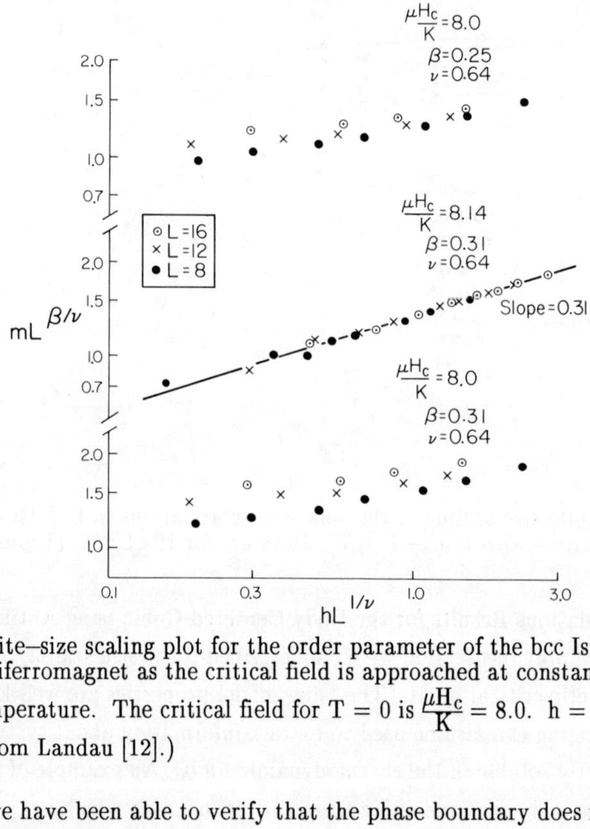

Fig. 5 Finite–size scaling plot for the order parameter of the bcc Ising antiferromagnet as the critical field is approached at constant temperature. The critical field for T = 0 is $\frac{\mu H_c}{K} = 8.0$. $h = (1-H/H_c)$. (From Landau [12].)

In this way we have been able to verify that the phase boundary does indeed show a very weak but clear maximum, see Fig. 6, which is not an artifact of the mean field theory which first predicted it. Note that the phase boundary obtained by finite size scaling is lower than that deduced from single simulations on much larger lattices of a single size and systematically slightly higher than that predicted by series expansions. The amount by which the phase boundary rises above the T=0 critical field is so small however that an accurate determination of this effect is possible <u>only</u> with the use of finite size scaling.

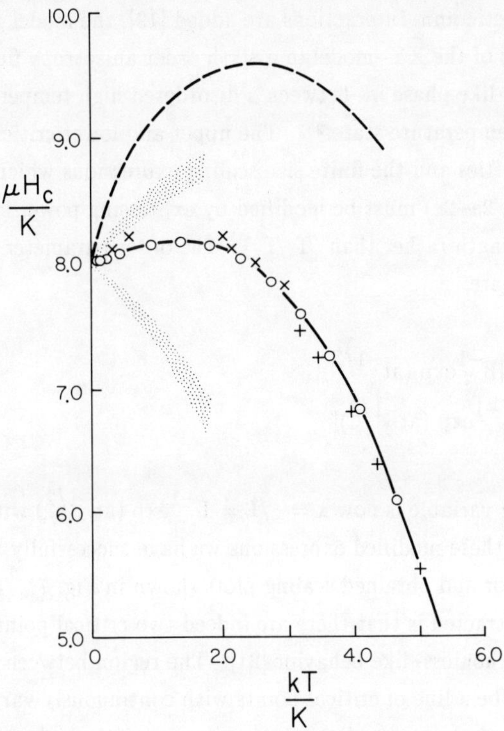

Fig. 6 Low–temperature portion of the critical field curve for the bcc Ising antiferromagnet. The dashed line is the mean field prediction, the series expansion results are given by + and the infinite–lattice Monte Carlo estimates by O. The thin solid line is merely intended as a guide to the eye. Earlier L = 25 Monte Carlo data (Ref. 17) are shown by x. The shaded area shows the crossover region between critical and noncritical behavior. (From Landau [12].)

2.4 Simulation Results for the Triangular Ising Antiferromagnet with Nearest– and Next–Nearest Neighbor Coupling

In zero magnetic field the triangular Ising antiferromagnet is relatively uninteresting in that it shows no phase transition at any non–zero temperature [18]. In the presence of a field an ordered ferrimagnetic state appears in which the system is divided into three interpenetrating nnn–sublattices and one of them is oppositely directed to the other two. This state is degenerate, of course, as there are six different states in which one sublattice is anti–parallel to the others.

When ferromagnetic nnn–interactions are added [19], the model joins the universality class of the XY–model in a sixth order anisotropy field, and possesses an XY–like phase in–between a disordered high temperature state and an ordered low temperature state.[20] The upper and lower critical points are essential singularities and the finite size scaling expressions which we have developed (Eqns. 2a–2c) must be modified by expressing power laws in terms of the correlation length rather than $(T-T_c)$. The order parameter m and ordering susceptibility χ^+ are:

$$m = L^{-b}\, \underline{X}\, [L^{-1} \exp\, (at^{-1/2})] \tag{5a}$$

$$\chi^+ T = L^{c}\, \underline{Y}\, [L^{-1} \exp\, (at^{-1/2})] \tag{5b}$$

where the scaling variable is now $x = \xi/L = L^{-1}\exp\,(at^{-1/2})$ with "a" being a constant. Using these modified expressions we have successfully analyzed the finite size behavior and obtained scaling plots shown in Fig. 7. The conclusion which may be extracted is that there are indeed two critical points with T_{c1} and T_{c2} Kosterlitz–Thouless–like behavior [21]. The region between T_{c1} and T_{c2} is thus expected to be a line of critical points with continuously varying critical exponents. Since each intermediate temperature is critical, the scaling form for the spin–spin correlation function reduces to a function of a single variable y = r/L, i.e., $\Gamma \sim r^{-\eta}\, G(0,y)$, and the asymptotic decay of the correlation should be a power law. This expectation is fulfilled, as shown in Fig. 8, moreover the exponent governing the decay does decrease with decreasing temperature. Although "vortex–equivalents" were also identified, the confirmation of the Kosterlitz–Thouless behavior is seen most clearly in the finite size scaling behavior.

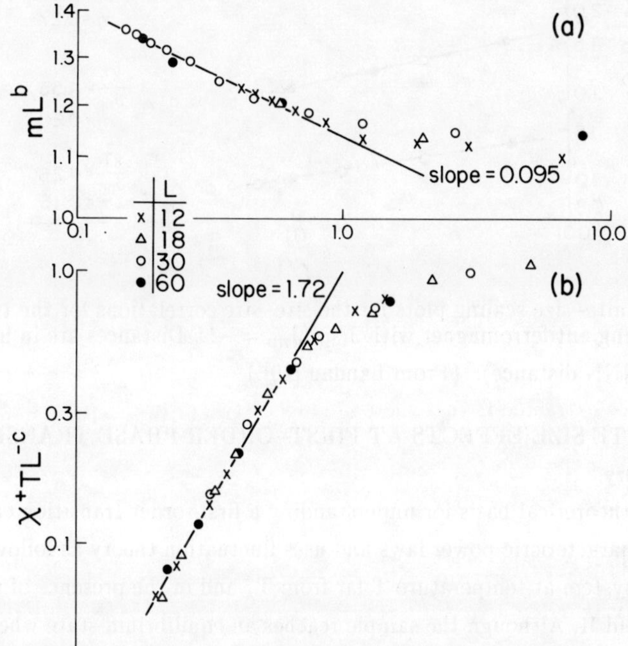

Fig. 7 Finite–size scaling plots in zero field for the triangular Ising antiferromagnet with $J_{nnn}/J_{nn} = -1$: (a) order parameter with $b = 0.095$ and $T_{c2} = 4.26\ J_{nn}/k$; (b) high–temperature ordering susceptibility with $c = 1.72$ and $T_{c1} = 4.89\ J_{nn}/k$. (From Landau [19].)

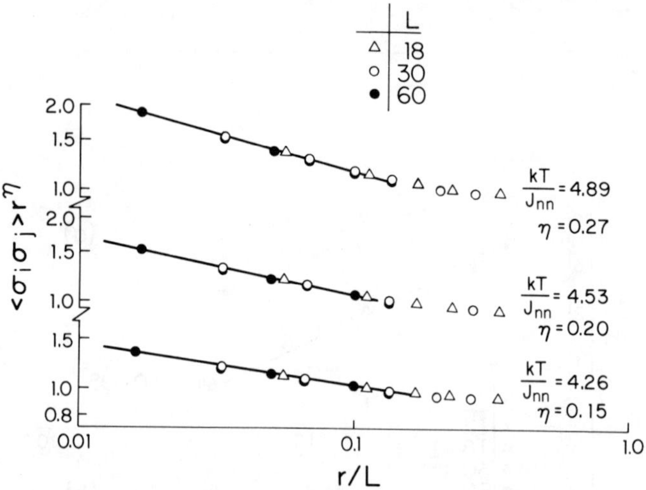

Fig. 8 Finite–size scaling plots for the site–site correlations for the triangular Ising antiferromagnet with $J_{nnn}/J_{nn} = -1$. Distances are in units of r_{NNN} (NNN distance). (From Landau [19].)

3. FINITE SIZE EFFECTS AT FIRST–ORDER PHASE TRANSITIONS

3.1 Theory

The theoretical basis for understanding a first–order transition cannot be related to characteristic power laws and uses fluctuation theory as follows. Consider a system at temperature T far from T_c and in the presence of a magnetic field H. Although the sample reaches an equilibrium state where the free energy is a minimum, spontaneous fluctuations about the equilibrium values still take place in quantities like the temperature $\langle T\rangle$ and magnetization $\langle s\rangle$ of the sample. Denoting these fluctuations by ΔT and Δs, respectively, and expressing all extensive quantities normalized per particle, we can expand the free energy about equilibrium and show (see, for example, Ref. 22) that the probability of a state which is not too far from the minimum is given by:

$$P_L(s,T) \sim \exp(-F_o L^d/k_B T)\exp\left(-\frac{CL^d(\Delta T)^2}{2k_B T^2}\right)\exp\left(-\frac{(\Delta s)^2 L^d}{2k_B T\chi}\right). \quad (6)$$

Here, k_B is the Boltzmann constant, F_o is the free energy at equilibrium, C is the specific heat at constant magnetization and χ is the isothermal susceptibility. We now restrict ourselves to first–order transitions involving (i) isothermal variations of H and (ii) temperature variations at H = 0. In case (ii), since $\Delta E = C\Delta T$ where E is the internal energy, we can convert to the energy distribution. Since we are interested only in changes in the free energy with respect to some reference point, $F_{ref}(T_{ref}, H_{ref})$, we can write:

$$P_L(s) \sim \exp(-F_o L^d/k_B T) \exp\{\frac{-(\Delta s)^2 L^d}{2k_B T\chi}\} \quad , \text{T constant} \tag{7a}$$

and

$$P_L(E) \sim \exp(-F_o L^d/k_B T) \exp\{\frac{-(\Delta E^2 L^d}{2k_B T^2 C}\} \quad , H = 0 \tag{7b}$$

Here $F_o = F_{ref} - s(H - H_{ref})$ in Eqn. (7a), while $F_o = F_{ref} - S(T - T_{ref})$ in Eqn. (7b) with S being the entropy. We draw attention to the fact that F, C and χ refer to the <u>infinite</u> lattice values. This is a valid assumption for $L >> \xi$ and the analysis is more complicated involving the scaling properties of these quantities. In fact, Binder [23] has shown that $P_L(s)$ itself scales at a critical point.

The application of Eqn. (7) to a first–order transition where two phases coexist is straightforward [23,24]. Since the free energies of the two phases are equal, the relevant distribution will exhibit two peaks in the vicinity of the transition, one representative of each phase. The approximation then is to take the distribution to be a superposition of Gaussians of the form in Eqn. (7), with F_{ref} being the free energy at the transition; it is then a trivial matter to calculate the moments of this distribution in order to extract thermodynamic properties. Rounding of the transition region is a natural consequence of this approach since the centers of the Gaussians will be determined by the response of each phase independently to the applied field, and the finite value of L in the free energy factors ensures a contribution from the minority phase for small deviations from the transition. The approach ignores interfacial effects which occur when the system possesses substantial domains of both phases, but this shortcoming becomes less relevant as system sizes increases. An important example of such a situation is the magnetic field–driven transition in the d=2 Ising square lattice at temperatures below the critical point. The Hamiltonian for this model is

$$\mathcal{H}_{\text{Ising}} = -J \sum_{(ij)} s_i s_j - H \sum_i s_i \quad , \quad \{s_i = \pm 1\} \tag{8}$$

where J>0 is the (ferromagnetic) coupling between nearest neighbor sites i and j. The model has a second–order transition in zero field at $k_B T_c / J \approx 2.269$. For $T<T_c$, the model has a first–order transition at H=0; when H is varied from negative to positive values the magnetization jumps from $-M_{sp}$ to $+M_{sp}$ as shown in Fig. 1. The two phases produce $\langle s \rangle > 0$ for H>0 and $\langle s \rangle < 0$ for $H \lesssim 0$. As the Hamiltonian is invariant under the transformation H, $\{s_i\} \to -H$, $\{-s_i\}$, the thermodynamic properties such as the susceptibility are identical in the two phases.

Taking $F_{ref} = F(T, H=0)$, with $\langle s \rangle = +M_{sp} + \chi H$ in one phase and $\langle s \rangle = -M_{sp} + \chi H$ in the other with $x = HM_{sp}L^d/k_BT$, from Eqn. (7a) we obtain the distribution function in the vicinity of the transition:

$$P_L(s) = A[e^x \exp\{-\frac{(s-M_{sp}-\chi H)^2 L^d}{2k_BT\chi}\} + e^{-x}\exp\{-\frac{(s+M_{sp}-\chi H)^2 L^d}{2k_BT\chi}\}] \tag{9}$$

where some of the constant factors have been adsorbed into the normalization constant A given by:

$$A = \{\frac{L^d}{2\pi k_BT\chi}\}^{1/2} \frac{1}{2\cosh(x)} \quad . \tag{10}$$

The first moment of the magnetization distribution is then:

$$\langle s \rangle_L = \chi H + M_{sp}\tanh(x) \quad . \tag{11}$$

For large L (i.e., $|x| >> 1$) we retrieve bulk behavior:

$$\langle s \rangle_L \to \chi H + M_{sp} \quad . \tag{12}$$

In the opposite limit, $|x|<<1$, we find

$$<s>_L \approx HM_{sp}^2 L^d/k_B T \quad , \tag{13}$$

i.e., the transition is rounded over a region of H of order

$$\Delta H = k_B T/M_{sp} L^d \quad . \tag{14}$$

instead of showing a sharp jump at $H = 0$. The susceptibility is then given by:

$$\chi_L = \frac{\partial <s>_L}{\partial H} = \chi + \frac{M_{sp}^2 L^d \mathrm{sech}^2(x)}{k_B T} \quad . \tag{15}$$

Thus, $\chi_L\,(H=0) \sim L^d$, instead of the δ–function singularity of the infinite system, as shown schematically in Fig. 1. It is also easy to verify that

$$<s^2>_L = M_{sp}^2 + \chi^2 H^2 + \frac{k_B T\chi}{L^d} + 2M_{sp}\chi H \tanh(x) \tag{16}$$

and that the fluctuation relation:

$$<s^2>_L - <s>_L^2 = k_B T \chi_L / L^d \tag{17}$$

holds. Another quantity of importance is the 4th–order cumulant

$$U_L = -[<s^4>_L - 3<s^2>_L^2 - 4<s>_L <s^3>_L + 12<s^2>_L <s>_L^2 - 6<s>_L^4]/(3<s^2>_L^2) \tag{18}$$

which shows characteristic behavior at the transition. It can be shown that for the Ising model:

$$U_L = 2\,\frac{\{1-\tanh^2(x)\}^2 - (2/3)\{1-\tanh^2(x)\}}{\{1+(\chi^2 H^2/M_{sp}^2) + (k_B T\chi/M_{sp}^2 L^d) + (2\chi H/M_{sp})\tanh x\}^2} \,. \tag{19}$$

For $L \to \infty$, $U_L \to 2/3$ at the first–order transition for H=0 and U_L vanishes for non–zero H. In contrast, at a second–order transition, U_L approaches a non–trivial value which lies between these limits [24]. The limiting value of the minimum in U_L thus serves both to identify the order of the transition and also to locate the field at which the transition occurs.

A more general scenario for a first–order transition involves states whose properties are not symmetric with respect to the transition. An example of such a possibility is the d=2 q–state Potts model which for all q > 4 exhibits a 1st–order transition [25] as the temperature changes. Although the theoretical presentation (and Monte Carlo data presented here) is in terms of the appropriate parameters for this model, we wish to emphasize that the general features of the theory are widely applicable to systems with discrete degrees of freedom. The Hamiltonian for the q–state Potts model is:

$$\mathcal{H}_{\mathrm{Potts}} = -J \sum_{(ij)} \delta_{s_i s_j} \qquad \{s_i=1, 2, \ldots, q\} \tag{20}$$

where J>0 is the interaction strength between "spins" on nearest neighbor sites i and j and δ is the Kronecker delta. The phase transitions in the d=2 Potts model have been extensively studied [27] and many features of the transitions are exactly known.

Here too we shall superimpose the Gaussians of the form of Eqn. (7b) in the transition region of this model but we note the following two complications: (a) the transition is asymmetric, and we must explicitly use C_+ and C_- for the specific heats of the two phases, respectively; (b) there are q degenerate ordered phases coexisting with a single disordered phase and the weighting of the two Gaussians must reflect this. We adopt the following notation: the subscripts + and – refer to quantities in the high– and low–temperature phases respectively. U denotes the internal energy so that $E_\pm=U_\pm(T_c)$ are the values at opposite ends of the discontinuity in internal energy at the transition. We define $\Delta F=F_+(T)-F_-(T)$ and $\Delta T=T-T_c$; then $\Delta F \approx -\{S_+(T_c)-S_-(T_c)\}\Delta T = -\frac{(E_+-E_-)\Delta T}{T_c}$ [10)] so that x (recall Eqn. (9)) has the symmetrized form:

$$x = -\frac{\Delta F L^d}{2 k_B T} \quad . \tag{22}$$

The energy distribution is then given by:

$$P_L(E) = A[\frac{a_+}{\sqrt{C_+}} P_L(+) + \frac{a_-}{\sqrt{C_-}} P_L(-)] \tag{23}$$

where $a_+ = e^x\sqrt{C_+}$, $a_- = qe^{-x}\sqrt{C_-}$ reflect the degeneracy of the ordered state explicitly. The functions $P_L(\pm)$ are given by

$$P_L(\pm) = \exp[-\frac{\{E-(E_\pm + C_\pm \Delta T)\}^2 L^d}{2k_B T^2 C_\pm}] \ , \tag{24}$$

and the normalization constant A is:

$$A = [\frac{L^d}{2\pi k_B T^2}] \frac{1}{(a_+ + a_-)} \ . \tag{25}$$

We have thus constructed the distribution such that $P_L(E_-) = qP_L(E_+)$.

The first and second moments of the energy distribution are then:

$$\langle E\rangle_L = \frac{(a_+E_+ + a_-E_-) + (a_+C_+ + a_-C_-)\Delta T}{a_+ + a_-} \tag{26}$$

and

$$\langle E^2\rangle_L = \frac{a_+(E_+ + C_+\Delta T)^2 + a_-(E_- + C_-\Delta T)^2}{a_+ + a_-} + \frac{k_B T^2}{L^d}\frac{(a_+C_+ + a_-C_-)}{a_+ + a_-} \ . \tag{27}$$

Using the fluctation–dissipation theorem and the first and second moments of the internal energy we obtain the specific heat

$$C_L = \frac{L^d}{k_B T^2}\{<E^2>_L - <E>_L^2\}$$
$$= \frac{a_+C_+ + a_-C_-}{a_+ + a_-} + \frac{a_+a_-L^d\{(E_+-E_-)+(C_+-C_-)\Delta T\}^2}{k_B T^2 (a_+ + a_-)^2} . \tag{28}$$

The identical expression results if we use the definition $C_L = d<E>_L/dT$.

For very large lattices sufficiently close to $T_c(\infty)$ the following (leading–order) scaling form can be derived for the specific heat from Eqn.(28):

$$\frac{C_L}{L^d} = \frac{(E_+-E_-)^2 q(C_-/C_+)^{1/2}}{k_B T_c^2(\infty)\{e^x + e^{-x} q(C_-/C_+)^{1/2}\}^2} . \tag{29}$$

Note that this assumes that the lattices are sufficiently large that the contribution from the first term in Eqn. (28) is negligible.

From Eqns. (26) and (28), we see that for $L \to \infty$, we recover the bulk behavior:

$$<E>_L \to E_\pm + C_\pm \Delta T, \quad C_L \to C_\pm$$

where the subscripts give the sign of ΔT. We define an effective transition temperature of the finite system $T_c(L)$ to be the location of the peak in the specific heat $C_{L|max}$, to leading order in ΔT the condition $dC_L/dT=0$ then leads to $a_+=a_-$. Thus, the transition temperature of the finite system corresponds to the two peaks having equal weights. For the q–state Potts model $T_c(L)$ is thus:

$$\frac{T_c(L)-T_c(\infty)}{T_c(\infty)} = \frac{k_B T_c(\infty)\ln\{q(C_-/C_+)^{1/2}\}}{(E_+-E_-)L^d} . \tag{30}$$

To leading order in ΔT, the energy at $T_c(L)$ is simply the mean of E_+ and E_- while the specific heat is:

$$C_{L|max} \approx \frac{C_+ + C_-}{2} + \frac{(E_+ - E_-)^2 L^d}{4 k_B T_c^2(\infty)} . \tag{31}$$

Equation (31) is also reflected in the behavior of $<E>_L$ for small L and ΔT. We find that for x<<1,

$$<E>_L \approx \frac{E_+ + E_-}{2} + \frac{(E_+ - E_-)^2 L^d (T - T_c(L))}{4 k_B T_c^2(\infty)} . \tag{32}$$

Thus the discontinuity in E is replaced by a smooth variation when L is finite, and the peak of the specific heat grows as L^d. The width of the transition region may be estimated by setting x ≈ 1 and thus yielding:

$$\frac{\Delta T_{round}}{T_c(\infty)} \approx \frac{2 k_B T_c(\infty)}{(E_+ - E_-) L^d} . \tag{33}$$

Therefore, the rounded region in the specific heat vanishes as L^{-d}. These features are in agreement with those found earlier for the Ising model except for the additional appearance of a shift in $T_c(L)$ induced by the asymmetry of the transition.

For the Potts model it is also useful to evaluate the behavior of the reduced 4th–order cumulant, V_L, of the energy,

$$V_L = 1 - \frac{<E^4>_L}{3 <E^2>_L^2} . \tag{34}$$

Away from the transition and for $L \to \infty$, $P_L(E)$ becomes a δ–function so that $V_L \to 2/3$; this result holds at second–order transitions as well; although $P_L(E)$ will be distinctly non–Gaussian for finite L, the width vanishes for $L \to \infty$. At a first–order transition on the other hand, the fact that $P_L(E)$ reduces to a sum of δ–functions indicates that V_L reaches a non–trivial limit at the transition: at $T = T_c(L)$, the minimum in V_L is of the form

$$V_{L|min} = 1 - \frac{2(E_+^4+E_-^4)}{3(E_+^2+E_-^2)^2} - \frac{4k_BT_c^2(\infty)D(q,E_+,E_-,C_+,C_-)}{L^d}. \tag{35}$$

where the full expression for the function D is given in Ref. 10. Note that the minimum of V_L reaches a non–trivial limit, the leading finite–size dependence again being of the order of L^{-d}. Of course, the location of $V_{L|min}$ also can serve as another definition of the $T_c(L)$.

3.2 Simulations results for the Ising square lattice below T_c

Monte Carlo results have been obtained for the Ising Hamiltonian in Eqn. (8) at the temperature of $k_BT/J=2.1$ using $4\leq L\leq 14$ with periodic boundary conditions.[9] At this temperature, $\xi \approx 3.7$ in the infinite sample so that the preceding theory should be valid. At this temperature, $M_{sp} \approx 0.87$,[26] so that a substantial gap separates the two peaks of $P_L(s)$ and quite long runs are needed to sample the entire distribution correctly, i.e., to insure that both peaks are visited sufficiently often that the relative sizes as well as individual shapes are accurately determined. Even for these relatively small lattice, satisfactory accuracy was obtained only by using runs of about $1.5\mathrm{x}10^6$ Monte Carlo steps/site (MCS).

Figure 9 shows the behavior of χ_L vs H for various L; a peak is seen at H=0 as expected from Eqn. (15). As L increases, the maximum grows while the transition region simultaneously narrows. Figure 10 shows the behavior of the cumulant U_L which vanishes for large H, reaches 2/3 at H=0 and has a minimum in between. As L increases, the minimum in U_L sharpens and approaches H=0 again exhibiting the vanishing of the transition region. The L–dependence of these features in χ_L and U_L are demonstrated in the log–log plots of Fig. 11. We see there that χ_L does indeed diverge as L^d while the width of the transition (represented by H_{min} and $H_{1/2}$ in Fig. 11) vanishes as L^{-d}. This behavior shows that the susceptibility has a δ–function singularity in the limit $L \to \infty$.

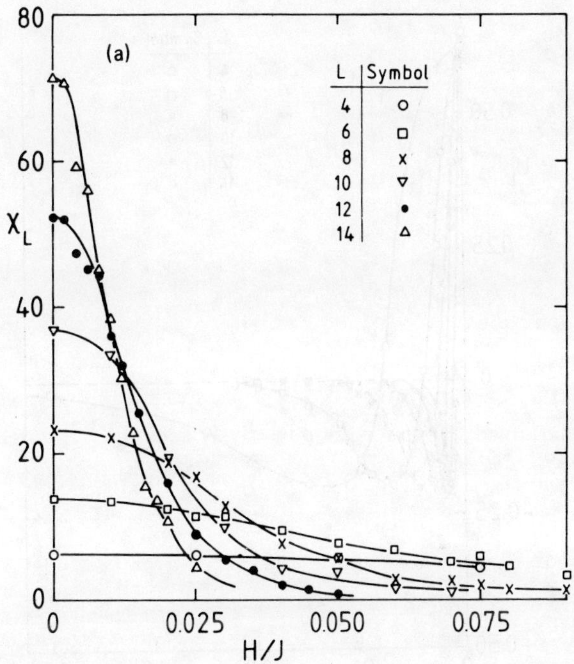

Fig. 9 Monte Carlo data for the susceptibility χ_L for LxL Ising square lattices with periodic boundary conditions plotted vs magnetic field H at kT/J=2.1 for various L. The infinite lattice critical temperature is kT_c=2.269 (from Binder and Landau [9]).

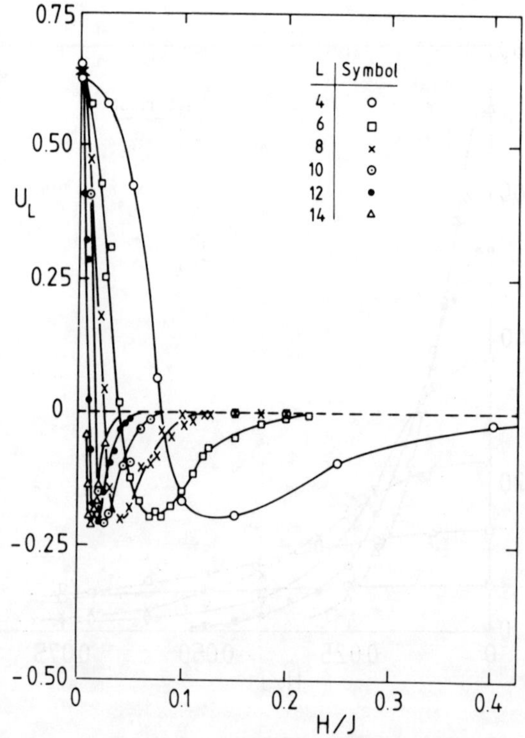

Fig. 10 Monte Carlo data for the reduced cumulant U_L for LxL Ising square lattices with periodic boundary conditions plotted vs magnetic field H at kT/J=2.1 for various L. The infinite lattice critical temperature is kT_c=2.269 (from Binder and Landau [9]).

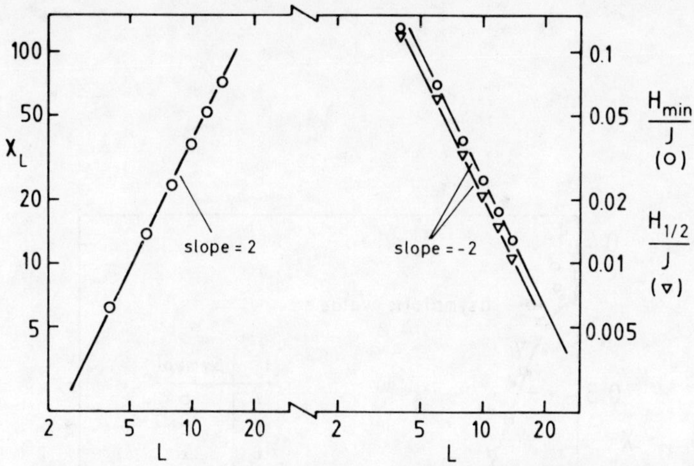

Fig. 11 Log–log plot of the susceptibility maximum of the Ising square lattice vs linear dimension L (left part) and of characteristic fields vs linear dimension (right part) at kT/J=2.1. Here H_{min} denotes the field where the cumulant U_L has its minimum, while $H_{1/2}$ is the field at which the susceptibility has fallen to half its maximum value (from Binder and Landau [9]).

From Eqn.(15), we see that, apart from a correction term which vanishes as L^{-d}, the scaled susceptibility χ_L/L^d can be represented as a function of the scaled variable, H/L^d. This scaling is tested in Fig. 12 and except for small systematic deviations near H=0 due to the correction term the data scale quite well. By examining the L–dependence of $\chi_L(H=0)$, one can also extract the discontinuity in the magnetization, M_{sp} [9]. Our Monte Carlo results are in full agreement with the theory of Section 3.1.

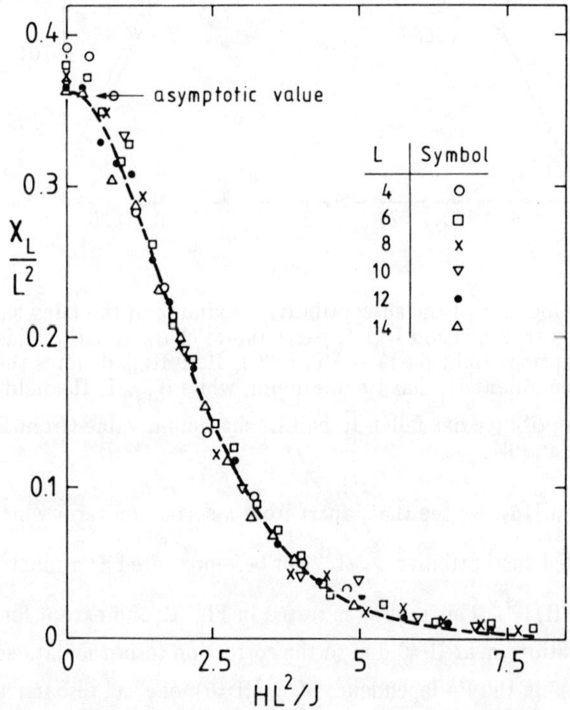

Fig. 12 Scaling plot for the susceptibilty χ_L of LxL square lattices at kT/J=2.1. The arrow indicates the asymptotic value $M^2_{sp}J/kT$ calculated from the exact solution of Yang [26] (from Binder and Landau [9]).

To show that first– and second–order transitions can be distinguished by such tests, we present in Figs. 13 and 14 the behavior of U_L and χ_L when the field is scanned at the critical temperature [11] of $k_BT_c/J \approx 2.269$. While the behavior of U_L at the critical temperature is not qualitatively different from Fig. 10, Fig. 13 shows that the cumulant converges to the value of 0.618, easily distinguishable from the value of 2/3 at the first–ordrer transitions. Figure 14 presents the L–dependence of the various quantities shown in Fig. 11 and clearly demonstrates that the slopes are distinctly different from 2 apart from the fact they are not equal: The slopes here reflect the critical exponents $\gamma/\nu = 1.75$ and $(\gamma+\beta)/\nu=1.875$ since finite–size scaling theory [3] predicts that $\chi_L=L^{\gamma/\nu}\,\tilde{\chi}(HL^{-(\gamma+\beta)/\nu})$ where $\tilde{\chi}$ is a scaling function.

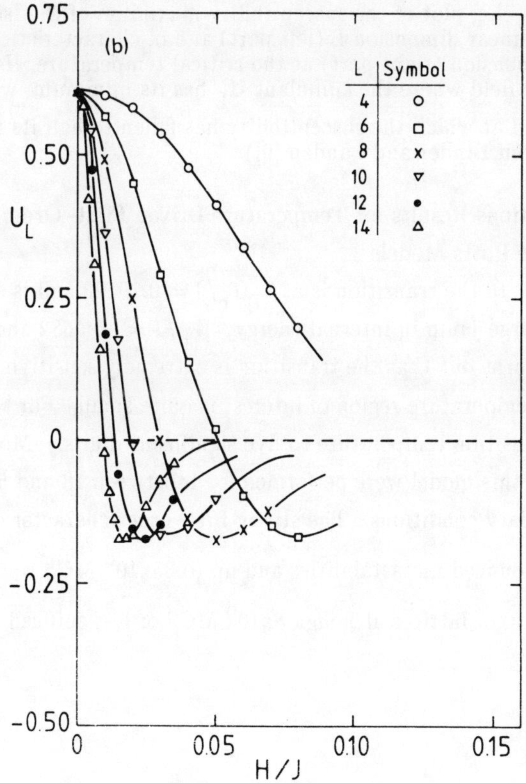

Fig. 13 Field dependence of the reduced cumulant U_L for the Ising square lattice at $k_B\,T = 2.1$ (from Binder and Landau [9]).

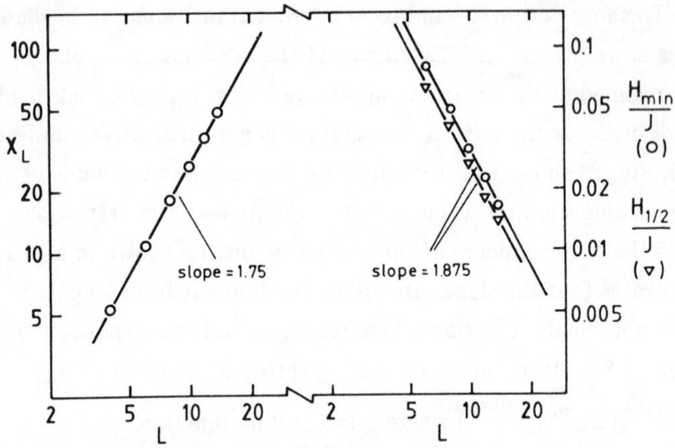

Fig. 14 Log–log plot of the susceptibility maximum of the Ising square lattice vs linear dimension L (left part) and of characteristic fields vs linear dimension (right part) at the critical temperature. Here H_{min} denotes the field where the cumulant U_L has its minimum, while $H_{1/2}$ is the field at which the susceptibility has fallen to half its maximum value (from Binder and Landau [9]).

3.3 Simulations Results for Temperature Driven First–Order Transitions in q–State Potts Models

For q = 10 the transition is at $k_BT_c/J = 0.70123$ and is strongly first order with a large jump in internal energy: $E_+/J = -0.9682$ and $E_-/J = -1.6643$. (It turns out that the transition is extremely sensitive to lattice size and that the temperature region of interest is quite small. For this reason we specify the transition temperature to five significant digits.) Monte Carlo simulations of this model were performed for L between 18 and 50 and with periodic boundary conditions. The strong first–order character of the transition produced pronounced metastabilities and up to 35×10^6 MCS were used per data point for the larger lattices although 8×10^6 MCS or less sufficed for the smaller ones.

In Fig. 15 we show data for the probability distributions for the internal energy of the q=10 Potts models; well above the transition a single gaussian peak is observed whereas quite close to $T_c(L)$ the peaks are no longer Gaussian. In Fig. 15 the data show a very strong overlap of peaks with only a shallow

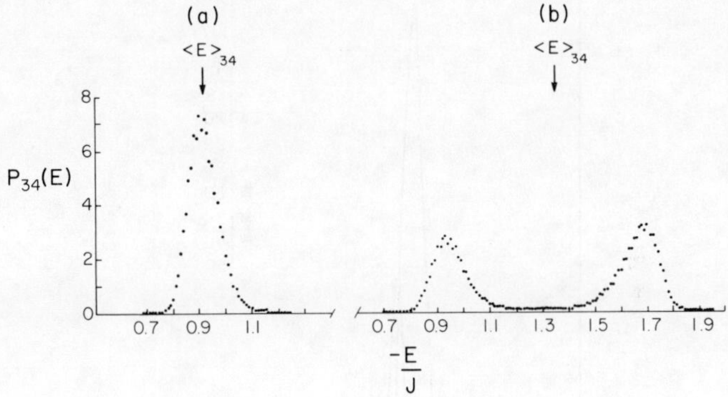

Fig. 15 Monte Carlo results for the probability distribution of the internal energy in d=2 10–state Potts models on 34×34 lattices with periodic boundary conditions. (a) kT/J=0.7012 (the specific heat peak is at $kT_c(L)/J$=0.7027), and (b) $kT/J \approx kT_c(L)/J$ (from Challa et al. [10]).

minimum. The formalism discussed earlier must be applied with care in this region.

Figure 16 shows the temperature dependence of the specific heat for different lattice sizes. As L increases the peaks grow while the transition region narrows and the peak location shift towards $T_c(\infty)$. The extraordinarily large peak values are a consequence of the approach to a δ–function latent heat which is present in the thermodynamic limit. The peak heights are plotted vs volume in Fig. 17; the data are well fitted by a straight line with slope of 0.250, in very good agreement with the exact slope of 0.254, thus verifying Eqns. 30–31.

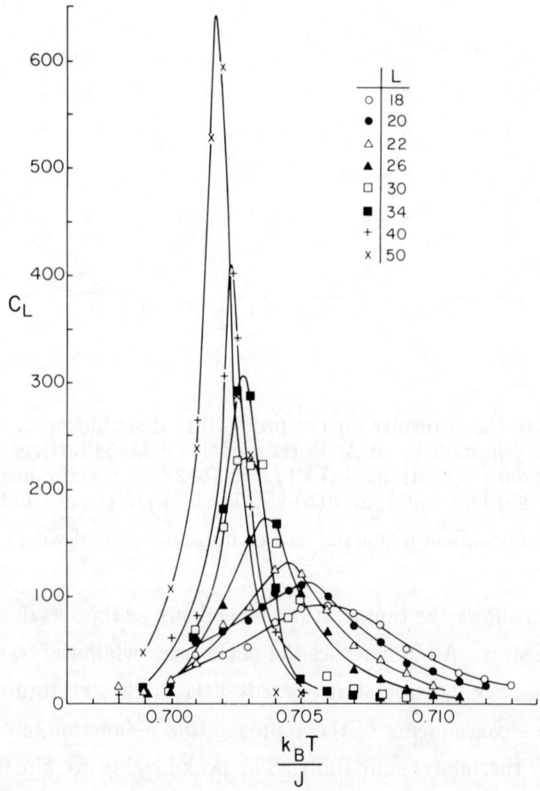

Fig. 16 Temperature dependence of the specific heat for the q=10 Potts model on LxL square lattices with periodic boundary conditions for various lattice sizes (from Challa et al. [10]).

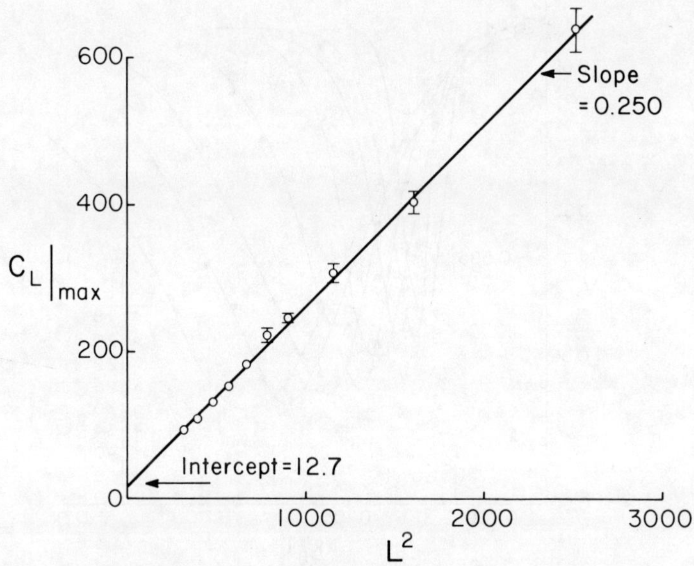

Fig. 17 Variation of the specific heat maximum for the q=10 Potts model with sample volume L^d.

The temperature dependence of V_L is shown for various L in Fig. 18 and also agrees with our expectations: with increasing L the minimum in V_L becomes sharper and seems to approach the non–trivial limit of 0.585 at $T_c(\infty)$. This result may be contrasted with the behavior of V_L at the 2nd order transition in the q=2 model with increasing lattice size, V_L rapidly approaches the trivial values of 2/3 for all temperature.

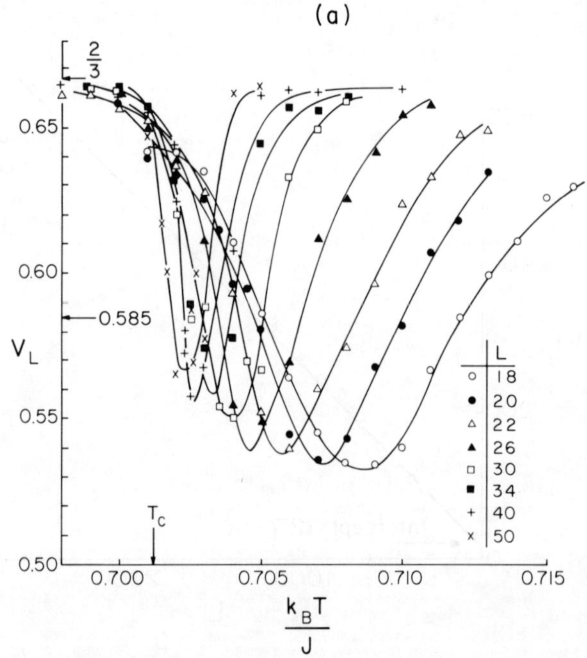

Fig. 18 Temperature variation of the reduced cumulant for the energy V_L for various lattice dimensions L: (a) results for the q=10 Potts model. The transition temperature and the trivial (2/3) and non–trivial (0.585) limits for $V_{L,min}$ are indicated (from Challa et al. [10]).

Denoting the locations of $(C_L)_{max}$ and $(V_L)_{min}$ by T_L^C and T_L^V respectively, we can extrapolate estimates vs L^{-d} to obtain $T_c(\infty)$. This is shown in Fig. 19 where we see that both definitions extrapolate very well to the exact value. Thus, combining this result for $T_c(\infty)$ with the slope of the line in Fig. 17 and using Eqn. (30), we estimate the discontinuity at the transition from Monte Carlo data.

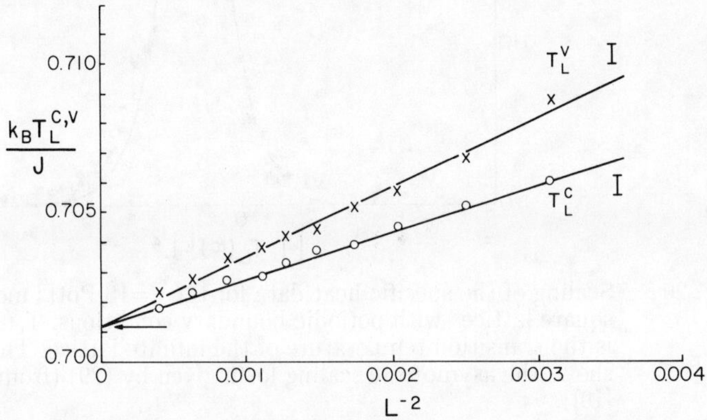

Fig. 19 Extrapolation of characteristic temperatures for the d = 2 q=10 Potts model vs inverse lattice volume L^{-d}. T_L^C is the location of the specific heat maximum and T_L^V is the location of the minimum of the reduced cumulant. The arrow shows the infinite lattice value. The maximum error bars are also indicated (from Challa et al. [10])

Recalling the definition of x (Eqn. (22)), we see that $C_L L^{-d}$ is a universal function of $\Delta T L^d$. Moreover, since C_-/C_+ can be obtained from Fig. 18 (see also Eqn. (29)), there are no adjustable parameters and Eqn. (29) can be directly tested. The results are shown in Fig. 20 and we find generally good agreement between theory and simulation. The small deviations near the peak result from the first term in Eqn. (28) which is not yet negligible for the values of L studied. The deviations at large values of scaling variable are due to remaining corrections from the term in $(C_+ - C_-)\Delta T$ in Eqn. (28) which is not included in Eqn. (29).

The phase transition in the q=5 Potts model is intriguing in that it exhibits only very small jumps in its properties at the transition and hence looks quite continuous in a finite system. In addition, simulations by Katznelson and

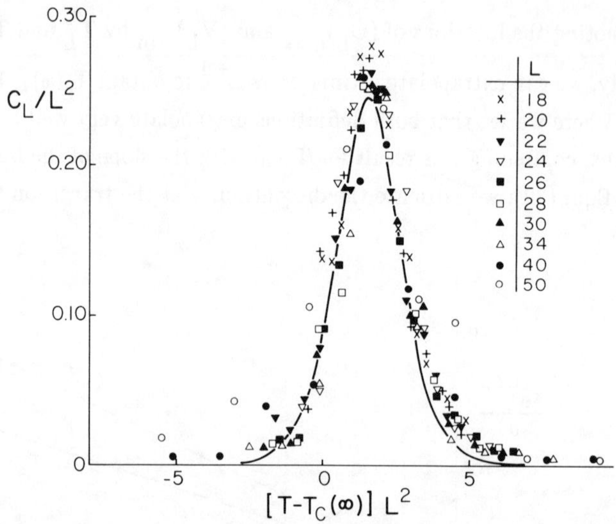

Fig. 20 Scaling of the specific heat data for the q=10 Potts model on LxL square lattices with periodic boundary conditions. $T_c(\infty)$ in the abscissa is the transition temperature of the infinite lattice. The solid curve shows the asymptotic scaling form given by (29) (from Challa et al. [10]).

Lauwers [28] seemed to show the existence of metastable peaks in the energy distribution which could not be explained within the context of the theory outlined in 3.1. More extensive Monte Carlo simulations of this model were carried out by Peczak and Landau [29] using L as large as 240 and periodic boundary conditions. Fluctuations were quite important and runs well in excess at 10^6 MCS were needed per data point in order to obtain reliable data. With sufficient statistics there are no "extra" peaks in the energy distribution but the two peaks which do occur are very close together, even in rather large systems; clearly interfacial effects will be much more pronounced than for q=10. Even for the largest lattices considered the peaks in the energy distribution overlapped appreciably and the size dependence was consistent with that of a system undergoing a second–order transition. For example, as we show in Fig. 21, the increase of the specific heat maxima is much slower than the expected L^2 dependence and could be well fitted by an effective exponent ratio $\alpha/\nu = 1.07 \pm 0.03$. A careful examination of small q=10 lattices (L<20) revealed similar behavior, and by comparing correlation lengths for the two models we

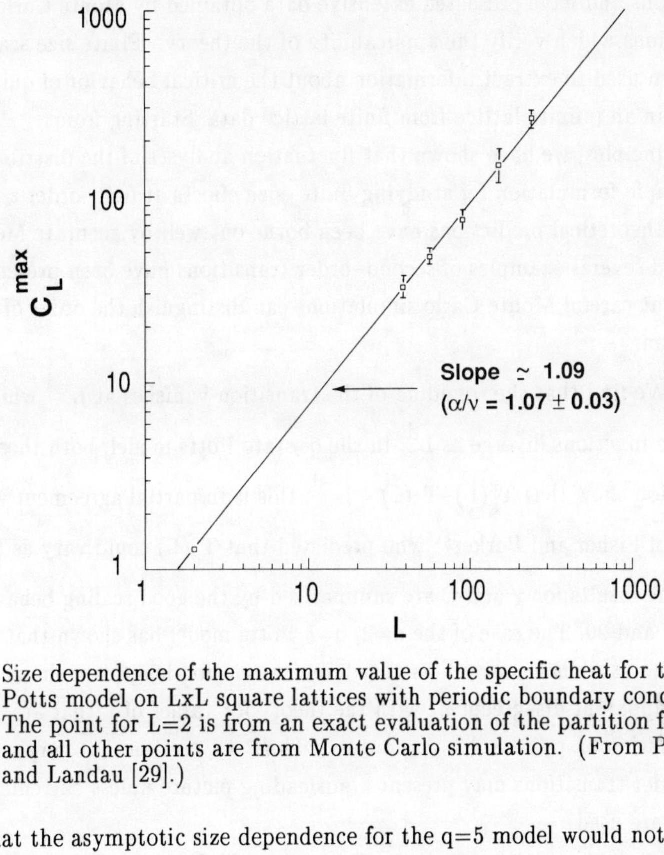

Fig. 21 Size dependence of the maximum value of the specific heat for the q=5 Potts model on LxL square lattices with periodic boundary conditions. The point for L=2 is from an exact evaluation of the partition function and all other points are from Monte Carlo simulation. (From Peczak and Landau [29].)

estimate that the asymptotic size dependence for the q=5 model would not be achieved until $L \gtrsim 2000$. Thus although it appears that the finite size behavior of the q=5 Potts model is not inconsistent with theory, it is also evident that if it were not previously known that the transition were first order it would have been quite easy to misidentify the order of the transition from the finite size behavior, particularly if the lengths of the Monte Carlo runs were too short. The scaling behavior is closer to critical scaling than to the thermodynamic scaling of Sec. 3.1 until quite large lattices are used.

4. CONCLUSIONS

We have reviewed the formalism for finite size scaling at 2nd order phase transitions and have presented extensive data obtained by Monte Carlo simulations which verify the applicability of the theory. Finite size scaling has also been used to extract information about the critical behavior of quite different models in an infinite lattice from finite lattice data. Starting from first–principles, we have shown that fluctuation analyses of the distributions can be a simple formulation for studying finite–size effects at first–order transitions. All the theoretical predictions have been borne out well by accurate Monte Carlo data and several examples of second–order transitions have been presented to show that careful Monte Carlo simulations can distinguish the order of the transition.

We find that the rounding of the transition vanishes at L^{-d} while the response functions diverge as L^{d}. In the q–state Potts model, both theory and simulation show that $T_c(L)-T_c(\infty) \sim L^{-d}$; this is in partial agreement with the results of Fisher and Berker[5] who predicted that $T_c(L)$ could vary as L^{-1} to L^{-d}. The results for χ and C are summarized by the good scaling behavior in Figs. 12 and 20. The case of the d=2, q=5 Potts model has shown that there are strong metastabilities as well as very large correlation lengths in the model that lead to apparent disagreement with the theory for values of L that are amenable to simulation on present day computers. Thus, we caution that very weak first–order transitions may present a misleading picture unless extremely large lattices are used.

Lastly, simulations are in progress to study finite size behavior at first–order transitions in systems with classical spin vectors; the effects of spin waves should modify the behavior discussed above [30].

Acknowledgements: Portions of this work were carried out in collaboration with K. Binder, M.S.S. Challa, and P. Peczak to whom the author is indebted. This research was supported in part by NSF grant DMR–8715740.

REFERENCES

1. M. E. Fisher, Proceedings of the International Summer School "Enrico Fermi," Course L. I. (Academic, New York, 1971); see also M. E. Fisher and M. N. Barber, Phys. Rev. Lett. 28, 1516 (1972).

2. D. P. Landau, Phys. Rev. B13, 2997 (1976); D. P. Landau, Phys. Rev. B14, 255 (1976).

3. For a review see M. N. Barber, in "Phase Transitions and Critical Phenomena", edited by C. Domb and J. L. Lebowitz, Vol. 8, p. 145 (Academic, New York) 1983.

4. Y. Imry, Phys. Rev. B21, 2042 (1980).

5. M. E. Fisher and A. N. Berker, Phys. Rev. B26, 2507 (1982).

6. J. L. Cardy and M. P. Nightingale, Phys. Rev. B27, 4256 (1983).

7. V. Privman and M. E. Fisher, J. Stat. Phys. 33, 385 (1983).

8. V. Privman and M. E. Fisher, J. Appl. Phys. 57, 3327 (1985).

9. K. Binder and D. P. Landau, Phys. Rev. B30, 1477 (1984).

10. M. S. S. Challa, D. P. Landau and K. Binder, Phys. Rev. B34, 1841 (1986).

11. L. Onsager, Phys. Rev. 65, 117 (1944).

12. D. P. Landau, Phys. Rev. B16, 4164 (1977).

13. J. M. Ziman, Proc. Phys. Soc. A64, 1108 (1951).

14. K. R. Subbaswamy and G. D. Mahan, Phys. Rev. Lett. 37, 642 (1976); E. Müller–Hartmann and J. Zittartz, Z. Phys. B27, 261 (1977); K. Binder and D. P. Landau, Phys. Rev. B21, 1941 (1980).

15. M. E. Fisher, Proc. Roy. Soc. A254, 66 (1960).

16. A. Bienenstock, J. Appl. Phys. 37, 1459 (1966); A. Bienenstock and J. Lewis, Phys. Rev. 160, 393 (1967).

17. T. E. Shirley, Phys. Lett. A42, 183 (1972).

18. G. H. Wannier, Phys. Rev. 79, 357 (1950); R. M. F. Houtappel, Physica 16, 425 (1950).

19. D. P. Landau, Phys. Rev. B27, 5604 (1983).

20. J. V. José, L. P. Kadanoff, S. Kirkpatrick, and D. R. Nelson, Phys. Rev B16, 1217 (1977).

21. J. M. Kosterlitz and D. J. Thouless, J. Phys. C6, 1181 (1973).

22. L. D. Landau and E. M. Lifshitz, "Statistical Physics", Vol 5 of Course of Theoretical Physics part I (Pergamon, New York) 1980.

23. K. Binder, Phys. Rev. Lett. 47, 693 (1981).

24. K. Binder, Zeit. Phys. B 43, 119 (1981).

25. R. J. Baxter, J. Phys. C6, L445 (1973).

26. C. N. Yang, Phys. Rev. 87, 410 (1952).

27. See the review by F. Y. Wu, Rev. Mod. Phys. 54, 235 (1982).

28. E. Katznelson and P. G. Lauwers, Phys. Lett. B186, 385 (1987); in "Non–Perturbative Methods in Quantum Field Theory, edited by Y. Norvath (World Scientific Publishing, Singapore, 1987)

29. P. Peczak and D. P. Landau, Phys. Rev. (1989), in press.

30. M. E. Fisher and V. Privman, Phys. Rev. B32, 447 (1985).

VI. MONTE CARLO STUDIES OF UNIVERSAL FINITE-SIZE SCALING AMPLITUDES

K. K. Mon

TABLE of CONTENTS

MONTE CARLO STUDIES OF UNIVERSAL FINITE-SIZE SCALING AMPLITUDES

K.K. Mon
Department of Physics and Astronomy
and Center for Simulational Physics
University of Georgia
Athens, Georgia 30602 U.S.A.

I. INTRODUCTION

Although the importances of finite-size effects in critical phenomena have been noted for many years[1], the subject continues to be a significant topic of current researchs . Recently, there are substantial progress in finite-size scaling theory and numerical simulation of finite system near and at criticality. In this contribution to the review , Monte Carlo techniques and applications to studies of universal finite-size scaling amplitudes will be considered. Alternative methods such as field-theoretical techniques[2], transfer-matrix[3], conformal invariance[4], Poisson summation[5], exact solutions[6] and others have produced important results on this subject . Reviews on their contributions will be discussed by other authors in this volume or elsewhere. The emphasis here will be on Monte Carlo techniques for physical systems for which these other methods have not been applied extensively . They include non-lattice models (such as Lennard-Jones fluids and solids) in two and three dimensions, large lattice systems in three dimensions with nearest-neighbor couplings and models with longer range interactions. In particular, we will restrict our discussions to the finite-size scaling amplitude of excess free energy density for systems with and without surfaces and interfaces for the three dimensional Ising lattice models. A short summary of finite-size scaling theory predictions for possible universality of the finite-size scaling amplitudes will follow in this section. In section 2 , a discussion of Monte Carlo techniques will be presented and a summary of published results can be found in section 3 . The review concluded in section 4 with some remarks .

The predictions and confirmations of universality for critical exponents and various thermodynamic scaling amplitude ratios are well known. They constitute one of the fundamental contributions of the theory of critical phenomena. The possibility of universality for finite-size scaling amplitudes are relatively more recent and their studies are motivated in part by developments of analytical and numerical techniques for theoretical investigations of model systems which are finite in one or more directions. Such theoretical studies have significance in physical and computer experiments involving fluids or magnetic systems in restricted geometry. This include thin fluid or magnetic films, capillary phenomena such as wettings and growths, properties of fluids in porous media , and analysis of simulation studies of finite model systems.[7-8]

Universality in the finite-size scaling amplitudes at criticality have been suggested by a variety of recent theoretical studies.[9-12] For two-dimensional systems some of the amplitudes can be obtained from applications of conformal invariance[4]. In general, universality has been proposed on the basis of scaling arguments[9]. In three-dimensions there are considerable analytical and numerical results, which support the universality for the finite-size scaling amplitudes.[13-27] Exact solutions exist only for the spherical model and ideal Bose gas [28]. For this reasons, numerical studies , such as Monte Carlo simulations play an important role. For the present discussions, we focus on the singular part of the free energy density for finite system at bulk criticality

Consider a hyper-cubic lattice system of lattice sizes L in d bulk dimensions with periodic boundary conditions in all d directions, for $d < d_>$, the upper critical dimension. At bulk criticality with zero field, the free density in units of $k_B T_c$ is predicted [9] to scale in the large L limit as,

$$f_L = f_\infty + U_o / L^d . \qquad (1)$$

U_o is universal. The second term is considered to be the singular part of the critical free energy density and vanishes in the thermodynamic limit. The possible universality of U_o is the finite-size scaling analog of the two scale-factor universality [29]for thermodynamic scaling.

The interfacial free energy density can be defined as the free energy differences per unit area of the cross-section (L^{d-1}) between a system with an interface and the identical system without an interface. For reviews, see references 30-32. An interface can be imposed by using anti-periodic boundary condition with the interfacial tensions given by ,

$$\tau \equiv (F_{ap} - F_{p})/(L^{d-1}) \tag{2}$$

or employing a strong wall potential which pin spins adjacent to the walls with,

$$\tau \equiv (F_{+-} - F_{++})/(L^{d-1}) \ . \tag{3}$$

An extension of the finite-size scaling arguments [18] would predict at bulk criticality ,

$$\tau_{L}/k_{B}T_{c} = \Omega \ / \ L^{d-1} \ . \tag{4}$$

Ω is universal.

Lattice systems with free surfaces can be studied by considering free boundary conditions. In these systems involving free boundary condition, the excess free energy due to the presences of free surfaces, edges, and corners can be systematically defined in reference to the finite system with full periodic boundary conditions. Two free surfaces are created by changing the boundary in one direction from periodic to free and four free surfaces with four edges are obtained by changing the boundary in two directions. Eight corners, twelve edges and six free surfaces are produced by setting all three directions to have free boundary conditions. The free energy differences between these three types of boundary conditions and the fully periodic boundary condition are denoted as F^{x}, F^{xy}, and F^{xyz}, indicating the free boundary conditions in the x, xy, xyz directions, respectively. The excess surface free energy per unit area can be expressed as $f^{s}(L)=F^{x}/2L^{d-1}$. The finite-size scaling predictions in units of $k_{B}T_{c}$ at bulk criticality[33,19] is,

$$f^{s}(L)=f^{s}_{\infty}+\Delta^{s}/L^{d-1} \quad , \tag{5}$$

where f^{s}_{∞} is the critical surface free energy density and Δ^{s} is the universal finite-size scaling amplitude. The edge excess free energy per unit length is obtained from F^{xy} by subtracting out the excess free energy of four free surfaces. The edge free energy density in units of k_BT_c finite-size scaling is proposed[25] to be,

$$f^{e}(L)=f^{e}_{\infty}+\Delta^{e}/L \ , \tag{6}$$

where f^{e}_{∞} is the critical edge free energy density and Δ^{e} is the universal finite-size scaling amplitude. The corner excess free energy per corner is obtained from F^{xyz} by subtracting out the contributions of six free surfaces and twelve edges. The finite-size scaling predictions[25] for corner excess free energy in units of k_BT_c was argued to be,

$$f^{c}(L)= u \ln(L) + c \quad . \tag{7}$$

u is universal but c is a non-universal constant.

Boundary conditions which combine free boundary and wall potentials are also possible. Examples of their considerations have been mostly in two bulk dimensions , where exact solutions are often possible.[33-37] In general, finite-size scaling amplitudes are sensitive to the shape or aspect ratios of the lattice sizes and other details of the boundary conditions. One important case is the semi-infinte slab geometry (strips in d=2). This has applications in wettings and fluids between closely spaced plates.[38,7] The finite-size scaling amplitudes than become a function of the geometry and is also an universal function.[33,19]

One of the important motivation and applications for calculating finite-size dependences of free energy density is to obtain the free energy density in the thermodynamic limit. A well know observation and technique is that the partition sum and hence the free energy for a sufficiently small lattice size can be evaluated numerically to arbitrary precisions.[39] This together with the understanding and ability to calculate the finite-size dependences constitute a method for obtaining the free-energy density for any finite system sizes and in the thermodynamic limit.[16,17] The free energy density in the thermodynamic limit are essential ingredients in the unambiguous determination of phase boundary. This is particularly relevent in discontinuous transitions. There, computer simulations with finite system sizes and limited simulation time often suffers from hystersis and other dynamical effects.[30] In those cases, the systems are not at a critical point, with the associated infinite correlation length and the free energy density approach the thermodynamic limit exponentially fast. The amplitudes and parameters governing this approach are not universal. Techniques used to study the finite-size dependences for system at criticality are generally also applicable to non-critical model systems.

2. MONTE CARLO TECHNIQUES FOR EVALUATING FINITE-SIZE SCALING AMPLITUDES

In this section, a summary of numerical techniques used in computer simulation studies of free energy and related quantities such as surface and interfacial tensions will be given , followed by a review of related recent progress in developing efficient and accurate method for routine applications.

To obtain the finite-size scaling amplitudes, the finite-size dependences of the related excess free energy density have to be calculated. In general the standard technique is to obtain the desired free-energy density for a range of system sizes and the difference between them is used to extract the finite-size dependences. We will now consider the calculation of the bulk free-energy density. The interfacial tensions and surface free energy will be discussed later. Because of its importances, a large number of methods have been proposed to calculate bulk free energy.[40-42] Most methods can be roughly divided into two types. One approach (labeled approach A) is to calculate free energy differences between the thermodynamic state of interests and a reference thermodynamic state. The standard method to calculate free-energy differences is thermodynamic integration of various derivatives of the free energy along a reversible path in the phase diagram.[40] A commonly used alternative technique is multistage sampling (umbrella sampling) and the free-energy difference is obtained from the relative overlap of distribtion functions.[43] The second type of approach (approach B) involves direct numerical evaluation of the partition function[39] or transfer matrix[3]. A recent proposal[44,45] combined Monte Carlo methods and transfer matrix in a variant of the Green's function Monte Carlo method.[46] There are also various stochastic methods[47-49] which provide estimates of the entropy or free energy. In this type of approach (approach B), the absolute free energy is obtained without the need for a suitable reference state.

These two types of methods (A and B) together, have been quite adequate in principle to calculate the bulk free energy to high accuracy, given the computational resources. Depending on the model and the type of informations desired, one method may be preferred over another. The study of finite-size dependence with these numerical methods will involves separate calculations of the free energy for different system sizes. There must be sufficient precisions, so that the difference (the finite-size dependence) can be analyzed and finite-size scaling amplitudes extracted.

The standard procedure to calculate the interfacial tension and surface free energy with these two types of methods has been to calculate separately the bulk free energy with an interface or surface and the same bulk free energy without an interface or surface, respectively . From the differences of the two bulk free-energies , the appropriate interfacial tension or surface free energy is obtained.[30-32] Thermodynamic integrations using the zero temperature ground state as the reference state have been used to obtained the interfacial tension for the three dimensional Ising model on a simple cubic lattice.[50] These calculations have the adventage that the bulk free-energies can also be obtained in addition to the interfacial tension or surface free energy. The disadventages include a need for high accuracy to extract a small quantity as a difference of two estimates. This is particularly relevent for critical interfacial tensions, where the value to be estimated vanishes in the thermodynamic limit.

Other approachs have also been proposed. One rather elegent technique obtains directly the interfacial free energy from sampling the (nonzero) minimum of the order-parameter distribution function.[51] This minimum is a consequence of the spontaneous formations of the interface from order-parameter fluctuations. The probability of such domain formations is directly related to the interfacial tensions and decreases exponentially with the intefacial free energy of the domain . For this reason, this method was proposed mainly for systems near criticality , where sampling efficiency are tractable with the small interfacial tensions. Both two and three dimensional Ising model were considered, with the data converging nicely to exactly known results for d=2 [52], and satisfactory results seem to have been obtained for moderate system sizes for the three dimensional simple cubic lattice.

Recent calculations of the finite-size scaling amplitudes for the three dimensional Ising model interfacial , surface and bulk free energy at bulk criticality employs extensions of standard methods and new approach. The free energy difference needed to extract the interfacial tensions and surface free energy density was estimated directly without first calculating the separate related bulk free energy.[18-21,53]

Critical interfacial tensions for the simple cubic and body-cenetred cubic lattice Ising model were obtained by extensions of multistage sampling (umbrella sampling) method to estimate the free energy differences between a lattice system with an interface and one without an interface.[18] The interfaces were imposed by appropriate boundary conditions. Most applications of multistage sampling have been concerned with free energy differences of systems with the same boundary conditions but different reduced Hamiltonian ($H/k_B T$). The system sizes remains the same. This permits estimates of the free energy differences of the same system at different temperatures or the differences between a given system and a reference system. The extensions needed to calculate interfacial tensions for lattice systems were the simple observation that differences in the boundary conditions can be formally incorporated into the Hamiltonians.[53] This simple observations though trivial, provided a straightforward scheme for implementing the multistage sampling method.[43] In the multistage sampling method, intermediate interpolating Hamiltonians have to be introduced to efficiently calculate the free energy difference of systems with very little overlaps in their distribution functions . This will always occur for large free energy difference and arises naturally for large system sizes. The larger the system size, the greater is the number of intermediate stages needed. High accuracy data also requires a large number of intermediate stages.

Interfacial tensions for the Ising model in both two and three dimensions at criticality were calculated by multistage sampling using the interpolating Hamiltonian ($H_i = \Gamma_i H_{ap} + (1-\Gamma_i)H_p$), which is a linear mixture of periodic boundary condition(H_p) and anti-periodic boundary condition (H_{ap}). Γ_i is a parameter between 0 and 1. The extensions to pinning wall boundary condition is , $H_i = \Gamma H_{+-} + (1-\Gamma) H_{++}$. These results provided the first numerical data in support of the the proposed extension of universal finite-size scaling amplitudes to critical interfacial tensions.[18] The number of stages needed was about six and 10^5 Monte Carlo steps per spins were used for a range of systems sizes , 4^3 to 16^3. Both the simple cubic and body centered cubic lattices were used.

More recent calculations of free energy differences for large lattice systems consider alternative method to the multistage sampling technique. As noted above, high accuracy data for large system often requires many intermediate stages. The optimum sequences of interpolating Hamiltonians are to be chosen for maximum overlaps of the related distribution functions, which cannot be accurately predicted without actually performing the simulations. This is far from ideal for routine applications. This problem was studied and the reliance on the multistage sampling was removed by adopting the "coupling-parameter" approach to calculating free energy differences.[41,42] The coupling-parameter technique has been used extensively in simulations of fluids [42] and is based on a thermodynamic integration along a reversible path defined by continuous variation of a parameter which takes the system from one thermodynamic state to another. For this reason, the method will be referred to as the parametric integration method. This method was first used to obtain the finite-size dependences of the critical free-energy density of the three dimensional Ising model to high accuracy [17] and will be discussed later in this section. To illustrate the applications of this method, a more recent applications to estimate the excess free energy of surfaces, edges, and corners will be discussed first.[21]

The excess free energy of interests involves the differences between a lattice system with free boundary conditions in one or more directions and the same system with the full periodic boundary conditions. They were denoted as, F^{x}, F^{xy}, and F^{xyz} with the related Hamiltonians, H^{x}, H^{xy}, and H^{xyz}. The directions with free boundary conditions are given by the superscrips. The system with full periodic boundary conditions is described by the Hamiltonian, H^{o}. One considers interpolating Hamiltonians, such as, $H^{x}(\Gamma)=(1-\Gamma)H^{o}+\Gamma H^{x}$ and $F^{x}=G^{x}(\Gamma=1)-G^{x}(\Gamma=0)$. $G^{x}(\Gamma)$ is the free energy of the system under the Hamiltonian $H^{x}(\Gamma)$. F^{x} can be obtained by an application of parametric integrations[42],

$$F^{x}=\int_{0}^{1} (dG^{x}(\Gamma)/d\Gamma)\, d\Gamma . \qquad (8)$$

The derivative is simply the internal energy differences under the two set of boundary conditions, which is the energy of the bonds being cut. This can be sampled efficiently by standard Monte Carlo simulations [8]. The integral must be estimated numerically and can be done accurately by Gauss-Legendre quadrature. The results for F^{xy} and F^{xyz} can also be derived in the same manner as for F^{x}. Both simple cubic and body centered cubic lattices at bulk criticality were considered for a range of lattice sizes (L≤20). Typically about 10^{5} Monte Carlo steps per spins for each point and a total of ten points in the

Gauss-Legendre quadrature were used. The success of the numerical quadrature depends on the integrand being smooth and non-singular. This is indeed the case for the Ising lattice models. Models with singular couplings, such as the Lennard-Jones fluids will demand special techniques, which is outside the scope of this review.[54]

The final topic for this section is a discussion of Monte Carlo method for calculations of the finite-size scaling amplitudes of the critical bulk free energy density. It was proposed recently that the finite-size dependence of the bulk free energy density can be directly sampled in Monte Carlo simulations.[16] This eliminate the need to calculate the bulk free energy density for a range of system sizes and than extract the finite-size dependence. The method is based on the observation that the free-energy density differences $(f_{rL}-f_L)$ between two different systems of sizes $(rL)^d$ (r is an integer) and L^d can be calculated directly. One considers a (rL^d) system under two sets of boundary conditions. The first is the usual periodic boundary conditions for a $(rL)^d$ system and denoted by the Hamiltonian H_{rL}. In the second set, one divide the $(rL)^d$ system into r^d separate L^d systems, each with the standard periodic boundary conditions, and denote the composite system by H_L. An effective Hamiltonian for the $(rL)^d$ system can be defined as,

$$H(r,L,\Gamma)=(1-\Gamma)H_L+\Gamma H_{rL} , \tag{9}$$

where Γ is the " finite-size " parameter with a value between 0 and 1. $H(r,L,\Gamma=0)$ denotes a composite system of r^d separate systems of sizes L^d and $H(r,L,\Gamma=1)$ denotes a $(rL)^d$ system. The finite-size dependence of the free-energy density $(f_{rL}-f_L)$ can now be related to the free-energy density differences between Γ=1 and Γ=0,

$$f_{rL} - f_{L} = g(r,L,\Gamma=1) - g(r,L,\Gamma=0), \tag{10}$$

where g(r,L,Γ) is the free-energy density of the system defined by H(r,L,Γ). Standard methods to calculate free-energy differences were first used with this technique. For the two-dimensional Ising model , the ratio method[55] was used for square lattice of up to 24x24. For the three-dimensional simple-cubic and body-centered cubic Ising model, multistage sampling were needed.[43] For system sizes of up to 12x12x12, six stages were used and only r=2 was considered. These first results were significant and provided the first estimates of the finite-size scaling amplitudes (U_o) for the Ising model in three dimensions. They were consistent with the universality predictions.[9] In a later calculations, the parametric integration method was adopted and allowed more efficient and routine calculations for system sizes of up to 32x32x32 and for a wide range of r.[17] It demonstrated that accurate free energy calculations for lattice systems can be done on a routine manner. In the next section, a summary of the reported finite-size scaling amplitudes for the three-dimensional Ising model will be given with comparisions to the universality predictions .

3. MONTE CARLO TEST OF THE UNIVERSALITY FOR THE FINITE-SIZE SCALING AMPLITUDES : THE THREE-DIMENSIONAL ISING MODEL

In this section, we collect together the Monte Carlo estimates for the various finite-size scaling amplitudes of the three-dimensional Ising model and comment on the theoretical predictions of universality. Although different Monte Carlo methods have been used to calculate free energy for the three dimensional Ising model (see for examples, references 40,44,47-49,56-58), there are only a few calculations on the finite-size dependences at criticality. A number of finite-size scaling amplitudes at bulk criticality have been calculated by the Monte Carlo methods discussed in section 2 for the bulk free energy density, interfacial tension, and excess free energy of free surfaces, edges, and corners . The amplitudes for both the simple cubic and body-centered cubic lattices have been considered as a minimum test of possible universality. Since there are no exact solution for the three-dimensional Ising model and physical measurements for these finite-size scaling amplitudes have not yet been reported, these Monte Carlo estimates assumes an important role in providing a test on theoretical predictions. In table I, a summary of Monte Carlo results for the finite-size scaling amplitudes, U_o, Ω, Δ^s, Δ^e, and u are given. They are the amplitudes for the bulk, interfacial, surface, edge, and corner free energy, respectively and have been defined in section 1. The critical couplings in units of $k_B T_c$ used are 0.221655 [59] for the simple cubic and 0.15737 [60] for the body-centered cubic lattice.

Table I. Critical free energies finite-size scaling amplitudes for the three-dimensional Ising model.

	Bulk	Interface	Surface	Edge	Corner
	U_o	Ω	Δ^s	Δ^e	u
SC	-0.625[a]	0.6[b]	0.196[c]	-0.052[c]	0.009[c]
	±0.005	±0.02	±0.008	±0.013	±0.005
	-0.657[d]		0.201[e]		
	±0.03		±0.008		
BCC	-0.643[d]	0.6[b]	0.209[c]	-0.048[c]	0.012[c]
	±0.04	±0.02	±0.005	±0.015	±0.003
			0.205[e]		
			±0.015		

References; a:[16],b:[18],c:[21],d:[17],e:[19]

From table I, it is quite satisfying that ,within the estimated statistical errors the Monte Carlo data are consistent with the theoretical predictions of universality for all the finite-size scaling amplitudes. [9,25] Further test with other physical systems in the three dimensional Ising universality class will be most interesting. Computer simulations studies with model fluids and fluid mixtures would be one possibility.

4. REMARKS

In this contribution , we have focused on recent progress in applying efficient method of calculating free energy differences to obtain the finite-size dependences of critical bulk free-energies, interfacial tensions, and free surface, edge, and corner free-energies in three bulk dimensions . One motivation has been the finite-size scaling theory predictions of universality for various finite-size scaling amplitudes.[9,25] Our summary of the Monte Carlo data for the three-dimensional Ising model provided clear support for the theoretical predictions. Extensions of these techniques to non-lattice models (such as Lennard-Jones fluids and solids) are possible with additional care needed to account for the singular nature of the short-range interactions. For example, in one recent preliminary calculations for a model Lennard-Jones mixture in two dimensions[61], the critical interfacial tension finite-size scaling amplitude has been estimated to be 1.58 in units of $k_B T_c$. This is in agreements with the exact calculations for the two dimensional Ising square lattice of 1.526....[36] Such calculations for models of fluids and mixtures in the three dimensional Ising universality class will be an important test of the finite-size scaling amplitude universality predictions . This is especially relevent in view of the rather small Monte Carlo estimates for these scaling amplitudes . Experimental measurements will be a very challenging and demanding project.

This work has been supported in part by the

Petroleum Research Fund of the American Chemical Society and the Advanced Computational Methods Center of the University of Georgia. The author is indebted to Professors David Jasnow, Peter Nightingale, and Dr. Pik-Yin Lai for collaborations . Numerous discussions and preprints of results prior to publication from Professor V. Privman have also been very helpful. He thanks them.

5. REFERENCES

1. M.E. Fisher, in: " Critical Phenomena ", Proceeding of 1970 E. Fermi Summer School of Physics, edited by M.S. Green (Academic, NY, 1971) Vol. 51. For a review see, M.N. Barber, in: " Phase transitions and Critical Phenomena ", edited by C. Domb and J.L. Lebowitz (Academic, NY, 1983) Vol. 8 .

2. For a review see , David Jasnow ,in: " Finite-Size Scaling and Numerical Simulation of Statistical Systems ",edited by V. Privman (World Scientific, 1989) .

3. For a review see , M.P. Nightingale ,in: " Finite-Size Scaling and Numerical Simulation of Statistical Systems ", edited by V. Privman (World Scientific, 1989) and in , J. Appl. Phys. 53, 7932 (1982).

4. For a review see , J.L. Cardy , in: " Phase Transitions and Critical Phenomena ", edited by C. Domb and J.L. Lebowitz (Academic, 1983) Vol. 7 .

5. For a recent example see , M.P. Gelfand and M.E. Fisher, Int. J. Thermophys, 9, 713 (1988).

6. For a review see , J. Rudnick ,in: " Finite-Size Scaling and Numerical Simulation of Statistical Systems ", edited by V. Privman (World Scientific,

1989)

7. For reviews in wetting, see: P.G. DeGennes, Rev. Mod. Phys. 57, 827 (1985).
D.E. Sullivan and M.E. Telo da Gama, in " Fluid Interfacial Phenomena ",edited by C.A. Croxton (Wiley, New York, 1986).
S. Dietrich, in " Phase Transitions and Critical Phenomena ", edited by C. Domb and J.L. Lebowitz (Academic, New York, 1988) Vol.12 .

8. For reviews see, " Monte Carlo Methods in Statistical Physics ", edited by K. Binder (Springer-Verlag, Berlin, 1987).

9. V. Privman and M.E. Fisher, Phys. Rev. B30, 322 (1984).

10. M.P. Nightingale and H.W.J. Blöte, J. Phys. A16, L657 (1983) .

11. J.L. Cardy, J. Phys. A17, L385 (1984).

12. K. Symanzik, Nucl. Phys. B190[FS3], 1 (1981).

13. H. Guo and D. Jasnow, Phys. Rev. B35, 1846 (1987).

14. J. Rudnick, H. Guo, and D. Jasnow, J. Stat. Phys. 44, 353 (1985).

15. E. Brèzin and J. Zinn-Justin, Nucl. Phys. B257[FS14], 867 (1985).

16. K.K. Mon, Phys. Rev. Lett. 54, 2672 (1985).

17. K.K. Mon, Phys. Rev. B39, 467 (1989).

18. K.K. Mon and David Jasnow, J. Stat. Phys. 41, 273 (1985).

19. K.K. Mon and M.P. Nightingale, Phys. Rev. B31, 6137 (1985).

20. K.K. Mon and M.P. Nightingale, Phys. Rev. B35, 3560 (1987).

21. Pik-Yin Lai and K.K. Mon, Phys. Rev. B (in press, 1989)

22. E. Eisenreigler, Z. Phys. B61, 299 (1985).

23. J. Rudnick, G. Gaspari and V. Privman, Phys. Rev. B32, 7594 (1985).

24. J.L. Cardy and I. Peschel, Nucl. Phys. B300[FS22], 377 (1988).

25. V. Privman, Phys. Rev. B38, 9261 (1988).

26. M.P. Nightingale and J.O.Indekeu, Phys. Rev. Lett. 54, 1824 (1985).

27. J.O. Indekeu , M.P.Nightingale, and W.V. Wang, Phys. Rev. B34, 330 (1986).

28. S. Singh and R.K. Pathria, Phys. Rev. B31, 4483 (1985); Phys. Lett. 55, 347 (1985); Phys. Rev. B32, 4618 (1985).

29. D. Stauffer, M. Ferer, and M. Wortis, Phys. Rev. Lett. 29, 345 (1972).

30. K. Binder, in: " Phase Transitions and Critical Phenomena ", edited by C.Domb and J.L. Lebowitz (Academic, New York, 1983), Vol. 8.

31. R.K.P. Zia, in: " Proceedings of the Twenty-Sixth Scottish Universities Summer School in Physics on Statistical and Particle Physics: Common Problems and Techniques, Edinburgh, 1983 ", edited by K.C. Bowler and A.J. McKane (Scottish Universities Summer School in Physics, Edinburgh, 1984).

32. D. Jasnow, in: " Phase Transitions and Critical Phenomena ", edited by C. Domb and J.L. Lebowitz (Academic , New York, 1986), Vol. 10 . D. Jasnow, Rep. Prog. Phys. 47, 1059 (1984).

33. H. Au-Yang and M.E. Fisher, Phys. Rev. B11, 3469 (1975).
H. Au-Yang and M.E. Fisher, Phys. Rev. B21, 3956 (1980).

34. M.E. Fisher and A.E. Ferdinand, Phys. Rev. Lett. 19, 169 (1967).

35. See for a review, D.B. Abraham in: " Phase Transitions and Critical Phenomena ", edited by C. Domb and J.L. Lebowitz (Academic, New York, 1986), Vol. 10.

36. D.B. Abraham and N.M. Svrakic, Phys. Rev. Lett.

56, 1172 (1986).

37. D.B. Abraham, L.F. Ko and N.M. Svrakic, Phys. Rev. Lett. 61, 2393 (1988).

38. M.E. Fisher and P.-G. de Gennes , C. R. Acad. Sci. Ser. B 287, 207 (1978).

39. See for example, K. Binder, Physica, 62, 508 (1972).

40. For reviews see, " Applications of Monte Carlo Methods in Statistical Physics ", edited by K. Binder (Springer-Verlag, Berlin, 1987).

41. For a review of free energy simulation see, M. Mezel and D.L. Beveridge, Ann. N.Y. Acad. Sci. 482, 1 (1986).

42. For a discussions of various techniques see, M.P. Allen and D.J. Tildesley, " Computer Simulation of Liquids"(Oxford University Press, New York, 1987).

43. J.P. Valleau and G.M. Torrie, in " Statistical Mechanics, Part A ", edited by B.J. Berne (Plenum, New York, 1977).

44. J. Kuti,in " Monte Carlo Methods in Quantum Mechanical Problems ", Vol. 125 of Nato Advanced Study Institute, Series B: Physics, edited by M. H. Kalos (Plenum, New York, 1984).

45. M.P. Nightingale and H.W.J. Blöte, Phys. Rev. Lett. 60, 1562 (1988).

46. For a review see, D.M. Ceperley and M.H. Kalos, in " Monte Carlo Methods in Statistical Physics ", edited by K. Binder (Springer-Verlag, Berlin, 1979).

47. For example, S.K. Ma, J. Stat. Phys. 26, 221 (1981). See reference 40 for extensive list of references.

48. Z. Alexandrowicz, J. Stat. Phys. 13, 231 (1975). H. Meirovitch, ibid. 30,681 (1983).

49. G. Bhanot, S. Black, P. Carter, and R. Salvador, Phys. Let. B 183,331 (1987). G. Bhanot, R. Salvador, S. Black, P. Carter , and R. Toral, Phys. Rev. Lett. 59, 803 (1987).

50. E. Bürkner and D. Stauffer, Z. Phys. B53, 241 (1983).
K.K. Mon, S. Wansleben, D.P. Landau, and K. Binder, Phys. Rev. Lett. 60, 708 (1988), 61, 902 (1988).

51. K. Binder, Phys. Rev. A25, 1699 (1982).

52. L. Onsager, Phys. Rev. 65, 117 (1944).

53. K.K. Mon and David Jasnow, Phys. Rev. A30, 670 (1984).

54. K.K. Mon, to be published.

55. C.H. Bennet, J. Comput. Phys. 22,245 (1976).

56. M.P. Nightingale and R.G. Caflisch, in: " Computer Simulation Studies in Condensed Matter Physics : Recent Developments", edited by D.P. Landau, K.K. Mon, and H.B. Schüttler (Springer Proceedings in Physics 33, Springer-Verlag, Berlin, 1988).

57. K. Binder, Z. Phys. B45, 61 (1981).

58. H. Meirovitch, Phys. Rev. B30, 2866 (1984).

59. G.S. Pawley, R.H. Swendsen, D.J. Wallace, and K.G. Wilson, Phys. Rev. B29, 4030 (1984).

60. M.J. George and J.J. Rehr, Phys. Rev. Lett. 53, 2063 (1984).

61. K.K. Mon, to be published.

VII. TRANSFER MATRICES, PHASE TRANSITIONS, AND CRITICAL PHENOMENA: NUMERICAL METHODS AND APPLICATIONS

M. P. Nightingale

TABLE of CONTENTS

(continued)

Transfer Matrices, Phase Transitions, and Critical Phenomena: Numerical Methods and Applications

M.P. Nightingale

Department of Physics,

University of Rhode Island, Kingston, R.I. 02881,

and

Theory Center and Laboratory of

Atomic and Solid State Physics,

Cornell University, Ithaca, N.Y. 14853, U.S.A.

INTRODUCTION

Transfer matrix techniques have been applied with great success in the study of statistical mechanical models, both in exact solutions and in numerical investigations. This paper is a review of numerical techniques, applied to these models in two and three dimensions. In many instances the statistical mechanical models have quantum mechanical analogs in one less spacial dimension. Many methods and theories apply equally to both of these in many ways equivalent formulations of the same physical problem. However, this review does not deal with the quantum mechanical models, even though this limitation is somewhat artificial; Bethe Ansatz methods are not part of this review either. For a previous review on the same subject see [1].

Closely related work that this review does not cover is the mean-field renormalization group or coherent anomaly method [2]. There are interesting connections between statistical mechanical models and cellular automata. For these the reader is referred to a paper by Rujan [3].

The layout of this paper is as follows. Section 1 is a review of the basic transfer matrix formalism. Section 2 covers finite-size scaling and phenomenological renormalization group methods, which are used to interpret the finite system data generated with the transfer matrix. Section 3 discusses applications, some in considerable detail; unavoidably the emphasis is on technique rather than physical interpretation of the models studied. The last section contains a discussion of the transfer matrix Monte Carlo method, which provides a way to directly calculate the free energy of systems for which the transfer matrix is otherwise intractable. This is the only section that contains material that has not been published before.

1 TRANSFER MATRIX METHODS

1.1 Introduction

In this section the transfer matrix formalism is reviewed. With the exception of Baxter's variational approximation at the end of this section, most of the results are already contained in the earliest papers on the subject and thus have been around for almost fifty years. Kramers and Wannier introduced the transfer matrix in 1941. In a series of two papers they reported their discovery of duality of the two-dimensional Ising model, proposed numerical and closed form approximations, and also used their formalism to derive high- and low-temperature series expansions of the free energy [4]. Ashkin and Lamb [5] used the transfer matrix to derive series expansions of correlation functions; their paper also contains further references to work predating Onsager's solution of the two-dimensional Ising model [6]. These remarks are not intended to begin a history of the subject. Their purpose is to provide the basic references; no attempt has been made to systematically provide references to further work that introduces variations on the basic theme, with apologies to authors whose originality may not be duly acknowledged.

The material in this section is presented from the perspective of numerical applications. Specifically, it is shown how the computation of thermodynamic quantities and correlation functions can be reduced to a small number of multiplications of an arbitrary vector by sparse matrices and evaluations of vector inner products, usually followed by numerical differentiation. With the ability to perform these numerical operations, systems with various boundary conditions can be handled; a notable exception is the case of periodic boundary conditions in the transfer direction, which requires the computation of a trace.

1.2 Definition of the transfer matrix

We shall consider a system of microscopic variables $s_1, s_2, \ldots s_N$ defined on a lattice. The nature of the s_i is immaterial at this point, but the method becomes impractical rapidly if the s_i can assume more than a few discrete values. Generalization of the formalism to continuous variables is no more difficult than replacing matrices by integral kernels and sums by integrals. In the case of continuous microscopic variables, however, the method can only be employed computationally in combination with a Monte Carlo or other truncation scheme.

To introduce a transfer matrix the first step is to group the sites of the lattice into a one-dimensional array of, say, M *slices*; the direction from one slice to the next is referred to as the *transfer direction*. For simplicity it is assumed that the slices are geometrically identical and fully cover the lattice without overlap. However, it should be noted that the method can be made much more powerful without these constraints, as discussed below. The microscopic configuration of the slice i is given by σ_i. The slices are to be chosen so that the total reduced Hamiltonian H (*reduced* in the sense that it is expressed in units $-k_{\mathrm{B}}T$) can be written as a sum of *two-slice Hamiltonians* $h(\sigma_i|\sigma_{i+1})$, each of which contains interactions within and between slices i and $i+1$:

$$H(s_1, \ldots, s_N) = \sum_{i=1}^{M} h(\sigma_{i+1}|\sigma_i), \tag{1}$$

assuming periodic boundary conditions $\sigma_{M+1} = \sigma_1$. In terms of two arbitrary slice configurations σ and τ and the partition (1) the transfer matrix $\mathbf{T}$ is defined by its elements:

$$T(\sigma|\tau) = \exp\, h(\sigma|\tau). \tag{2}$$

Whether or not it is possible to write a given Hamiltonian in the slicewise additive form (1) depends on the choice of the variables σ_i. In some cases the additivity is already realized if one chooses the σ_k to be just local, ordered subsets of the original microscopic variables s_i. In other cases, however, slice configurations specify nonlocal information, for example the state of connectivity at the surface of a semi-infinite system. The latter is the case for transfer matrices such as have been used in the finite-size scaling studies of ordinary and directed percolation [7,8], ordinary and directed lattice animals [9,10] (treated in some detail as an example below), the random resistor problem [11], and the q-state Potts [12] and n-vector [13] models with continuous q and n.

We shall assume that the transfer matrix has a complete set of right and left eigenvectors, denoted respectively by $\psi_0, \psi_1, \ldots$ and $\tilde{\psi}_0, \tilde{\psi}_1, \ldots$ with components $\psi_0(\sigma), \psi_1(\sigma), \ldots$ and $\tilde{\psi}_0(\sigma), \tilde{\psi}_1(\sigma), \ldots$, and with eigenvalues $\lambda_0 > |\lambda_1| \geq \ldots$†. For finite systems, the dominant eigenvalue is positive and nondegenerate, as a consequence of the Perron-Frobenius theorem [14].

For a given Hamiltonian H there are many different choices of two-slice contributions h, all of which satisfy Eq. (1). In particular, one can choose an arbitrary function g, defined on the states of a slice, and define $\hat{h}$ by

$$\hat{h}(\sigma|\tau) = g(\sigma) + h(\sigma|\tau) - g(\tau). \tag{3}$$

Clearly, Eq. (1) remains valid if h is replaced by $\hat{h}$; in the case of free boundary conditions [that is if the summation in Eq. (1) runs only up to M-1] this transformation constitutes a change of boundary conditions. Upon exponentiation, Eq. (3)

† Note that the tilde does not denote the transpose.

can written as

$$\hat{T}(\sigma|\tau) = \gamma(\sigma)T(\sigma|\tau)/\gamma(\tau), \tag{4}$$

with $\gamma = \exp g$, which defines a transfer matrix $\hat{\mathbf{T}}$, a similarity transform of the original matrix $\mathbf{T}$.

In some cases Eq. (3) can be exploited to define a symmetric two-slice Hamiltonian h:

$$h(\sigma|\tau) = h(\tau|\sigma), \tag{5}$$

in which case the transfer matrix is also symmetric. The physics required for this is that the system be symmetric under reflection along the transfer direction. This, however, is not sufficient: each slice configuration must also be invariant under this transformation. A system with next-nearest neighbor interactions along the transfer direction requires slices in two dimensions consisting of two neighboring columns of lattice sites; such a system violates the latter condition, even if it satisfies the first. An example is the ANNNI model if the transfer direction is chosen along the direction of possible spacial modulations, which for their description require a subdominant complex eigenvalue, *i.e.*, a non-symmetric transfer matrix. As a second example, a system with helical boundary conditions lacks reflection symmetry altogether. In both of these instances the slice Hamiltonian can be chosen so as to satisfy:

$$h(\sigma|\tau) = h(\tilde{\tau}|\tilde{\sigma}), \tag{6}$$

where $\tilde{\sigma}$ (or $\tilde{\tau}$) denotes the slice configuration obtained from σ (or τ) by inversion, as defined precisely in an explicit example discussed below (see Section 1.6). If Eq. (6) holds, the transfer matrix $\mathbf{T}$ and its transpose $\mathbf{T}^\dagger$ are related as follows:

$$T^\dagger(\sigma|\tau) = T(\tau|\sigma) = T(\tilde{\sigma}|\tilde{\tau}). \tag{7}$$

For any vector ψ with components $\psi(\sigma)$ one can define a conjugate vector ψ' with components $\psi'(\sigma)$ via by the relation:

$$\psi'(\sigma) = \psi(\tilde{\sigma}). \tag{8}$$

It is simple to show that if ψ is a right eigenvector of the transfer matrix $\mathbf{T}$ and if Eq. (6) holds, then the transpose of conjugate vector ψ', defined via Eq. (8), is a left eigenvector and *v.v.*

1.3 Free energy and correlation lengths

The conditional partition function $Z_r(\sigma_r)$ of a system of r slices with slice r in a given state σ_r is defined as:

$$Z_r(\sigma_r) = \sum_{\sigma_{r-1}} \sum_{\sigma_{r-2}} \cdots \sum_{\sigma_1} \exp\Big[\sum_{i=1}^{r-1} h(\sigma_{i+1}|\sigma_i)\Big] Z_1(\sigma_1), \tag{9}$$

where $Z_1(\sigma_1)$ is determined by the (free, mean-field, fixed,...) boundary conditions. An immediate consequence of the additivity property (1) of the Hamiltonian and definition (2) is the following recursion relation:

$$Z_r(\sigma_r) = \sum_{\sigma_{r-1}} T(\sigma_r|\sigma_{r-1}) Z_{r-1}(\sigma_{r-1}). \tag{10}$$

A convention suggested by Eq. (10) is to interpret the transfer matrix multiplication as an extension of the array of slices by one slice to the *left*. Referring back to Eq. (6), inversion turns a slice configuration to be added if the system is grown towards the left into an equivalent slice growing to the right.

It follows from Eq. (10) that for large r one has:

$$Z_{r+1}(\sigma) = \lambda_0^r \Big[z_0\psi_0(\sigma) + z_1\psi_1(\sigma)e^{-\kappa_1 r} + z_2\psi_2(\sigma)e^{-\kappa_2 r} + \ldots \Big], \tag{11}$$

where $z_0, z_1, \ldots$ are the expansion coefficients of Z_1, and the (possibly complex) correlation lengths $1/\kappa_i$, measured in units of the lattice spacing, are defined as:

$$\kappa_i = \ln(\lambda_0/\lambda_i). \tag{12}$$

As an immediate implication of Eq. (11), the reduced free energy F per slice is given by

$$F = -\ln \lambda_0, \tag{13}$$

in the limit $r \to \infty$.

It was assumed that the distance between neighboring slices, used in the definition of the transfer matrix, is one lattice spacing. This distance may be less than one, *e.g.*, in the case of the Kramers-Wannier transfer matrix defined in Section 1.6, or greater than one, so that a rescaling factor may be required in Eq. (12). Similarly, the free energy defined in Eq. (13) is the free energy per fraction of a slice if neighboring slices overlap.

The subdominant inverse correlation length κ_n has a simple physical interpretation, if the system has symmetries with which the leading coefficients $z_1, \ldots, z_{n-1}$ can be made to vanish. For example, the Baxter (symmetric eight-vertex) model [15], in its Ising-spin formulation [16], has spin inversion symmetry for the lattice as whole, as well as for its two alternating-site sublattices separately, symmetries that lead to low-temperature phases with spontaneous magnetization and polarization. The coefficient z_0 vanishes if Z_1 is odd under either of these symmetries; z_1 and z_2 vanish if Z_1 is even under inversion of the magnetization and polarization respectively. This yields the interpretation of the three leading correlation lengths as the magnetization, polarization, and energy autocorrelation lengths, in order of decreasing magnitudes as it turns out.

It should be noted that correlation lengths as defined above may be infinite in the thermodynamic limit in regions where there are coexisting phases; generally, one expects the largest eigenvalue to be p-fold degenerate asymptotically, if there are p coexisting phases. However, symmetrized transfer matrices, for example, may have "physically incomplete" spectra for which this is not the case.

1.4 Correlation functions

Being the asymptotic conditional partition function, the eigenvector component $\psi_0(\sigma)$ is the relative probability for the leftmost slice of a semi-infinite system to be in state σ; $\tilde{\psi}_0(\sigma)$ is the same for the rightmost slice of a system extending infinitely far to the left. With a minor generalization of the arguments, one shows that $\tilde{\psi}_0(\sigma)\psi_0(\sigma)$ is the relative probability to be in a state σ for a slice in the middle of an infinite system. Consequently, given the dominant right and left eigenvectors one can compute bulk and surface correlations within one slice. That is, suppose that O is a function defined on the configurations of a slice, then the thermal average of O at the left surface ($\langle O\rangle_{\rm l}$), the right ($\langle O\rangle_{\rm r}$), and in the bulk ($\langle O\rangle$) are given by:

$$\langle O\rangle_{\rm l} = \sum_\sigma O(\sigma)\psi_0(\sigma)/\sum_\sigma \psi_0(\sigma); \quad \langle O\rangle_{\rm r} = \sum_\sigma O(\sigma)\tilde{\psi}_0(\sigma)/\sum_\sigma \tilde{\psi}_0(\sigma) \tag{14}$$

$$\langle O\rangle = \sum_\sigma O(\sigma)\tilde{\psi}_0(\sigma)\psi_0(\sigma)/\sum_\sigma \tilde{\psi}_0(\sigma)\psi_0(\sigma); \tag{15}$$

Similarly, the computation of the correlation function G_{12} of two functions O_1 and O_2, associated with two slices, a distance r apart, can also be reduced to repeated multiplication of a vector by a matrix (and computation of an innerproduct) by recursive evaluation of the expression:

$$G_{12}(r) = \sum_{\sigma,\tau} \tilde{\psi}(\sigma)O_1(\sigma)\mathbf{T}^r(\sigma|\tau)O_2(\tau)\psi(\tau)/\sum_{\sigma,\tau} \tilde{\psi}(\sigma)\mathbf{T}^r(\sigma|\tau)\psi(\tau), \tag{16}$$

where the vectors $\tilde{\psi}$ and ψ are determined by the left and right boundary conditions respectively; the special choice of dominant left and right eigenvectors yields the correlation function of an infinitely long system.

1.5 Modulated and random systems

So far, it was assumed that in the transfer direction the system is translationally invariant over one unit of the distance between the slices. Generalization is straightforward to cases where there is such symmetry only over a multiple of P such units. In this case, instead of one transfer matrix, one introduces as many different single-step matrices as there are different pairs of nearest-neighbor slices; a P-step transfer matrix is defined as the product of the single-step transfer matrices over one period of the system. To be precise, suppose that instead of Eq. (1) the Hamiltonian can be written as a sum involving two-slice Hamiltonians $h_0, h_1, \ldots, h_{P-1}$:

$$H(s_1, \ldots, s_N) = \sum_{r=1}^{M} h_{r \bmod P}(\sigma_{r+1}|\sigma_r), \tag{17}$$

leading to transfer matrices $\mathbf{T}_r$, $r = 0, 1, \ldots, P-1$, with elements:

$$T_r(\sigma|\tau) = \exp h_r(\sigma|\tau), \tag{18}$$

The reduced free energy per slice is derived from the largest eigenvalue Λ_0 of the P-step transfer matrix $\mathbf{T} = \mathbf{T}_0\mathbf{T}_1 \ldots \mathbf{T}_{P-1}$ in terms of the average quantity $\overline{\lambda}_0$ by:

$$F = -\ln \overline{\lambda}_0, \quad \overline{\lambda}_0 = \Lambda_0^{1/P}. \tag{19}$$

As an illustration, consider the calculation of the structure factor of the two-dimensional Ising model [17] †. The microscopic variables are Ising variables $s_{i,j} = \pm 1$ with $i = 1, 2, \ldots, M$ and $j = 1, \ldots, L$. There are nearest-neighbor interactions K and a coupling with a magnetic field of strength B. For the purpose of calculating the structure factor at a wavelength commensurate with the lattice spacing and with a wavevector in the transfer direction, a spacially modulated magnetic field of amplitude μ and of wavenumber $q = 2\pi/P$ is added to the Hamiltonian. That is, the reduced Hamiltonian reads:

$$H = \sum_{r=1}^{M}\sum_{j=1}^{L}\Big[K s_{r,j+1} + K s_{r+1,j} + B + \mu\cos(qr)\Big] s_{r,j}, \tag{20}$$

where periodic boundary conditions are assumed: $s_{r,L+1} = s_{r,1}$ and $s_{M+1,j} = s_{1,j}$. The slices are chosen to be columns $\sigma_r = (s_{r,1}, s_{r,2}, \ldots, s_{r,L})$ with the two-column Hamiltonian:

$$\begin{aligned} h_r(\sigma|\tau) = \sum_{j=1}^{L}\Big[\frac{1}{2}K(2s_jt_j + s_js_{j+1} + t_jt_{j+1}) + \Big(B + \mu\cos(qr)\Big)s_j + \\ + \Big(B + \mu\cos(q(r+1))\Big)t_j\Big], \end{aligned} \tag{21}$$

† Reference [18] describes an alternative method.

with $\sigma = (s_1, \ldots, s_L)$ and $\tau = (t_1, \ldots, t_L)$. This defines the transfer matrices $\mathbf{T}_r$ via Eq. (18).

The structure factor $S(q)$ of an infinite strip is defined as

$$S(q) = \frac{1}{L} \sum_{r=-\infty}^{\infty} \sum_{j=1}^{L} \langle s_{01} s_{rj} \rangle \cos(qr), \tag{22}$$

and can be calculated from the free energy per slice F, given in Eq. (19), as:

$$S(q) = \frac{1}{L} \partial^2 F / \partial \mu^2 |_{\mu=0}. \tag{23}$$

A further generalization allows the application of the transfer matrix technique to random systems. Technically, these are not different from the previous example generalized to the calculation of the structure factor at a wavelength incommensurate with the lattice: there is no translation symmetry at all in the transfer direction so that there are infinitely many transfer matrices $\mathbf{T}_r$, corresponding to $P \to \infty$ in Eq. (17). The free energy per slice in this case can be obtained from the relation:

$$F = -\ln \overline{\lambda}_0, \quad \overline{\lambda}_0 = \lim_{P\to\infty} (|\mathbf{T}_1 \mathbf{T}_2 \ldots \mathbf{T}_N \psi| / |\psi|)^{1/P}. \tag{24}$$

Here $|\cdot|$ denotes any norm defined on the vector space. The limit in Eq. (24) is independent of the vector ψ for almost all choices. Furthermore, in the case of random systems the limit is independent of the realization of the random interactions, which define the random transfer matrices $\mathbf{T}_r$ [19]. In other words, Eq. (24) states that the free energy is a self-averaging quantity: it can be calculated with any desired accuracy from a single realization of a sufficiently big random system.

A more general method that can be used instead of Eq. (24) to calculate thermodynamic functions or correlation functions of random systems is the following: *(i)* for any given realization of the random interactions compute free energy and correlations from Eqs. (10,14,15,16), all of which can be generalized trivially to random transfer matrices; *(ii)* average these results over sufficiently many realizations.

For random magnets with up-down symmetry the inverse correlation length κ_1 can be argued to be given by [20]:

$$\kappa_1 = \ln(\overline{\lambda}_0/\overline{\lambda}_1), \quad \overline{\lambda}_1 = \lim_{P\to\infty} (|\mathbf{T}_1\mathbf{T}_2 \ldots \mathbf{T}_N \psi^{(-)}|/|\psi^{(-)}|)^{1/P}, \tag{25}$$

where $\psi^{(-)}$ is an arbitrary initial vector that is odd under up-down symmetry.

1.6 Sparse-matrix factorization and examples

All transfer matrices introduced above are dense matrices: their elements vanish only if there are infinite repulsions. However, use of matrices in this form in numerical applications is highly inefficient. For example, multiplication of a general vector by the transfer matrix of the Ising model, as given by Eqs. (18) and (21), requires on the order of 2^{2L} elementary arithmetic operations. The number of operations performed to extend the system by a full column of L new sites can be reduced to roughly $L2^L$ by use of sparse matrices. The simplest example is provided by the following Kramers-Wannier transfer matrix [4]:

$$T(\sigma|\tau) = \exp[K(s_1t_L + \frac{1}{2}s_1s_2 + \frac{1}{2}t_Lt_{L-1})] \prod_{i=2}^{L} \delta_{s_i t_{i-1}}, \tag{26}$$

with $\sigma = (s_1, \ldots, s_L)$ and $\tau = (t_1, \ldots, t_L)$, which generates an Ising model with Hamiltonian (20) without magnetic fields ($B = \mu = 0$) and with helical boundary conditions. The sparseness of this matrix, as manifested by the Kronecker delta functions, reflects the fact that, for example, the variables s_2 and t_1 specify one and the same spin. Note that the matrix satisfies Eq. (7) with inverted configurations defined by $\tilde{\sigma} = (\sigma_L, \sigma_{L-1}, \ldots, \sigma_1)$ and $\tilde{\tau} = (\tau_L, \tau_{L-1}, \ldots, \tau_1)$; actually, Eq. (26) is a similarity transform [*cf.* Eqs. (3) and (4)] of the original Kramers-Wannier matrix. The eigenvalue spectrum of the transfer matrix (26) was obtained by Newell [21], as a special case of his exact solution of the Ising model on a triangular lattice.

Somewhat more elaborate techniques are required to obtain sparse matrices in case of periodic boundary conditions, or further-than-nearest-neighbor interactions [22,23,24]. The basic idea is that the sum in Eq. (1), which expresses the total Hamiltonian as a sum of two-slice Hamiltonians, does not exploit additivity ***within*** slices, which is present in many cases, as illustrated by Eq. (21). This additivity of the Hamiltonian translates into multiplicativity, decomposing the transfer matrix into factors each of which extends the lattice typically by no more than one site at a time. This is done by introduction of slices that overlap and, in contrast with the case of the helical boundary conditions, need not be geometrically identical, so that the decomposition may involve different factors. To accommodate overlap of the slices on the one hand the state space of the system is formally expanded to include unphysical states that arise if each slice varies independently. On the other hand, unphysical states are eliminated, because transfer matrix elements between physically incompatible neighboring slice configurations vanish, the mechanism that creates sparse matrices.

As an example, consider the Ising model with periodic boundary conditions. The transfer matrix defined via Eqs. (18) and (21) can be factored into:

$$\mathbf{T} = \mathbf{V}^L \mathbf{H}^L \mathbf{V}^L, \tag{27}$$

where the matrices $\mathbf{V}$ and $\mathbf{H}$ are defined as follows, in the special case $B = \mu = 0$:

$$V(\sigma|\tau) = e^{K s_1 s_2/2} \prod_{i=1}^{L} \delta_{s_{i+1} t_i}, \tag{28}$$

and

$$H(\sigma|\tau) = e^{K s_1 t_1} \prod_{i=2}^{L} \delta_{s_{i+1} t_i}, \tag{29}$$

with $s_{L+1} = s_1$. The matrix $\mathbf{V}$ adds just half a vertical bond, $\mathbf{H}$ adds a full horizontal bond in addition to a new site; both matrices perform a translation in the periodic, vertical direction over one lattice spacing.

A next example illustrates a similar decomposition for a transfer matrix defined to study lattice animals [10]: the polymers formed by the bonds connecting occupied nearest neighbor sites of a square-lattice gas. Particles and nearest-neighbor bonds have a fugacity x and y respectively. The correlation function with information about the geometric size of these clusters is the generating function g_r defined by:

$$g_r(x,y) = \sum_{N,B} \omega(N,B) x^N y^B, \tag{30}$$

where $\omega(N,B)$ is the number of different, connected animals that live on columns 0 through r, have N sites and B bonds and form a polymer connecting columns 0 and r.

To define a transfer matrix, the configuration of an arbitrary column is defined as $\sigma = (s_1, s_2, \ldots, s_L)$ with *(i)* $s_i = 0$, if site i is empty, *(ii)* $s_i = 1$, if site i is occupied, but not connected to column 0, and *(iii)* $s_i = 2$ if site i is occupied and connected to column 0. A transfer matrix is introduced to calculate recursively the conditional correlation function $g_r(x, y; \sigma_r)$, which is defined as in Eq. (30), with the additional constraint that only those animals are counted that leave column r in state σ_r:

$$g_r(x,y;\sigma_r) = \sum_{\sigma_{r-1}} T(\sigma_r|\sigma_{r-1}) g_{r-1}(x,y,\sigma_{r-1}). \tag{31}$$

The inverse size of a representative animal is proportional to $\ln \lambda_0$ with λ_0 the largest eigenvalue of $\mathbf{T}$. This transfer matrix can be written as:

$$\mathbf{T} = (\mathbf{PV})^L (\mathbf{PH})^L, \tag{32}$$

where $\mathbf{V}$ is a matrix that adds a vertical bond in a column:

$$V(\sigma|\tau) = \begin{cases} 1, & \text{if } \sigma = \tau \text{, and sites 1 and 2 are not both occupied} \\ y, & \text{if sites 1 and 2 are both occupied, and } \sigma \text{ is the state} \\ & \text{obtained from } \tau \text{ by adding a bond between these sites} \\ 0, & \text{otherwise.} \end{cases} \tag{33}$$

The matrix **H** adds a new, empty or occupied site to the lattice and deals with the corresponding horizontal bond:

$$H(\sigma|\tau) = C_{s_1 t_1} \prod_{i=2}^{L} \delta_{s_i t_i}, \tag{34}$$

where,

$$\mathbf{C} = \begin{pmatrix} 1 & 0 & 1 \\ x & xy & 0 \\ 0 & 0 & xy \end{pmatrix}, \tag{35}$$

Note that the matrix element C_{01} vanishes to protect animals from having their extremities severed. Finally, the matrix **P** cyclically shifts the columns over one lattice spacing:

$$P(\sigma|\tau) = \prod_{i=1}^{L} \delta_{s_{i+1} t_i}, \quad s_{L+1} = s_1. \tag{36}$$

Derrida and Herrmann do not use sparse matrices and describe an algorithm to construct all elements of a transfer matrix symmetrized with respect to translation [10].

None of the sparse matrices introduced above have translation symmetry: building up a lattice one bond or site at a time destroys translation invariance, at least at intermediate steps of the process. Therefore, sparse matrix decomposition most of the time is incompatible with the block-diagonalization method that exploits translation symmetry [25]. However, in matrix-vector multiplications the latter method produces savings at best equal to the square of the number of sites in a slice, which (except for small systems) is less efficient than the sparse-matrix method, the savings of which are exponential in the system size. For example, the transfer matrix for the two-dimensional $Z(n)$ (n-state clock) model is a matrix of order n^{L-1}, if the $Z(n)$ symmetry is exploited. Block diagonalization yields a matrix that requires $t_b = n^{2(L-1)}/L^2$ operations per matrix-vector multiplication; this number is $t_s = Ln^L$ for the sparse-matrix method. The largest eigenvalue of the transfer matrix for the $Z(5)$ model with cylindrical boundary conditions was calculated by sparse-matrix factorization for strips of 11 sites [26]. In this case one has $t_s = 10^8$ and $t_b = 8 \times 10^{11}$. It is worth noting in this context that the conjugate gradient method (see Section 1.7), used in this calculation of the largest eigenvalue, converges to double precision in roughly 15 iterations, starting from an infinite temperature initial vector consisting of all ones, and takes less than an hour of CPU time on an IBM 3090.

1.7 Methods to compute eigenvalues

With Eq. (10) the calculation of the free energy is reduced to a fixed number of transfer matrix multiplications, in the case of free boundary conditions; as follows from Eq. (13), the computation of the dominant eigenvalue is required for systems that are infinite in one direction. There are numerous methods to calculate the dominant transfer matrix eigenvalue; the relative merits of these methods in the context of statistical mechanics have not been studied systematically. The following paragraph describes what seems to be current practice, which, it should be noted, leaves room for improvement.

The simplest, most generally applicable method to compute the dominant eigenvalue, and also one of the slowest, is the power method, which is based on Eq. (11): almost all initial vectors (*i.e.* provided $z_0 \neq 0$) multiplied repeatedly by the transfer matrix will ultimately converge to the dominant eigenvector. For symmetric matrices a popular method is the Lanczos method [27,28]. This method has the advantage of yielding several leading eigenvalues simultaneously; eigenvectors can also be obtained, but probably not with very high accuracy. Another possibility for symmetric matrices is the conjugate gradient method [29]. This method has the disadvantage of only producing one eigenvalue at a time together with the corresponding eigenvector (at either end of the spectrum); it is of course always possible to use a projection technique to obtain a few additional eigenvalues. The advantage is that to compute the eigenvalue one can use $R(\psi)$, the

Rayleigh quotient:

$$R(\psi) = \sum_{\sigma,\tau} \psi(\sigma) T(\sigma|\tau) \psi(\tau) / \sum_{\sigma} \psi(\sigma) \psi(\sigma). \tag{37}$$

The result is a variational estimate of the eigenvalue; that is, if ψ differs from an eigenvector to order ϵ then $R(\psi)$ differs to order ϵ^2 from an eigenvalue.

It is impossible to tell from the literature what techniques are used for non-symmetric matrices. Of course, the power method still works; Blöte and Nightingale [12 (1982)] proposed a generalization of this method that is much faster.

In this context it should be noted that Eq. (8) is useful for approximation methods. First of all, the standard Rayleigh quotient can be modified to yield $\tilde{R}(\psi)$, a variational but not necessarily a lower-bound estimate of the dominant eigenvalue:

$$\tilde{R}(\psi) = \sum_{\sigma,\tau} \psi(\tilde{\sigma}) T(\sigma|\tau) \psi(\tau) / \sum_{\sigma} \psi(\tilde{\sigma}) \psi(\sigma) \tag{38}$$

Also, the relation (7) between the transfer matrix and its transpose allows the efficient implementation of the unsymmetric Lanczos algorithm [27], by choosing initial vectors that satisfy Eq. (8), a relation that is maintained during the iteration process, so that the algorithm effectively becomes as simple as in the symmetric case.

1.8 Derivatives

In addition to the free energy, thermodynamic functions are required in most applications. They can be obtained from the free energy by standard, numerical differentiation. The resulting loss of numerical accuracy poses no serious problem in most cases: eigenvalues can readily be calculated to one part in 10^{14}, in which case second-order derivatives can be computed with an accuracy typically better than 10^{-7} [12(1982),30]. In principle, derivatives can be calculated without loss of numerical accuracy, but at the expense of a substantially increased programming effort: one expands the transfer matrix in a power series, truncated at the appropriate order, and then performs term by term the matrix multiplications, inner products and the like. This is a numerical implementation of the series expansion methods used in the forties [4,5] and is described in some detail by Glaus [31], who attributes it to Yeomans and Derrida [32].

Perturbation theory is an attractive alternative for first-order derivatives, if left and right dominant eigenvectors are known. However, it should be noted that this is at the expense of the variational property which holds, if the Rayleigh quotient [Eqs. (37) or (38)] is used to find the eigenvalue.

1.9 Baxter's variational approximation

Baxter [33,34,35] introduced trial vectors that can be used with transfer matrices to variationally calculate the partition function per site of lattice models. For two-dimensional systems there is an infinite sequence of trial vectors, which in the limit converge to the exact result; at the lowest level one has the Kramers-Wannier approximation in the special case of the Ising model [4]. The main idea can be described as follows.

Consider the *column-to-column* transfer matrix of an Ising model on a square lattice with periodic boundary conditions in the vertical direction:

$$T(\sigma|\tau) = \exp\Big[\frac{1}{2}\sum_{j=1}^{L}\Big(K(2s_jt_j + s_js_{j+1} + t_jt_{j+1})\Big)\Big]. \tag{39}$$

It was shown above that the element $\psi_0(\sigma)$ of the dominant eigenvector is the partition function of a semi-infinite strip of rows of lattice sites, conditional on the fixed, but arbitrary state σ of the surface column. Suppose we truncate the lattice to M, rather than infinitely many columns. The partition function of this finite system can be expressed in terms of four $2^M \times 2^M$ *row-to-row* transfer matrices $\mathbf{F}^{(a_0,b_0)}$ with elements:

$$F^{(a_0,b_0)}(a_1,a_2,\ldots,a_M|b_1,b_2,\ldots,b_M) = \exp\Big[\frac{1}{2}\sum_{j=0}^{M-1}\Big(K(2a_jb_j + a_ja_{j+1} + b_jb_{j+1})\Big)\Big], \tag{40}$$

here the a_i and b_i are Ising variables; a_0 and b_0 play the role of arbitrary, fixed surface magnetic fields. The partition function $Z_{M+1}(\sigma)$ of the (M+1)×L lattice can be expressed as:

$$Z_{M+1}(\sigma) = \mathrm{Tr}\ F^{(s_1,s_2)}F^{(s_2,s_3)}\ldots F^{(s_L,s_1)}, \tag{41}$$

which is a good approximation for the eigenvector of the transfer matrix if M, the truncation length, is greater than the correlation length. In other words, this vector has a Rayleigh quotient which is a lower bound that converges to the exact partition function per site for M→ ∞. Baxter has shown how one can do better yet: consider *all elements* of the matrices $\mathbf{F}^{(a,b)}$ as parameters of a trial vector in a variational calculation of the partition function per site of the infinite system. For $M = 0$ these matrices reduce to numbers, and the Kramers-Wannier and Bethe approximations are reproduced, depending on whether the method is applied to one single or to two independent, interpenetrating square lattices. Baxter [34,35] gives algorithms, which can be understood in terms of corner transfer matrices, to solve the general variational problem, either numerically or by power series expansion.

Convergence properties of this method for the two-dimensional Ising model close to T_c , the critical temperature, were studied numerically by Tsang [36]: the deviation from T_c of the critical point ($T_{c,M}$) of the M-th-level approximation numerically appears to be given by $T_c - T_{c,M} = O(\exp(-a\sqrt{M}))$ with a constant $a > 0$. The approximation to the spontaneous magnetization as a function of temperature (T) behaves as $(T_{c,M} - T)^\beta$, where β is an effective exponent which crosses over from the exact Ising value $\frac{1}{8}$ to the classical value $\frac{1}{2}$ very close to $T_{c,M}$;

the cross-over occurs on the scale of $T - T_{c,M}$. Convergence is surprisingly more rapid than the lower bound suggested by finite-size scaling according to which corrections would be proportional to a negative power of M.

Further applications to two-dimensional systems include the following. Baxter and Tsang [37] used the method to calculate the entropy of the hard hexagon model, prior to its exact solution [35]. To illustrate the numerical accuracy of this approximation we quote the numerical result: the entropy per site is 0.333 242 741 976 k_B. Baxter and Enting [38] generated low-temperature series expansions for the Ising model on a square lattice in a magnetic field. Malakis [39] applied the method to the triangular Ising model with two- and three-spin interactions. Pearce and Seaton studied the phase diagram of interacting hard squares [40].

For three-dimensional systems the variational method is intractable, except at the lowest level ($M = 0$) for which Baxter and Forrester applied it to the three-dimensional Ising model [41]. Solution of the variational problem by series expansion correctly reproduces the first 14, 19, and 23 terms of the low- temperature series of the free energy respectively for the simple cubic, body centered cubic, and face centered cubic models. Numerical solution of the variational problem yields estimates of critical temperatures which for these lattices are in error by one or two percent.

2 PHENOMENOLOGICAL RENORMALIZATION AND FINITE-SIZE SCALING

2.1 Introduction

In Section 1 the transfer matrix technique was reviewed. It was shown how the formalism can be used to compute thermodynamic functions and correlation functions without any but numerical approximations. The systems that one can deal with are finite or one-dimensionally infinite at best, so that they have no phase transitions. To extract from finite system data the critical points and scaling exponents, which describe the phase transitions of the infinite systems, finite-size scaling theory, introduced by Fisher and Barber [42], and phenomenological renormalization, developed by Nightingale [22,43], have been employed with great success.

In this section, firstly, the main finite-size scaling relations that play a role in transfer matrix applications are heuristically derived from standard renormalization group theory; secondly, results of conformal invariance for finite-size scaling in two dimensions are listed. The reader interested in more theory is referred to Barber's review [44], and to collections of reprints edited by Cardy [45] for finite-size scaling and by Itzykson, Saleur, and Zuber on conformal invariance [46].

2.2 Renormalization group for finite systems

Consider a system with n interaction parameters $\vec{K} = (K_1, K_2, \ldots, K_n)$. Suppose that under a real-space renormalization transformation rescaling space by a factor b, these interactions transform into $\vec{K}^{(b)}$ given by the renormalization group equations

$$\vec{K}^{(b)} = \vec{K}^{(b)}(\vec{K}). \tag{42}$$

A non-trivial fixed point $\vec{K}^*$ defines a universality class characterized by scaling exponents, which are found by linearization of the renormalization group equations at the fixed point, *i.e.* expansion of Eq. (42) in $\Delta\vec{K} = \vec{K} - \vec{K}^*$ as:

$$\Delta\vec{K}^{(b)} = \mathbf{R}^{(b)}\Delta\vec{K} + O(\Delta\vec{K}^2), \tag{43}$$

where $\mathbf{R}^{(b)}$ is the $n \times n$ matrix of derivatives evaluated at the fixed point:

$$R_{ij}^{(b)} = \partial K_i^{(b)} / \partial K_j. \tag{44}$$

From the matrix $\mathbf{R}^{(b)}$, the scaling exponents y_i and scaling fields u_i to linear order can be obtained by diagonalization:

$$\mathbf{R}^{(b)} = \mathbf{E}\mathbf{D}^{(b)}\mathbf{E}^{-1}, \quad \vec{u} = \mathbf{E}^{-1}\Delta\vec{K}, \tag{45}$$

with a diagonal matrix $\mathbf{D}^{(b)}=\mathrm{diag}(b^{y_1}, b^{y_2}, \ldots)$. A scaling field or exponent is called relevant if $y_i > 0$, irrelevant if $y_i < 0$, and marginal if $y_i = 0$.

For a finite system the renormalization transformation maps a system of size L to one of size L/b; in other words, the inverse system size $1/L$ is an additional relevant scaling field, with scaling exponent 1. In general, scaling fields are nonlinear combinations of all parameters, in this case the interactions $\vec{K}$ and inverse system size $1/L$; we assume that the first and the latter do not mix [47].

Under renormalization, the free energy per lattice cell f transforms according to the equation:

$$f(\vec{K}, L) = g^{(b)}(\vec{K}) + b^{-d} f(\vec{K}^{(b)}, L/b) \tag{46}$$

for a system of spacial dimensionality d. Inverse correlation lengths κ_i, measured in units of the lattice spacing, rescale according to:

$$\kappa_i(\vec{K}, L) = b^{-1}\kappa_i(\vec{K}^{(b)}, L/b). \tag{47}$$

Again following standard arguments [48,49], one derives that the finite-size scaling behavior of the singular part of the free energy f_{sing} and the correlation length κ_i are expressed in terms of scaling functions A and Ω as follows:

$$f_{\mathrm{sing}}(\vec{K}, L) = L^{-d}\mathrm{A}(L^{y_1}u_1, L^{y_2}u_2, \ldots), \tag{48}$$

$$\kappa_i(\vec{K}, L) = L^{-1}\Omega_i(L^{y_1}u_1, L^{y_2}u_2, \ldots), \tag{49}$$

where the u_i are the *nonlinear* scaling fields with the corresponding scaling exponents y_i; furthermore,

$$f_{\mathrm{sing}} = f - f_{\mathrm{reg}},$$
$$f_{\mathrm{reg}}(\vec{K}, L) = \sum_{i_1=0}^{\infty}\sum_{i_2=0}^{\infty} \ldots \gamma^{(b)}_{i_1 i_2 \ldots} u_1^{i_1} u_2^{i_2} \ldots / (1 - b^{-d+i_1 y_1 + i_2 y_2 + \cdots}), \tag{50}$$

that is, $f_{\rm reg}$ is the solution of Eq. (46) obtained by power series expansion in the scaling fields u_i so that the $\gamma^{(b)}_{i_1 i_2 \ldots}$ are the expansion coefficients of $g^{(b)}$.

Finally, a relation similar to Eqs. (48) and (49) can be written down for correlation functions. An example is the scaling form of the correlation function G at distance r of two local densities conjugate to a relevant scaling field u_i:

$$G(r, \vec{K}, L) = L^{-2x_i}\Gamma(r/L, L^{y_1}u_1, L^{y_2}u_2, \ldots), \quad x_i + y_i = d \tag{51}$$

Here we introduced the *critical dimension* x_i of the density conjugate to the scaling field u_i with *scaling exponent* y_i.

2.3 Universality of critical finite-size amplitudes

Privman and Fisher [50] argued that the scaling functions A and Ω introduced in Eqs. (48) and (49) are universal apart from a scale factor for each scaling field except $1/L$. A simple way to arrive at this result is dimensional analysis: if all scaling fields, including the inverse system size, are transformed according to $u_i \to m_i u_i$ and $1/L \to m_0/L$ the change of the scaling functions A and Ω, as given in Eqs. (48) and (49), does not depend on m_0.

For anisotropic systems the amplitudes depend on the degree of anisotropy, which destroys universality. However, universality can be reintroduced by an anisotropic spacial rescaling chosen so that in the infinite system rotation symmetry of the correlation functions is restored [51,52,53,54].

2.4 Phenomenological renormalization

In principle, the scheme outlined above can be run backwards, in the sense that so-called phenomenological renormalization group equations for n interaction parameters are computed as follows [22,43]: *(i)* compute n distinct inverse correlation lengths κ_i and free energies f for system sizes L and L/b; *(ii)* solve the n simultaneous Eqs. (47) for $\vec{K}^{(b)}$, the renormalized couplings; and *(iii)* compute $g^{(b)}$, the "internal free energy" per Kadanoff block, from Eq. (46). Actual calculations are restricted to a small number of couplings; hence, they fall short of the ideal: the independence of $K^{(b)}$ and $g^{(b)}$ of the size of the systems from which they were calculated.

McMillan [55] developed the domain wall renormalization method along these same lines. He considered a square-lattice Ising model with random nearest neighbor interactions K with average $\bar{K}$ and variance $\tilde{K}$. To obtain two equations to implicitly determine the renormalized parameters $\bar{K}^{(b)}$ and $\tilde{K}^{(b)}$ he used the average domain wall energy and its variance, instead of employing two inverse correlation lengths. By studying the renormalization group flow in this two-parameter space he was able to obtain various critical fixed points: *(i)* a non-random fixed point describing conventional ferromagnetic criticality; *(ii)* a zero-temperature fixed point associated with the transition from spin-glass to ferromagnetism; *(iii)* a bicritical fixed point on the critical curve connecting the other two fixed points. The last fixed point is completely unstable, whereas the first two are unstable only in one direction.

In a related development Binder [56] introduced a modified version of these ideas based on rescaling block distribution functions which is suitable for implementation in a Monte Carlo context.

There are only very few applications in which the internal free energy $g^{(b)}$, as calculated with phenomenological renormalization, has been used to derive the singular part of the free energy from finite system data. Nightingale's application [43(1976)] is the calculation of the amplitude of the logarithmic singularity in the specific heat of the two-dimensional Ising model; from strips of maximum width 16 a result can be obtained that agrees to within 0.4% with the exact result. For the chiral three-state clock model, Stella *et al.* [57] calculated the amplitudes $A_\pm$ in the singular, temperature dependent part of the free energy:

$$f_{\mathrm{sing}}(T) \approx A_\pm |T - T_c|^{d/y_T}, \tag{52}$$

using the standard real-space renormalization recipe [48] to extract the amplitude. The authors quote an accuracy of 10% for their estimates. This calculation is more demanding than that of the amplitude of the Ising model: for the clock model the function $g^{(b)}$ has to be summed over a full renormalization group trajectory, while the logarithmic singularity requires only first- and second-order derivative of quantities at T_c.

2.5 Ordinary critical points

In most applications one uses a single or at most two different correlations lengths, the free energy, and their derivatives. Numerical estimates of critical point and exponents derive from the leading finite-size scaling behavior of various quantities rather than from actual solution of the phenomenological renormalization group equations. First, we discuss a system with only two thermodynamic fields. This suffices for the generic case of an ordinary critical point with critical behavior characterized by two relevant scaling fields and their scaling exponents. Specifically, we shall consider a ferromagnetic system with a temperature T and magnetic field h and with associated scaling fields u_T and u_h, and scaling exponents y_T and y_h.

Denote by $f^{(k,l)}$ derivatives of the singular part of the reduced free energy at the critical point, where the superscript (k,l) indicates the orders of the derivatives with respect to temperature and magnetic field, in that order; $\kappa^{(k,l)}$ stands for the same derivatives of the inverse correlation length. It follows from the discussion above that the dominant finite-size behavior of these quantities is given by:

$$f_{\text{sing}}^{(k,l)}(L) \approx m_T^k m_h^l \tilde{\mathrm{A}}^{(k,l)} L^{ky_T+ly_h-d}, \tag{53}$$

$$\kappa_i^{(k,l)}(L) \approx m_T^k m_h^l \tilde{\Omega}^{(k,l)} L^{ky_T+ly_h-1}, \tag{54}$$

where we explicitly introduced the thermal and magnetic metric factors m_T and m_h which render the amplitudes $\tilde{\mathrm{A}}^{(k,l)}$ and $\tilde{\Omega}^{(k,l)}$ universal. Suppressed in Eqs. (53)

and (54) are the correction-to-scaling terms, which determine convergence properties of the the numerical procedures outlined below. Sources of these terms are the usual: irrelevant scaling fields and nonlinearities of the scaling fields [12(1982)].

In Eqs. (53) and (54), $k = l = 0$ represents a case of special interest: genuinely universal amplitudes are found. For two dimensions the theory of conformal invariance predicts the values of these amplitudes, as discussed below. Very little numerical work has been done to verify the theoretical predictions in instances in which the metric factors explicitly occur (i.e. $k \neq 0$ or $l \neq 0$). Burkhardt and Derrida [58] calculated the ratio $\tilde{A}^{(4,0)}/\left(\tilde{A}^{(2,0)}\right)^2$ using conformal invariance for Ising strips and found excellent agreement of this ratio and numerical transfer matrix results. For a related study of the structure factor for this model see Bartelt and Einstein [18]. Blöte and Nightingale [30] calculated $\tilde{A}^{(2,0)}/(\tilde{\Omega}^{(0,1)})^2$ for a spin-1 Ising model and spin-$\frac{1}{2}$ models with next-nearest neighbor interactions. The result agrees with the value $2/\pi$, obtained exactly for the square, nearest neighbor spin-$\frac{1}{2}$ Ising model, to within about 0.1%. Ratios $\tilde{A}^{(0,2)}/\tilde{\Omega}^{(0,2)}$, for which no exact result is not known in the spin-$\frac{1}{2}$ case, are the same to within 0.5%.

Eqs. (53) and (54) can be employed for numerical purposes in many ways. To estimate the location of a critical temperature T_c from the inverse correlation length of systems of linear dimension L_1 and L_2 one solves:

$$L_1\kappa_1(T_c, L_1) = L_2\kappa_1(T_c, L_2), \tag{55}$$

There is a convenient alternative, which is particularly useful to determine a Kosterlitz-Thouless transition: the right-hand side of Eq. (55) can be set to $\tilde{\Omega}^{(0,0)}$, the value of which in two dimensions in many cases is known from universality or from conformal invariance. Incorrect constants can also be used; however, corrections in that case will be $O(L^{-y_T})$, rather than $O(L^{-y_T+y_0})$, where y_0 is the most relevant, irrelevant exponent.

Once T_c has been determined, exponents can then be found from derivatives of the inverse correlation length, for example:

$$y_T = 1 + \ln\Big(\kappa_1^{(1,0)}(L_1)/\kappa_1^{(1,0)}(L_2)\Big)\,/\,\ln(L_1/L_2). \tag{56}$$

Of course, also derivatives of the free energy can be used for this purpose [59]; however, the presence in some case of the regular background term usually leads to somewhat less smooth convergence of the estimates with system size. More attractive in this respect are correlations calculated *e.g.* at a fixed value of r/L (either with Eq. (15) for correlations in a direction perpendicular to the the strip, or Eq. (16) for the parallel ones). The most accurate results are obtained if estimates are calculated for a sequence of lattice sizes and subsequently (repeatedly) extrapolated assuming power-law corrections [12(1982),51].

2.6 Phase diagrams and first-order transitions

The case of a ferromagnet, discussed above, is representative of a class of special systems in which symmetry determines the location of the coexistence line. In this case, just the critical point remains to be determined, if it is not known from duality. For these systems the method has been most successful. The problem of estimating the position of the coexistence curve in the generic case, which lacks symmetry, has received less attention. In general, to determine the number of coexisting phases, to gain insight into the symmetries of coexisting phases and to estimate the location of coexistence lines, one has to rely on degeneracies or near-degeneracies in the eigenvalue spectrum of the transfer matrix, finite-size scaling behavior of the spectrum, and transformation properties of eigenvectors.

Wood and Osbaldestin [60] proposed to obtain approximations to the full phase diagram by locating the surface defined by the solution to Eq. (55) generalized to more than one parameter. Employing the same equation with a sequence of subdominant transfer matrix eigenvalues, it is possible to count the number of coexisting phases, as follows from the expected properties of the eigenvalue spectrum: for $p-$phase coexistence the first p-1 of these surfaces are calculated from eigenvalues that in the ordered phase and in the thermodynamic limit are degenerate with the largest eigenvalue. The surfaces obtained from the p most dominant eigenvalues contain the coexistence surface up to the critical point; the next ones will only contain the critical points. In practice contours of equal residual are plotted, which also works if no real solutions to Eq. (55) exist. The method was applied to the q-state Potts model with symmetry breaking fields for $q = 2, 3, 4$ and 5. Lavis and Quinn used this method to obtain the phase diagram of a two-dimensional Ising ferrimagnet on a triangular lattice [61].

A different approach was used by Bartelt, Einstein and Roelofs [62], who introduced a method that at the same time deals with the problem to calculate discontinuities in quantities such as the magnetization, energy, or particle density at first-order phase transitions: no such discontinuities are present in the thermodynamic quantities calculated as derivatives of the free energy of systems finite in all but possibly one direction. A way to address this difficulty is based on the method Yang used for the exact calculation of the spontaneous magnetization of the Ising model, a method later modified by Uzelac and applied to finite systems by Uzelac and Julien, and Hamer [63]. Bartelt *et al.* [62] used these ideas to locate the phase coexistence line of a square-lattice gas. They also calculated the discontinuity in the particle density across this line. This application will be discussed in detail in Section 3.2.

In principle, the phase diagram with critical points and phase boundaries can also be obtained from the phenomenological renormalization group flow diagram; the only application in this spirit that I am aware of is McMillan's domain wall renormalization group study of the the two-dimensional random Ising model, mentioned in Section 2.4 (also see Section 3.8). For applications to surface adsorption phase diagrams see Section 3.7.

2.7 Multicritical points

To find a critical point the procedure is to solve Eq. (55) with one correlation length in a one-dimensional subspace. A simple way to locate a tricritical point is, also in this case, to use just two values of system size L_1 and L_2, and to solve the analog of Eq. (55) for an additional inverse correlation length κ_2. Assuming that the value of one thermodynamic field at the critical point is known from symmetry, this suffices to determine the remaining fields. In general, in an n-dimensional parameter space, once a multicritical point has been located, the matrix $\mathbf{R}^{(b)}$ of the linearized renormalization group equations can be found in terms of the matrix of derivatives of the inverse correlation lengths, which can be computed directly from eigenvalues of the transfer matrix:

$$\vec{\kappa}(\vec{K},L) = \vec{\kappa}(\vec{K}^*,L) + D(L)\Delta\vec{K} + O(\Delta\vec{K}^2), \quad \vec{\kappa} = (\kappa_1,\ \kappa_2,\ldots,\kappa_n) \tag{57}$$

Substitution in Eq. (47) of this expression together with the linearized renormalization group Eqs. (43) yields:

$$\mathbf{R}^{(b)} = b\mathbf{D}^{-1}(L/b)\mathbf{D}(L), \tag{58}$$

from which the linear scaling fields and scaling exponents follow with Eq. (45). The matrix $\mathbf{R}^{(b)}$ can also be evaluated at ordinary critical points, in which case it can be used, *e.g.*, to determine irrelevant scaling exponents and the corresponding linear scaling fields. To obtain one additional exponent and scaling field, or to accelerate convergence for the leading exponent [64], an alternative is to use *one* inverse correlation length and *three* system sizes [10,65,66,67]. However, this method seems to converge less rapidly, and the reader is referred to the literature for details. In the case of one correlation length and a sequence of system sizes that successively scale into each other with the same rescaling length b, expression (58) can be readily generalized [66], if different scaling lengths enter, the situation becomes somewhat messy [10].

Starting from the exact eigenvalues of the transfer matrix, Barber [66] studied the convergence of various approximation schemes for the anisotropic square Ising model. If the coupling strengths are given by K_1 and K_2, where the transfer direction is along the bonds of strength K_2, the coordinates of the points of the estimated critical curve are given by $(1, K_2/K_1)\beta(L, L/b)$ for results derived from strips of widths L and L/b. Here β_c is given by:

$$\beta_c(L, L/b) = K_{1,c} - \frac{\pi^3 b(b+1)(1+\rho^2)\rho K_{1,c}}{192b^2(K_{1,c}+\rho K_{2,c})} L^{-3} + O(L^{-5}), \qquad (59)$$

where $\sinh 2K_{1,c} \sinh 2K_{2,c} = 1$ is the exact critical curve, and $\rho = \sinh 2K_{1,c}$. If the thermal exponent is calculated by taking derivatives at the critical curve, in the direction defined by variations of the inverse temperature, the estimate of the thermal exponent is given by

$$y_T(L, L-2) = 1 - \frac{1}{4}\pi\sqrt{(1+\rho^2)}\frac{K_{1,c}-\rho K_{2,c}}{K_{1,c}+\rho K_{2,c}} L^{-1} + O(L^{-2}), \qquad (60)$$

where the corrections vanish to all orders in the isotropic case. If the 2×2 matrix $\mathbf{R}^{(b)}$ is calculated at the critical line from three system sizes ($L, L/b$, and L/b^2) two exponents are found: a thermal one precisely equal to unity, and a marginal exponent y_1, associated with anisotropy and given by:

$$y_1(L, L/b, L/b^2) = -1 + \frac{\pi^2(1+3\rho^2)(1-b^3)(1+b)}{48L^2 \ln b} + O(L^{-4} \qquad (61)$$

As shown by Barber, the same approach with three system sizes also works well for the thermal and symmetry breaking field exponents of the the Baxter-Wu model and some of the associated linear scaling fields; for the symmetric eight-vertex model the marginal exponent is not reproduced convincingly [66].

Derrida and Herrmann [10] studied the tricritical point of lattice animals both ways: *(i)* with three system sizes and one correlation length, and *(ii)* with two correlation lengths and two sizes. Herrmann [65] studied the tricritical Ising metamagnet in the same way. The errors in the results appear to be typically

0.1% for the critical couplings and the leading scaling exponent, and 1% for the second relevant thermal critical exponent.

2.8 Conformal invariance

For systems in two dimensions with conformally covariant critical behavior there are predictions for the values of the universal amplitudes $\tilde{A}^{(0,0)}$ and $\tilde{\Omega}^{(0,0)}$. We shall omit the superscripts, but it is understood in this section that the scaling functions are evaluated at the critical point.

An infinite system can be mapped onto a strip with periodic boundary conditions by a conformal transformation. As Cardy [68] showed, this implies for an isotropic system that:

$$\kappa_i \approx \Omega/L, \quad \tilde{\Omega} = 2\pi x_i, \tag{62}$$

which relates the unknown amplitude in Eq. (49) to the critical dimension of the corresponding inverse correlation length, a relation that can also be derived in other ways [69].

The theory of conformal invariance also predicts the amplitude of the singular part of the free energy at the critical point [70,71]:

$$\tilde{A} = \begin{cases} -\pi c/6 & \text{for periodic boundary conditions,} \\ -\pi c/24 & \text{for free or fixed boundary conditions,} \end{cases} \tag{63}$$

where c is the conformal anomaly number (the value of the central charge in the Virasoro algebra).

As shown by Cardy [72], the eigenvalue spectrum of the transfer matrix can be predicted in detail from conformal invariance in the case of unitary theories. Friedan, Qiu and Shenker [73] showed that for these theories with $c < 1$ the values of c are restricted to

$$c = 1 - \frac{6}{m(m+1)}, \quad m = 3, 4, 5, \ldots \tag{64}$$

Primary operators and their conformal descendants in the case of a unitary theory have anomalous dimensions given by

$$x(p,q,r;p',q',r') = h_{pq} + h_{p'q'} + r - r', \quad r,r' = 0,1,2,\ldots \tag{65}$$

where the h_{pq} and $h_{p'q'}$ are given by the Kac formula [74]:

$$h_{pq} = \frac{[p(m+1) - qm]^2 - 1}{4m(m+1)}, \text{ with } m \geq 3, \text{ and } 1 \leq q \leq p \leq m-1 \tag{66}$$

Cardy showed that for these theories the spectrum of the transfer matrix can be determined completely, in the sense that he was able to predict which of the allowed operators actually occur and what the degeneracy of the corresponding eigenstates is. The eigenstate of the transfer matrix of a strip of width L associated with an operator $x(p,q,r;p',q',r')$ has a crystal momentum P measured in units $2\pi/L$:

$$P = (h_{pq} + r) - (h_{p'q'} + r') \pmod{L}, \tag{67}$$

where non-integral values of P are associated with systems with various possible choices of anti-periodic boundary conditions. For systems with free boundary conditions and various combinations of infinite boundary fields Cardy [71] discusses the spectrum of the transfer matrix, and gives detailed results for the Ising model and the three-state Potts model, also see Burkhardt and Burkhardt and Guim [75]. The most extensive numerical verification of these predictions was performed for systems in the one-dimensional quantum Hamiltonian limit. For two-dimensional systems see references [69], [62], and [54], the latter two of which are discussed in more detail below. For transfer matrix studies of the non-unitary n-vector model the reader is referred to Batchelor and Blöte [76], who report exact Bethe-Ansatz and numerical results. Results for this model are also reported by Blöte and Nienhuis [77] and by Saleur [78]. Blöte and den Nijs [79] studied corrections-to-scaling at Ising critical points for the antiferromagnetic Ising model in a magnetic field, the Blume-Capel model, and the $q = 2 + \epsilon$-state Potts model.

3.1 Introduction

This section discusses applications and elaborations of the theory discussed in Sections 1 and 2. Some subsections are quite detailed, such as the the discussion of the calculations for the square-lattice gas, the magnetic hard-square and the restricted solid on solid models are discussed in detail; others mainly serve as guides to the literature.

3.2 Square-lattice gas

Bartelt, Einstein and Roelofs [62] and Kim and Pearce [54] studied a square-lattice gas with nearest neighbor repulsion and next-nearest neighbor attraction. In terms of the site occupancy variable n_i (assuming a value 1 if site i is occupied, 0 otherwise) the Hamiltonian (in units $-k_{\mathrm{B}}T$) reads:

$$H = \mu \sum n + K_1 \sum{}' n_i n_j + K_2 \sum{}'' n_i n_j, \tag{68}$$

where the first summation is over single sites, the second and third summations are over nearest and next-nearest neighbor pairs of sites respectively. Two separate cases are considered *(i)* $K_1 = -\infty$; and *(ii)* $K_2/K_1 = -\frac{1}{2}$, with $K_2 > 0$ (attractive) in both cases. At low temperatures a low density, disordered phase coexists with two $c(2 \times 2)$ phases of high density, each of which preferentially occupies one of the alternating-site sublattices. For temperatures above the tricritical temperature T_{t}, the triple line in the temperature-chemical potential plane turns into a critical line at which the $c(2 \times 2)$ phase disorders. For $K_1 = -\infty$ the triple line and the particle density, among other things, are known for the more general model with anisotropic next-nearest neighbor interactions, exactly solved by Baxter and Baxter and Pearce [80]. See Huse [81] for a detailed interpretation of the multicritical behavior of this model in the context of these exact results.

The method to compute the location of the triple line and the densities of the coexisting phases is based on the following behavior of the three dominant eigenvalues of the column-column transfer matrix [62]. These are: firstly, the largest and third largest eigenvalues, λ_{1+} and λ_{2+}, with normalized eigenvectors, $|1+>$ and $|2+>$, that are even under a unit translation in the column direction; secondly, the second largest eigenvalue λ_{1-} with an odd eigenvector. In the thermodynamic limit, the particle density of the stable phase has a discontinuity at the triple line and is given by $<1+|\hat{\rho}|1+>$, where $\hat{\rho}$ is the number density of a column; on the other hand, the metastable phase [82] is expected to have a density $<2+|\hat{\rho}|2+>$ with an equal but opposite discontinuity, so that $<1+|\hat{\rho}|1+>$ connects continuously with $<2+|\hat{\rho}|2+>$ at the triple line. For a strip of finite width, both quantities vary smoothly across the incipient triple line, but increasingly rapidly as the system becomes larger; points of intersection,

$$<1+|\hat{\rho}|1+>=<2+|\hat{\rho}|2+>, \tag{69}$$

provide estimates of the coexistence curve. The densities $<1+|\hat{\rho}|1+>$ and $<2+|\hat{\rho}|2+>$ are proportional to the slopes of λ_{1+} and λ_{2+} *vs.* the chemical potential μ; in other words, λ_{1+} and λ_{2+} develop discontinuities in slope, and λ_{1+} and λ_{2+} are expected to cross at this line in the thermodynamic limit. The smooth behavior of the derivatives of finite systems can be turned into discontinuous behavior by applying nearly degenerate perturbation theory, so that the densities $\rho_{\pm}$ at coexistence are approximated by the eigenvalues of the 2×2 matrix of $\hat{\rho}$ sandwiched between $|1+>$ and $|2+>$, *i.e.*

$$\rho_{\pm}=<1+|\hat{\rho}|1+>+<1+|\hat{\rho}|2+>=<2+|r|2+>\pm<1+|\hat{\rho}|2+> \tag{70}$$

The problem with Eq. (70) is the converse of the original difficulty: for a finite system $<1+|\hat{\rho}|2+>$ does not vanish, even if the infinite system has a continuous density, such as at the tricritical point and along the critical line at higher temperatures. Bartelt *et al.* use data of three strips of increasing width and fit to the form $b + aL^{-x}$ to locate the tricritical point as the point at which $< 1+|\hat{\rho}|2+ >$ extrapolates to zero. Although somewhat *ad hoc*, this can be expected to work for the following reason: for temperatures above T_t one has $x \to 1$ for large L, while at the tricritical point $x \to 1/5$. Close to T_t the three-point fit can therefore be expected to over-extrapolate, causing a change of sign in the estimate of $< 1+|\hat{\rho}|2+ >$ close to T_t. For the largest system sizes considered ($L = 12, 14,$ and 16) this yields an estimate $0.99985 T_t$, where T_t is the exact tricritical temperature . Also the results of the calculation of the triple line and the discontinuity in density agree very well with the exact results for this model. Similar calculations were performed for the case without infinite repulsions.

The quantity $< 1+|\hat{\rho}|2+ >$ must peak at the incipient triple line, an alternative criterion to locate the triple line. This is argued as follows. On the one hand, Eq. (70) implies $\rho_+ - \rho_- = 2| < 1+|\hat{\rho}|2+ > | \neq 0$; on the other hand, in the vicinity of the triple line $| < 1+|\hat{\rho}|2+ > |$ must become small: it follows from second-order perturbation theory that κ, the compressibility, is given approximately by:

$$\kappa = \frac{\partial \rho}{\partial \mu} \propto \frac{(< 1+|\hat{\rho}|2+ >)^2}{\lambda_{1+} - \lambda_{2+}}. \tag{71}$$

Since on approach of the triple line for large systems κ is expected to remain finite, the numerator has to become small: as was argued above, the denominator becomes exponentially small.

Bartelt *et al.* [62] verified, to within numerical accuracies of typically a fraction of 0.1%, that at tricriticality the leading seven inverse correlations lengths –five primary operators and two conformal descendants– have the amplitudes expected from conformal invariance, as given by Eqs. (62) and (65) with $m = 4$. With Eq. (63), the finite-size correction to the free energy yields an estimate of the central charge $c = 0.700\,03$, to be compared with $c = \frac{7}{10}$ from Eq. (64). Kim and Pearce [54] generalized this calculation to the case of the anisotropic hard-square model. They conjectured the anisotropy factor that restores universality, and confirmed this conjecture and by establishing its consistency with the conformal invariance prediction $c = \frac{7}{10}$ for the central charge and the leading critical dimension $x = \frac{3}{40}$. The results of their calculations are very accurate: agreement with the exact values of typically 5 to 7 significant digits for the central charge and of 3 to 5 digits for the exponent. The calculations were done up to strip sizes of width $L = 48$ as follows. Eigenvalues of the transfer matrix can be written as polynomials of order L for systems of size L. The zeros of these polynomials can be found from the inversion relation that the transfer matrix is known to satisfy. In addition, the authors present exact results confirming $c = 1$ for the symmetric eight-vertex model. For the hard-hexagon model, which is in the three-state Potts universality class, they establish consistency, to within an accuracy better than 0.1%, of a conjectured anisotropy factor with a value of $c = \frac{4}{5}$, obtained from Eq. (64) with $m = 5$.

3.3 Magnetic hard-square model

Pearce and Kim [83] study a lattice gas of magnetic hard-square particles on a simple quadratic lattice; the hard core leads to nearest-neighbor exclusion. In addition, the particles have spin-$\frac{1}{2}$ (Ising) degrees of freedom, and anisotropic next-nearest neighbor interactions. The state of the system at site i is given by σ_i: σ_i vanishes if site i is empty, and assumes values ± 1 if the site is occupied by a spin-up or spin-down particle. The variables around a face in counterclockwise fashion are given by $(\sigma_1, \sigma_2, \sigma_3, \sigma_4)$; the Boltzmann weight of this face is given by:

$$w(\sigma_1, \sigma_2, \sigma_3, \sigma_4) = (z/2)^{(\sigma_1^2+\sigma_2^2+\sigma_3^2+\sigma_4^2)/4}\ e^{L\sigma_1^2\sigma_3^2+M\sigma_2^2\sigma_4^2+J\sigma_1\sigma_3+K\sigma_2\sigma_4}, \tag{72}$$

where the interaction along the diagonals can be taken ferromagnetic $(J, K \geq 0)$ without loss of generality. Exact solutions have been obtained on four manifolds in the five dimensional parameter space.

The model has two order parameters: the difference in density $R = \rho_1 - \rho_2$ of the sublattice densities ρ_i; and the sum of the sublattice magnetic densities

m_1 and m_2, i.e. the total magnetization m. The phases found, either from exact solutions or variational approximations [40], are: *(i)* fluid ($R = 0$, $m = 0$); *(ii)* paramagnetic solid ($R \neq 0$, $m = 0$); *(iii)* ferromagnetic solid ($R \neq 0$, $m \neq 0$). There are no indications of the existence of a ferrofluid phase ($R = 0$, $m \neq 0$) in this model.

In their study of the finite-size behavior of the gaps in the eigenvalue spectrum of the transfer matrix, the authors restrict themselves to isotropic models $L = M$, $J = K$ and one particular manifold T, which reduces to the curve $e^L = (1 - \alpha^2)^{\frac{1}{2}}/\alpha^2$, $z = \alpha^4$, where $\alpha = \tanh J$. For $0 < a < \frac{1}{2}$ (TI) there is a line of three-phase coexistence; the range $\frac{1}{2} < a < 1$ is a multicritical regime (TII), to which the discussion is limited from here onwards.

The magnetic hard-square model has a column-to-column transfer matrix of order $2^\ell + (-1)^\ell$ for a strip of width ℓ, and is shown to satisfy an inversion identity, which the authors numerically solve, as described briefly at the end of Section 3.2, to calculate dominant eigenvalues for systems up to width $\ell = 32$. Thus, they obtain the finite-size corrections to the exactly known free energy and an estimate of the central charge c, which is found to be equal to unity (with accuracies ranging from one part in 10^5 to 10^7): an indication of continuously varying critical exponents [84]. Amplitudes of the size dependence of the inverse correlation lengths, as found from the ratios of the eigenvalues, yield conjectured exact scaling dimensions, which are identified with various primary operators of the model. The results are as follows.

Define $\Delta = (1 - 2\alpha^2)/2\alpha^2 = [-3 + \surd(1 + 4e^{2L})]/4$, and $\Delta = \cosh\lambda$ in TI $(\infty > \Delta > 1;\ \infty > \lambda > 0)$, and $\Delta = \cos\lambda$ in TII $(1 > \Delta > -\frac{1}{2};\ 0 < \lambda < 2\pi/3)$; note that $\Delta = 1$ separates the multicritical and first-order segments. In the sector spanned by transfer matrix eigenvectors that are even under spin inversion, four singlets are found with critical dimensions of spin-wave operators as present in the

Gaussian, Baxter symmetric eight-vertex, and Ashkin-Teller models [85], *viz.*

$$x_n = n^2 x_e, \tag{73}$$

with $n = 1, 2, 3,$ or 4 and where

$$x_e = 1/9(2 - y), \quad y = 2\lambda/\pi. \tag{74}$$

The corresponding operators are the sublattice density difference $(\sigma_1^2 - \sigma_2^2)$, the total energy $(\sigma_1^2\sigma_3^2 + \sigma_2^2\sigma_4^2)$, the sublattice energy difference $(\sigma_1^2\sigma_3^2 - \sigma_2^2\sigma_4^2)$, and the total density $(\sigma_1^2 - \sigma_2^2)$. In the sector odd under spin-inversion there are doublets with gaps that translate into critical dimensions

$$x_{-1} = \frac{1}{8} \text{ and } x_{-2} = \frac{9}{8}, \tag{75}$$

associated with magnetization $(\sigma_1 \pm \sigma_2)$, and correlation $[\sigma_1\sigma_3(\sigma_1 + \sigma_3) \pm \sigma_2\sigma_4(\sigma_2 + \sigma_4)]$. The numerical estimates agree with the above conjectures typically to within one part in 10^4.

Finally, in addition to these primary zero-spin operators, the numerical results are consistent with the presence of a doublet and quartet of states belonging to the respective conformal descendants with $(r, r') = (1, 0)$ and $(0, 1)$ of the dominant operators in the even and the odd sectors (*i.e.* the sublattice density difference and magnetization.)

3.4 Restricted solid-on-solid (RSOS) model

Models with a wealth of critical phenomena was solved by Andrews, Baxter, and Forrester [86]; the free energy and critical exponents were obtained exactly. The models are defined for integral $r = 4, 5, \ldots$ on a square lattice in which site i has associated with it an integral height variable $1 \leq l_i \leq r - 1$. The Boltzmann weight of each configuration is a product of weights associated with the faces of the lattice. A face (l_1, l_2, l_3, l_4), with the variables around a face ordered counterclockwise, starting with l_1 at the at the bottom left, has a Boltzmann weight $W(l_1,l_2,l_3,l_4)$ that vanishes unless nearest neighbor heights differ by 1. For $r = 4$ the only allowed configurations are those with one sublattice frozen in $l_i = 2$, so that the model is a transcription of an anisotropic Ising model in zero field; the $r = 5$ case is the hard-square model discussed in Section 3.2. For each value of $r \geq 4$ the model has been solved exactly on a two-dimensional manifold, which contains a critical line parameterized by v. On the critical line the non-zero weights are given by

$$W(l,l+1,l,l-1) = W(l,l-1,l,l+1) = \alpha_l, \tag{76}$$

$$W(l+1,l,l-1,l) = W(l-1,l,l+1,l) = \beta_l, \tag{77}$$

$$W(l+1,l,l+1,l) = \gamma_l, \quad W(l-1,l,l-1,l) = \delta_l, \tag{78}$$

where †

$$\alpha_l = \sin(\eta+v)s_{l+1}s_{l-1}/s_l^2, \quad \eta = \pi/2r, \quad s_l = \sin^{\frac{1}{4}}(2\eta l) \tag{79}$$

$$\beta_l = \sin(\eta-v)s_{l+1}s_{l-1}/s_l^2, \tag{80}$$

$$\gamma_l = \sin\Big((2l+1)\eta - v\Big)\sin(2\eta)/s_l^2 s_{l+1}^2, \tag{81}$$

$$\delta_l = \sin\Big((2l-1)\eta + v\Big)\sin(2\eta)/s_{l-1}^2 s_l^2. \tag{82}$$

The weights are invariant under the "up-down" transformation $l_i \to r - l_i$. Also, they are symmetric under reflection about the diagonals of a face. Finally, the model is spacially isotropic if $v = 0$. The manifold on which Andrews *et al.* solved the model consists of four regimes, labeled I, II, III, and IV. The model is disordered in regime I and condenses into a modulated commensurate $(r-2)\times 1$ phase in regime II. In regime III there are $r-2$ homogeneous phases of which

---

† This follows from Andrews *et al.* equations (1.2.7) and (1.2.12) with $\rho' = 1/2p^{\frac{1}{4}}$, $w_0 = 0$ for $p \to 0$, $K = \pi/2$, upon transformation according to (1.2.14) with $g_l = \sin^{-\frac{1}{4}}(2\eta l)$, $x = 1$. The latter yields Boltzmann weights manifestly isotropic for $v = 0$.

there are $r-3$ in regime IV. The segment of the critical line with $-\eta < v < \eta$ separates regimes III and IV.

The $III - IV$ critical points have been interpreted by Huse [87] to be in the universality class of generic multicritical points associated with $(r-2)$-phase co-existence. He also conjectured that the thermal field that varies as one approaches the critical point on the manifold of the exact solution in regime IV, does not couple (except for $r=4$, the Ising case) to the leading thermal field. Instead, this field governs the cross-over from $(r-2)$- to $(r-3)$-phase criticality in addition to coupling to an irrelevant field. The critical dimension x_{th} of the cross-over field is given by

$$x_{\text{th}} = \Big((2r-3)^2 - 1\Big)/2r(r-1), \tag{83}$$

corresponding to the case $m = r-1$, $p = m-1$, and $q = m-2$ of Eq. (66). The irrelevant field has critical dimension x_{irr}

$$x_{\text{irr}} = (2r+2)/(r-1), \quad r \geq 5 \tag{84}$$

which is $m = r-1$, $p = 3$, $q = 1$ in Eq. (66). (Note that the relation between Huse's correction exponent $\theta^{(n)}$ and the thermal and irrelevant exponents y_{th} and y_{irr} is $\tau^{(n)} = -y_{\text{th}}/y_{\text{irr}}$.)

In addition, there are $r-3$ magnetic critical dimensions $x_{\text{mag}}^{(k)}$, $k = 1, \ldots, r-3$, given by

$$x_{\text{mag}}^{(k)} = ((k+1)^2 - 1)/2r(r-1), \tag{85}$$

cf. Eq. (66) with $m = r-1$, $p = q = 2, 3, \ldots, m-1$.

The I-II critical points, interpreted as the continuous melting transitions of a $(r-2) \times 1$ commensurate phase, were conjectured by Huse to be in the universality class of the chiral $(r-2)$-state clock model. Apparently the chiral symmetry breaking field vanishes at the critical points, even though there does not seem to be an obvious symmetry that would explain this. Taking the absence of this symmetry breaking for granted, the conclusion is that the respective universality classes are those of the $(r-2)$-state clock [alias $Z(n)$ with n=$r-2$] models: Ising for $r = 4$; three-state Potts for $r = 4$; a particular point of the Ashkin-Teller line of continuously varying exponents for $r = 6$; and, finally, for $r > 6$ the universality class of the special multicritical point which is at the boundary of first-order and the two-stage Kosterlitz-Thouless melting. From the exact solution one can derive the following thermal (x_{th}) and magnetic ($x_{\text{mag}}^{(k)}$, $k = 1, \ldots, r-3$) critical dimensions:

$$x_{\text{th}} = 4/r, \tag{86}$$

$$x_{\text{mag}}^{(k)} = k(r-2-k)/r(r-2). \tag{87}$$

Whereas the exponents of the *III-IV* critical points are a subset of the set given in Eq. (66)) with conformal anomaly number $c < 1$, the exponents of the *I-II* critical points do not fit into this scheme. However, Zamolodchikov and Fateev [88] have constructed a self-dual conformal field theory with $Z(n)$ symmetry. Indeed, the conformal anomaly number is found to be greater than unity and is given by

$$c = 2(n-1)/(n+2). \tag{88}$$

The theory has the following thermal $x_{\text{th}}^{(k)}$ and magnetic $x_{\text{mag}}^{(k)}$ critical dimensions:

$$x_{\text{th}}^{(k)} = 2k(k+1)/(n+2), \quad k = 1, 2, \ldots, \leq n/2 \tag{89}$$

$$x_{\text{mag}}^{(k)} = k(n-k)/n(n+2), \quad k = 1, 2, \ldots, n-1 \tag{90}$$

of which Eqs. (86) and (87) are special cases with $n = r - 2$ and respectively $k = 1$ and $k = 1, 2, \ldots, n-1$.

Wills [89] studied the restricted solid-on-solid models by numerical calculation of dominant eigenvalues of the transfer matrix. Since the site variables, the column heights, can assume r different values, this might at first sight seem to be a formidable task; however, the order of the transfer matrix of a strip of width L

behaves as $O(r2^{L-2})$ for increasing L or r as a consequence of the restriction on the differences in height of neighboring columns. Estimates for the central charge c were obtained at the isotropic *III-IV* critical point ($v = 0$) by employing Eq. (63) and making two-point extrapolations to the finite system data. The results for values in the range $4 \leq r \leq 12$ agree with Eq. (66)), with Huse's identification $m = r - 1$. The numerical uncertainties range from 0.02% to 0.3%. In addition, Wills calculated the next-to-leading critical dimension, using Eq. (62). The results agree Eq. (83) for $r = 4$ and Eq. (85) for $r > 5$ (with $k = 1$ for odd r, and $k = 2$ for even r respectively) with absolute errors on the order of a few times 10^{-3}. At the *I-II* critical points the RSOS model is anisotropic ($v \neq 0$). As was mentioned before, finite-size amplitudes (*e.g.* of the dominant finite-size correction term in the free energy or the inverse correlation length) in such cases are not universal; however, the ratios of these amplitudes are universal †.

Wills computed the ratio $c/x_{\text{mag}}^{(k)}$ for r ranging from 4 to 9 for values of the anisotropy parameter v chosen so as to optimize convergence with increasing system size. The computations were done using a diagonal-to-diagonal transfer matrix for which the direction of the phase modulations coincides with the transfer direction. The results agree, typically with an accuracy of 5%, with the ratios obtained for clock models from Eqs. (88) and (90) (with $n = r - 2$ and $k = 1$ for r odd, and $k = 2$ for r even) with values of r in the range 4 to 9.

† The author suggests that anisotropy introduces logarithmic finite-size corrections by generalization of Cardy's results for the 4-state Potts model [90]. This does not appear to be justified: in the case of the Potts model the corrections derive from a marginally *irrelevant* operator, whereas anisotropy is expected to be *truly* marginal.

3.5 Kosterlitz-Thouless transitions

Identification of a phase transition as an infinite-order, Kosterlitz-Thouless transition and in particular estimation of the transition temperature poses a serious challenge to all numerical methods. For example, use of Eq. (55) may fail to produce a meaningful result for the transition temperature, as can be seen as follows. Define an effective exponent y_{eff} by the relation [23]:

$$y_{\text{eff}} = \frac{\ln\Big(\kappa(T, L-1)/\kappa(T, L)\Big)}{\ln\Big(L/(L-1)\Big)}. \tag{91}$$

In the thermodynamic limit the following behavior of y_{eff} is expected: the correlation length remains finite in the disordered phase so that $y_{\text{eff}} \to 0$; in the massless phase $y_{\text{eff}} \to 1$; in an ordered phase the dominant eigenvalues of the transfer matrix are expected to be degenerate up to an exponentially small correction so that $y_{\text{eff}} \to \infty$. Unless there is such an ordered low-temperature phase there is no guarantee that Eq. (55) has a solution, which is equivalent to $y_{\text{eff}} = 1$. Even if there is a solution, the temperature found in this way may have nothing to do with the Kosterlitz-Thouless temperature and only signify that a correction-to-scaling term that changes sign. The main advantage of computing the effective exponent y_{eff} is that the results can be extrapolated as a function of system size, so that the massless phase will show up as a plateau at which this exponent y_{eff} assumes a value of unity as a function of temperature.

Alternatively, one can exploit predictions for the universal amplitudes (for isotropic systems, or ratios thereof otherwise) as given in Eq. (62) to obtain estimates of the temperature at which a Kosterlitz-Thouless transition takes place. For example, for isotropic systems, such as the two-dimensional XY-model, the temperature at which the inverse correlation length has a finite-size amplitude of $\pi/4$ can be identified as the transition temperature. This follows from Eq. (62) and the fact that the critical dimension of the vortices that drive the transition has the value $\frac{1}{8}$. If the model under investigation is well understood, critical dimensions of other operators can usually also be predicted [85,23] and used in a similar way. Furthermore, a value of the conformal anomaly number c of unity and continuous variation of critical exponents are signatures of a Kosterlitz-Thouless phase.

Kitani and Oguchi [91] studied the antiferromagnetic triangular Ising model with ferromagnetic next-nearest neighbor interactions. This model is expected to be in the universality class of the six-state clock [i.e. $Z(6)$] model. In the model studied, the spin on each site interacts with all of its six nearest neighbors, but with only four of its six next-nearest neighbors: in one direction the bonds are removed. This truncation has the advantage that, if the transfer matrix is chosen to operate along this special direction, the computation requires a matrix of order 2^L for a lattice of width L, rather than double-column transfer matrix of order 2^{2L}, as needed when all next-nearest neighbor bonds are present. The price paid for this simplification is anisotropy. The presence of a massless phase extending over a finite temperature range is investigated employing Roomany-Wyld [92] finite-size approximants to the β-function. The results, based on calculations of strips of maximum width 9, suggest the presence of a massless phase and are comparable with the results for other models in the same universality class, *viz.* the antiferromagnetic three-state Potts model on a square lattice [23,93], and the triangular antiferromagnetic Ising model in a field studied by Saito and Igeta [94]. The finite-size scaling behavior of the spin autocorrelation function yields further evidence and allows determination with estimated accuracy of 10% of the upper and lower Kosterlitz-Thouless temperatures, which bound the intermediate massless phase. Calculated values of the magnetic critical exponent $x_{\rm h}$ fall in the expected range $\frac{1}{18} \leq x_{\rm h} \leq \frac{1}{8}$ [85].

Related to systems with quasi-long-range order are quasi-crystals. The reader is referred to work by Kawamura, by Henley, and by Widom *et al.* for details [95].

3.6 Commensurate-incommensurate transitions

Models with floating phases such as ANNNI [96], chiral clock [32,97,57] and other anisotropic models [98] models have been studied extensively with transfer matrix methods. Since these models can display anisotropic scaling, one of the quantities that has been used to study their behavior is the effective exponent y_{eff} introduced in Eq. (91); the value of this exponent is expected to be $\frac{1}{2}$ for a Pokrovsky-Talapov transition [99]. Another technique that has been used is called *wavevector scaling* [97]: to identify and to locate a commensurate-incommensurate transition the complex inverse correlation length (12) is used. Both real and imaginary parts of this quantity satisfy the usual scaling relation (47) for the inverse correlation length with minor modifications. Firstly, the wavevector of the commensurate phase has to be subtracted from the imaginary part. Secondly, the imaginary part may remain finite and non-scaling, while the real part goes to zero with increasing system size. There are various possible scenarios [97]: *(i)* as the temperature decreases, both real and imaginary parts scale as the inverse system size at (approximately) the same temperature, an indication of a transition from a modulated disordered to a commensurate ordered phase; *(ii)* if the real part scales at a higher temperature than the imaginary part, the transition takes place in two steps: a transition to a floating phase followed by a commensurate-incommensurate transition at a lower temperature.

3.7 Surface adsorption

Two-dimensional models of physi- and chemisorbed atoms or molecules at surfaces have also been studied with transfer matrix techniques. An incomplete list of this work follows.

Phase diagrams of the following systems were investigated: H adsorbed on Fe(110) [100]; Si on W(110) [101]; Cl on Ag(100) [102]; Se on Ni(100) [103]; H on Ni(111) [104] H on Pd(100) [105]. Rikvold *et al.* [106] calculated adsorption isotherms applied to the problem of sulfur poisoning of H on Pt(111); for a more theoretical study of the phase diagram of the spin-1 triangular Ising model used in this study see Collins *et al.* [107]. Bartelt and Einstein investigated the question of the order of $p(2 \times 2)$ order-disorder transition on a triangular lattice gas [108]. Bartelt *et al.* studied a model of oxygen ordering in $YBa_2Cu_3O_z$ [109].

3.8 Random systems

Morgenstern and Binder [110] and Fernández [111] introduced the transfer matrix method to study properties of Ising models with random nearest neighbor bonds, described by *e.g.*. a reduced Hamiltonian

$$H = \sum K_{ij} s_i s_j, \quad (s_i = \pm 1), \tag{92}$$

where the K_{ij} are independent random variables associated with the bonds of the lattice. In the $\pm J$ model the interactions have probability density $p\delta(K_{ij} - K) + (1-p)\delta(K_{ij} + K)$, with $K > 0$ so that $p = 1$ and $p = 0$ respectively are the pure ferromagnetic and antiferromagnetic cases; the Gaussian case has probability density $\exp(-(K_{ij} - \bar{K})^2/2\tilde{K}^2)/\sqrt{(2\pi\tilde{K})}$.

Independent realizations of bond strengths are considered, and for each of these finite systems the partition function is calculated recursively with Eq. (10), using free boundary conditions in the transfer direction and periodic ones transversely. The free energy follows and thermodynamic quantities can then be calculated as numerical derivatives. Subsequently, these data are averaged, typically over 100 independent bond configurations. An advantage of the method is that true equilibrium is guaranteed, in sharp contrast with standard Monte Carlo methods. A disadvantage is the limitation to small systems, since the time and storage complexity of the calculations is proportional to the dimension of the transfer matrix.

Clearly, the ground state energy of any given realization can be calculated directly by this method; finding the ground state spin configurations is somewhat more involved, but this can be done as follows [110]. Select the n (roughly 100) states of lowest intra-column energy of the first column of the lattice. Add an additional column in each possible configuration and keep the lowest n configurations of the system with this additional column; repeat this last step until the desired system size is obtained. Compare with the ground state energy to check if a groundstate has been found; by comparing with the partition function it can be determined in principle whether all ground states have been found. If necessary, the procedure is repeated with greater n.

An example of an interesting quantity in the study of random Ising models is the *associate susceptibility* [110]. A quenched average is calculated of the quantity

$$\chi^{11} = (Nk_BT)^{-1} \sum s_i^{(l)} s_j^{(l)} \langle s_i s_j \rangle, \tag{93}$$

where the summation is over all pairs of sites of a lattice of N sites; $\langle \cdot \rangle$ denotes the thermal average, and $s_i^{(l)}$ and $s_j^{(l)}$ are the values of the spin variables at sites i and j in ground state number l. This quantity can be obtained as the second-order derivative with respect to a magnetic field $h^{(l)}$ of equal magnitude throughout the system and aligned at each site with the local ground state spin. The correlation function $< s_i s_j >^2$ can be obtained from the expression

$$< s_i s_j >= \Big(Z(+) - Z(-) \Big)^2 / Z^2, \tag{94}$$

where $Z(+)$ denotes a restricted partition function summed over states $s_i s_j = \pm 1$, alternatively Eq. (15) may be used.

In all applications mentioned above the main result was the absence of spin-glass order in two-dimensional models at temperatures above zero, and possibly quasi-long-range order (i.e. power-law decay of correlations) at zero temperature. As a matter of fact, even in three dimensions no indications of spin-glass order were found by study of the decay of correlations with this method [112]; of course, the limitations imposed by system size in this case are severe: the calculations were restricted to sizes $L \times L \times M$ with $L = 4$ and $4 \leq M \leq 10$. Besides studies of the $\pm J$ model with and without an applied uniform magnetic field, applications of this method include investigations of various Ising models with quenched dilution, such as models with ferromagnetic nearest and antiferromagnetic next-nearest neighbor interactions, and the antiferromagnetic triangular model [113].

Also, the two-dimensional Ising model in a random field was studied with this method [114]. The cross-over behavior of the specific heat from the zero-field to the random-field model was studied by Binder [115]. Glaus [31] studied this model using the transfer matrix technique in combination with the method of Droz and Malaspinas [17] for calculation of the static structure factor [see Eq. (22)]. Note that the latter is the sum of two contributions: one is a self-averaging susceptibility, which can be calculated from one single long strip; the other term is not self-averaging and has to be calculated from independent realizations of (shorter) strips.

Cheung and McMillan [20] introduced a method to calculate the free energy and in particular the correlation length for strips of essentially infinite length with Eqs. (24) and (25). They calculated the exponent ν, which describes the divergence of the correlation length at zero temperature:

$$\xi(T) \sim T^{-\nu}. \tag{95}$$

Since the original estimate of ν by Cheung and McMillan similar calculations were done by Bray and Moore [116], using scaling of domain wall energies, and by Huse and Morgenstern [117], who again used the correlation length. Substantial increases in the estimate of ν have resulted as time went on.

Ozeki and Nishimori [118] investigated the $\pm J$ Ising model on a simple square lattice. The phase diagram in the plane of the coupling K and probability p was studied using finite-size scaling properties of the susceptibility $\chi(K,p)$ and the correlation length ξ_L of strips of width L. In terms of the susceptibilities and the correlation lengths two effective exponents $a(K,p)$ and $b(K,p)$ are defined:

$$\chi_L(K,p) \propto L^{a(K,p)}, \tag{96}$$

and

$$\xi_L(K,p) \propto L^{b(K,p)}. \tag{97}$$

These exponents a and b were computed from least-squares fits of the data for three subsequent, even strip widths in the range 6 through 14. To locate the phase boundaries they solved either one of the equations

$$a_L(K,p) = 2y_h - 2 = \frac{7}{4}, \tag{98}$$

or

$$b_L(K,p) = 1, \tag{99}$$

which provides two distinct criteria to determine the phase boundary. Eq. (98) can be used successfully to locate a critical point at the boundary of a disordered phase where the susceptibility remains finite and an ordered phase where χ_L diverges sufficiently rapidly, that is with an effective exponent in excess of $\frac{7}{4}$. Indeed, for a ferromagnetic phase one expects the susceptibility to diverge as the "volume" of the system (that is the total number of lattice points) or even exponentially for infinitely long strips. In other words, the specific value $\frac{7}{4}$ associated with the Ising model, should affect only the rate of convergence. Eq. (99), on the other hand, is of more general validity and is not predicated on the assumed onset of ferromagnetic order.

In the $\pm J$ Ising model there is the Nishimori line, $\tanh(K) = 2p - 1$, along which the internal energy can be calculated exactly [119]. The boundary of the ferromagnetic phase, as found by Ozeki and Nishimori from Eq. (98) in the disorder-temperature plane, follows the exact tangent obtained by Domany [120] for weak disorder (p close to 1) and then curves, and intersects the Nishimori line at the bicritical point $T = 0.961 \pm 0.009$, $p = 0.889 \pm 0.002$ and remains at almost constant disorder at low temperatures. The boundary between the para- and ferromagnetic phases, as estimated using the correlation length criterion Eq. (99), coincides with the one obtained from the susceptibility in the high-temperature range; in contrast, below the Nishimori line the two criteria yield different results. At any temperature below the Nishimori line the correlation length seems to diverge as a function of decreasing disorder before the susceptibility does. Ozeki and Nishimori interpret this as an indication of the existence of an intermediate phase, the *random antiphase* state. For temperatures below the multicritical point the correlation length criterion definitely converges more slowly than above, but the existence of such a phase in this model, which is inconsistent with the renormalization group fixed point structure as found by McMillan [55] remains controversial.

4 THE TRANSFER MATRIX MONTE CARLO METHOD

4.1 Introduction

Currently, exact numerical calculations are limited to systems for which the order of the transfer matrix does not exceed a few times 10^7, except in special cases such as mentioned above. For transfer matrices of higher order, or for transfer kernels for systems with continuous degrees of freedom in general, a Monte Carlo method related to the Green's function Monte Carlo method [121] can be used to perform stochastically the matrix multiplication in Eq. (10), and to find the largest eigenvalue iteratively with a stochastic implementation of the power method. In this section, firstly, the Monte Carlo algorithm is described; secondly, it is shown how the freedom of Eqs. (3) and (4) can be used to obtain a process with a strongly reduced variance, as illustrated explicitly with two examples: the three-dimensional Ising and planar rotator models.

4.2 Weighted random walkers

In the Monte Carlo process a vector is represented by a sequence of r random walkers $R_\iota = (\sigma_\iota, W_\iota)$, $\iota = 1, \ldots, r$, where σ_ι defines the configurations of a slice, as discussed in Section 1.2. Random walker ι has a statistical weight $W_\iota \geq 0$. The number of random walkers r fluctuates as described below, but is maintained within a few percent of a constant target value r_0; the weights are kept in the range $b_l < W_\iota < b_u$, with $b_l \approx \frac{1}{2}$ and $b_u = 2$. Elements of the transfer matrix are written as

$$T(\sigma|\tau) = P(\sigma|\tau)D(\tau), \quad D(\tau) = \sum_\sigma T(\sigma|\tau) \tag{100}$$

where $D(\tau)$ is a normalization constant, introduced so that $P(\sigma|\tau)$ can be interpreted as a probability (density) for a transition from state τ to σ. Note that performing the sum over all σ for a system with helical boundary conditions is simple: the transfer matrix can very naturally be chosen to be a sparse matrix.

A Monte Carlo run consists of a sequence of M_0 sweeps over all random walkers at times labeled by $t = 1, \ldots, M_0$. The sweep at time t consists of two steps that are designed to stochastically perform a transfer matrix multiplication. Affixing primes to indicate variables at time $t + 1$, there are the following steps, following Nightingale and Blöte [122]:

(i) For $\iota = 1, \ldots, r$ change $R_\iota = (\sigma_\iota, W_\iota)$ to $R'_\iota = (\sigma'_\iota, W'_\iota)$ according to the transition probability (density) $P(\sigma'_\iota|\sigma_\iota)$, with $W'_\iota = D(\sigma_\iota)W_\iota/c'$. If $\hat\lambda_0$ is a running estimate of the eigenvalue to be calculated (τ_0), choose $c' = \hat\lambda_0 r/r_0$ to maintain the number of random walkers close to r_0 in the next step.

(ii) From the random walkers R'_ι construct a new set using each old random walker precisely once, as follows:

(ii.1) If $W'_\iota > b_u$ add two random walkers $(s'_\iota, \frac{1}{2}W'_\iota)$ to the new set.

(ii.2) For pairs $(\sigma'_\iota, W'_\iota)$ and $(\sigma'_\kappa, W'_\kappa)$ with $W'_\iota < b_l$ and $W'_\kappa < b_l$ add $(\sigma'_\lambda, W'_\iota +$

W'_κ) to the new set of random walkers, where $\sigma'_\lambda = \sigma'_\iota$ or $\sigma'_\lambda = \sigma'_\kappa$ with relative probabilities W'_ι and W'_κ.

(ii.3) If $b_l < W'_\iota < b_u$, or if R_ι is left unpaired in step *(ii.2)*, add R'_ι to the new set of random walkers.

The algorithms commonly used in Green's function or diffusion Monte Carlo employ random walkers of unit weight. An advantage of the algorithm described above is that number of random walkers that are duplicated (or discarded), Δr, is kept smaller on average by the use of weighted random walkers. This reduces deleterious correlations among the random walkers of the same generation. To be more quantitative, let ΔD^2 be the variance of $D(\sigma)$. If ΔD is small, then $\Delta r \propto \Delta D$, while for an algorithm in which all random walkers are of equal weight the results is $\Delta r \propto \sqrt{\Delta D}$ [123].

The random walkers represent a stochastic vector $\mathbf{\Phi}$ with components

$$\Phi(\tau) = \sum_{\iota=1}^{r} \Big(W_\iota \delta(\tau, \sigma_\iota) \Big), \tag{101}$$

where the delta function is a Kronecker or Dirac function.

Denote the stochastic vector realized in sweep t by $\mathbf{\Phi}_t$. The main result is the following relation, obtained by p-fold iteration:

$$< \prod_{b=1}^{p} c_{t+b} \mathbf{\Phi}_{t+p} >= \mathbf{T}^p \cdot \mathbf{\Phi}_t, \tag{102}$$

where $< \cdot >$ denotes the average over the stochastic process defined above, starting from the fixed vector $\mathbf{\Phi}_t$. With Eq. (102) one constructs an estimator [124] of the dominant eigenvector $\mathbf{\Psi}_0$ of the transfer matrix:

$$\mathbf{\Psi}_0(p) = \frac{1}{M_0} \sum_{t=1}^{M_0} \Big(\prod_{b=0}^{p} c_{t-b} \mathbf{\Phi}_t \Big). \tag{103}$$

This expression is the iterate of order p in the power method with the average of $\mathbf{\Phi}_t$ as the initial vector; as such, this estimator has a systematic bias for any

$p < \infty$. The dilemma is that as the bias decreases with increasing p, the variance of this quantity amplifies. As a compromise, following reference [122], one can make the following *ad hoc* choice: the largest p for which the estimated value of the autocorrelation of c_t at lag p still exceeds its statistical noise by a factor of order unity. An estimate of the dominant eigenvalue of the transfer matrix derived from Eq. (103) is:

$$\lambda_0 \approx W_0(p+1)/W_0(p), \tag{104}$$

where W_0 is obtained from $\Psi_0(\sigma|p)$, the components of $\mathbf{\Psi}_0(p)$, by summation over σ, which –with Eq. (101)– takes the form of a summation over all random walkers in one generation.

4.3 Expectation values

Eq. (103) provides an estimator for the dominant right eigenvector of the transfer matrix, and can be employed in conjunction with Eq. (14) to yield estimates of surface correlations. Results for the three-dimensional Heisenberg model indicate that for small systems this method has smaller variance than results produced with a standard Metropolis algorithm; for large systems the situation is reversed. There has not yet been any systematic study to determine where the cross-over in efficiency occurs, in particular when variance reduction (see Section 4.4) is used.

Not only for the surface expectation values can be computed; for the bulk this can be done as follows [125]. In terms of a transfer matrix $\mathbf{T}$ one defines a transfer matrix $\mathbf{U}$ with a "memory" of m slices:

$$U(\sigma_1, \sigma_2, \ldots, \sigma_m|\tau_1, \tau_2, \ldots, \tau_m) = T(\sigma_1|\tau_1) \prod_{i=2}^{m} \delta_{\sigma_i \tau_{i-1}} \tag{105}$$

The dominant right eigenvector element $\chi(\sigma_1, \ldots, \sigma_m)$ of this matrix is proportional to the joint probability of finding the leftmost m slices respectively in the configurations $\sigma_1, \ldots, \sigma_m$. Given the algorithm for the Monte Carlo multiplication by $\mathbf{T}$, implementation of the multiplication by $\mathbf{U}$ requires generalization of the original one-slice random walkers to walkers with m slices. The only change in the algorithm is that the delta functions in Eq. (105) are implemented by a shift operation: for each random walker slice i of the current generation is transformed into slice $i+1$ of the next, for $i = 1, \ldots, m-1$, and current slice m is deleted. In this way, calculation of an expectation value for slice m (counted from the left surface) becomes the calculation of a generalized surface expectation; the result converges to a bulk expectation value for m sufficiently large. In practice, it is not necessary to store all the details of every slice, but only the values of the required functions of their configurations; the shift operation is then applied only to these values.

4.4 Variance reduction

An approximate left eigenvector, say $\vec{\gamma}$, can be used to reduce the variance of the eigenvalue estimates obtained with the Monte Carlo method described above, as is well-known [121], by application of the method described above to a similarity transform of the original transfer matrix $\mathbf{T}$. This new matrix $\mathbf{T}$ is defined [see Eq. (4)] by:

$$\hat{T}(\sigma|\tau) = \gamma(\sigma) T(\sigma|\tau) / \gamma(\tau). \tag{106}$$

In terms of $\hat{\mathbf{T}}$, a stochastic process can be defined via Eq. (100), which involves a diagonal matrix $\hat{\mathbf{D}}$. Ideally, if $\vec{\gamma}$ is an exact left eigenvector, the dependence on σ of its diagonal elements $D(\sigma)$ disappears, and with it all statistical noise; in practice, the variance is merely reduced.

4.5 Applications and implementation: three dimensional models

As applications we discuss results for the 3-dimensional Ising and planar rotator models with helical boundary conditions. For the Ising model on a simple cubic lattice the transfer matrix for an $n_x \times n_y \times \infty$ system, with slices of $N = n_x \times n_y$ sites each, is defined as:

$$T(s_1, \ldots, s_N | t_1, \ldots, t_N) = \prod_{i=2}^{N-1} \delta_{s_i, t_{i-1}} \exp\big(K s_1 (t_1 + t_{n_x} + t_N)\big) \tag{107}$$

Trial vectors that capture some of the essential physics and do so without considerably slowing down the computation are of the following form:

$$\tilde{\gamma}(s_1, \ldots, s_N) = \exp \sum_{i,j} K_{ij} s_i s_j, \tag{108}$$

The K_{ij} are variational parameters determined by the Monte Carlo scheme introduced by Umrigar, Wilson and Wilkins [126], i.e. by minimization of the variance of

$$\hat{D}(\sigma_\iota) = \sum_\tau \tilde{\gamma}(\tau) T(\tau|\sigma_\iota)/\tilde{\gamma}(\sigma_\iota), \tag{109}$$

where the variance is approximated by a weighted sum over the states σ_ι of the random walkers of one generation, during the initial stage of the Monte Carlo run. For temperatures close to or above the critical temperature this produces adequate estimates of the variational parameters with as few as one or two hundred random walkers; at lower temperatures this number increases, but in all cases discussed below roughly one thousand random walkers is sufficient.

The interactions K_{ij} have to be truncated for the trial vector (108) to be of practical use. It is clear that at or above the critical temperature the strength of the K_{ij} will decrease as the distance d_{ij} between sites i and j increases. It is less clear how the distance between these sites should be defined. The following scheme has been found to work satisfactorily.

The transfer matrix defined in Eq. (107) generates a system with a screw dislocation, as a consequence of the helical boundary conditions. The surface has a step which terminates at the point where the next site will be added. There are bonds, which have strength K, between all nearest-neighbor surface sites, except across the step. The right eigenvector of the transfer matrix is proportional to the probability to find the spins at this surface in a particular configuration. The left eigenvector describes the surface of the complementary semi-infinite lattice that has not yet been grown. In the latter, the order of the sites is reversed and the surface bonds are missing. The interactions K_{ij} that feature in the left trial vector (108) are mediated by the spins in the underlying bulk. The distance d_{ij} between sites i and j will by definition be the length of the shortest path connecting these sites via sites exclusively in the bulk. By setting $K_{ij} = 0$ in Eq. (108) if $d_{ij} > d_0$, a sequence of trial vectors is generated, parameterized by the maximum path length d_0: for $d_0 = 2$ only pairs of sites across the step contribute; for $d_0 = 3$ all pairs of nearest neighbor sites are included, in addition to a few odd ones across the seam.

This scheme makes sense only for relatively small values of d_0. Its main strength is that it can deal successfully with the vagaries of the surface defects. For larger values of d_0 the Euclidian distance between sites seems to be a more reasonable criterion to cut off the interactions; also, short-range many-spin interactions may be more important than long-range two-spin ones. A more serious shortcoming is that, the whole scheme in essence is a high- rather than a low-temperature approximation; nevertheless, it also works at low temperatures.

Table 1 shows results for the free energy of the Ising model [127] at three different coupling strengths (above, at, and below T_c) for four different system sizes ($L \times L \times \infty$, with L=4, 6, 8, and 10). The table shows the following trends: firstly, performance of the method improves with increasing temperature; secondly, the variance reduction scheme works at all temperatures, but again best at high temperature; and, thirdly, the standard errors seem to be overestimated in some cases.

With the ability to calculate the free energy, one can calculate the surface tension as the difference in free energy of two systems: one with ferromagnetic and the other with antiferromagnetic interactions, with linear dimensions chosen so as to force an interface in the latter, *i.e.* for $L \times L \times \infty$ systems with helical boundary conditions L has to be even.

The finite-size analysis described in Section 2 can be applied [122] to obtain an estimate of the critical coupling and the thermal exponent y_T for the three-dimensional Ising model. The results are: $K_c = 0.22162 \pm 0.00002$ and $y_T = 1.584 \pm 0.004$ using data with $L = 6$ through 12; and $K_c = 0.22167 \pm 0.00004$ and $y_T = 1.584 \pm 0.014$ if the $L = 6$ are omitted. To be compared with the results from *(i)* high-temperature series expansions $K_c = 0.22164 \pm 0.00002$ and $y_T = 1.587 \pm 0.004$ [128]; *(ii)* Monte Carlo renormalization group calculations [129] $K_c = 0.221654 \pm 0.000006$ and $y_T = 1.590 \pm 0.010$.

The transfer matrix for the planar rotator model is defined by a slight modification of Eq. (107): the Ising variables s_i and t_i are replaced by two-component vectors of unit length; inner products replace scalar products where appropriate. Again from a finite-size analysis of the interface free energy, the critical coupling can be derived together with the thermal exponent y_T . The results are [130]: $K_c = 0.454102 \pm 0.000022$ and $y_T = 1.490 \pm 0.003$ using data with $L = 6$ through 12; and $K_c = 0.454135 \pm 0.000038$ and $y_T = 1.486 \pm 0.004$ if the $L = 6$ are omitted. These results are roughly an order of magnitude more accurate than transfer matrix Monte Carlo results obtained without the variance reduction for the three-dimensional Heisenberg model [122]. They are to be compared with high-temperature series result $K_c = 0.4539 \pm 0.0012$ and renormalization group result $y_T = 1.495 \pm 0.004$ [131]. Finally, for applications to the two-dimensional Ising and $Z(5)$ models see reference. [26]

Table 1. Transfer matrix Monte Carlo results for the largest eigenvalue (λ_0) and standard errors σ for three-dimensional $L \times L \times \infty$ Ising lattices with helical boundary conditions at various coupling strengths K calculated with various trial vectors, labeled by d_0 as defined in the text. All calculations were done with 5000 random walkers and 2500 flips per spin.

K	L	λ_0	σ	L	λ_0	σ	d_0
0.1000	4	2.03088021	0.00000250	6	2.03082595	0.00000210	0
		2.03087796	0.00000060		2.03082558	0.00000048	2
		2.03087790	0.00000024		2.03082524	0.00000011	3
		2.03087777	0.00000015		2.03082526	0.00000008	4
	8	2.03082540	0.00000120	10	2.03082283	0.00000110	0
		2.03082429	0.00000027		2.03082478	0.00000027	2
		2.03082448	0.00000009		2.03082459	0.00000007	3
		2.03082457	0.00000005		2.03082457	0.00000003	4
0.2216	4	2.188328	0.000110	6	2.179887	0.000078	0
		2.188304	0.000026		2.180110	0.000046	2
		2.188295	0.000011		2.180163	0.000014	3
		2.188302	0.000005		2.180163	0.000009	4
	8	2.178206	0.000083	10	2.177155	0.000064	0
		2.178124	0.000034		2.177480	0.000032	2
		2.178159	0.000015		2.177432	0.000017	3
		2.178156	0.000007		2.177424	0.000009	4
0.3000	4	2.546603	0.000290	6	2.545337	0.000410	0
		2.546676	0.000069		2.546056	0.000085	2
		2.546577	0.000017		2.546103	0.000020	3
		2.546570	0.000013		2.546098	0.000014	4
	8	2.546041	0.000230	10	2.545190	0.000390	0
		2.546066	0.000080		2.546045	0.000120	2
		2.546085	0.000019		2.546086	0.000026	3
		2.546074	0.000017		2.546096	0.000019	4

ACKNOWLEDGMENTS

This work was funded at the University of Rhode Island by the National Science Foundation (NSF) under contract No. DMR-87-04730, and was generously supported by the Cornell National Supercomputing Facility, a resource of the Center for Theory and Simulation in Science and Engineering at Cornell University, which is funded in part by the NSF, New York State, and the IBM Corporation. It is a great pleasure to thank Drs. Piali De, Vladimir Privman, Karl Runge, Cyrus Umrigar and Ben Widom for their valuable comments on the manuscript.

REFERENCES

1. M.P. Nightingale, *J. Appl. Phys.* **53**, 7927 (1982).
2. J.O. Indekeu, A. Maritan, and A.L. Stella, *J. Phys. A* **15**, L291 (1982); J.O. Indekeu, A. L. Stella, and L. Zhang, *J. Phys. A* **17**, L341 (1984); J.O. Indekeu, A. Maritan, and A.L. Stella, *Phys. Rev. B* **35**, 305 (1987); J.O. Indekeu and G. Menu, *J. Phys. A* **21**, L523 (1988); M. Suzuki, M. Katori, and X. Hu, *J. Phys. Soc. Jpn.* **56**, 3092 (1987); M. Katori and M. Suzuki, *ibid.*, p. 3113; X. Hu, M. Katori, and M. Suzuki, *ibid.*, p. 3865; X. Hu and M. Suzuki, *ibid.*, **57**, 791 (1988).
3. P. Rujan, *J. Stat. Phys.* **49**, 139 (1987).
4. H.A. Kramers and G.H. Wannier, *Phys. Rev.* **60**, 252 (1941); *ibid.* p. 263.
5. J. Ashkin and W.E. Lamb, *Phys. Rev.* **64**, 159 (1943).
6. L. Onsager, *Phys. Rev.* **65**, 117 (1944); B. Kaufman, *Phys. Rev.* **76**, 1232 (1949).
7. B. Derrida, *J. Phys. A* **14**, L5 (1981);
8. B.Derrida and J. Vannimenus, *J. Physique Lett.* **41**, L473 (1980); W. Kinzel and J.M. Yeomans, *J. Phys. A* **14**, L163 (1981); both reprinted in [45].
9. B. Derrida and L. De Seze, *J. Physique* **43**, 475 (1982), reprinted in [45]; J.P. Nadal, B.Derrida, and J. Vannimenus, *J. Physique* **43**, 1561 (1982).
10. B. Derrida and H.J. Herrmann, *J. Physique* **44**, 1365 (1983).
11. B. Derrida and J. Vannimenus *J. Phys. A* **15**, L557 (1982), reprinted in [45].
12. H.W.J. Blöte, M.P. Nightingale, B. Derrida, *J. Phys. A* **14**, L45 (1980); H.W.J. Blöte and M.P. Nightingale, *Physica* **112A**, 405 (1982).
13. H.W.J. Blöte and M.P. Nightingale, *Physica* **119A**, 1 (1984).
14. See for example F.R. Gantmacher, *The Theory of Matrices II* (Chelsea Publishing Company, 1974).
15. R.J. Baxter, *Ann. Phys.* **70**, 193 (1972).
16. F.Y. Wu, *Phys. Rev. B* **4**, 2312 (1971); L.P. Kadanoff and F. Wegner, *Phys. Rev. B* **4**, 2312 (1971).

17. M. Droz and A. Malaspinas, *J. Phys. C* **16**, L231 (1983).
18. N.C. Bartelt and T.L. Einstein, *J. Phys. A* **19**, 1429 (1986).
19. H. Furstenberg and H. Kersten, *Ann. Math. Statist.* **31**, 457 (1960); H. Furstenberg, *Trans. Am. Math. Soc.* **108**, 377 (1963).
20. H.-F. Cheung and W.L. McMillan, *J. Phys. C* **16**, 7027 (1983); *ibid.* p. 7033.
21. G.F. Newell, *Phys. Rev.* **79**, 876 (1950).
22. M.P. Nightingale, *Proc. Koninklijke Nederlandse Acad. van Wetenschappen* **B82**, 245 (1978).
23. M.P.M. den Nijs, M.P. Nightingale, and M. Schick, *Phys. Rev. B* **26**, 2490 (1982).
24. M.T. Batchelor, *J. Aust. Math. Soc. B* **28**, 462 (1987).
25. F.H. Ree and D.A. Chestnut, *J. Chem. Phys.* **45**, 3983; L.K. Runnels and L.I. Combs, *ibid.*, 2482; W. Kinzel and M. Schick, *Phys. Rev. B* **23**, 3435 (1981).
26. M.P. Nightingale, *Proceedings of the Third International Conference on Supercomputing*, Boston, Massachusetts, USA, May 1988, edited by L.P. Kartashev and S.I. Kartashev, vol. I, p.427; M.P. Nightingale and R.G. Caflisch, in *Computer Stimulation Studies in Condensed Matter Physics*, edited by D.P. Landau, K.K. Mon, and H.-B. Schüttler, *Springer Proceedings in Physics* 33 (Springer-Verlag, Berlin, 1988) p.208.
27. J.H. Wilkinson, *The Algebraic Eigenvalue Problem* (Clarendon Press, Oxford).
28. L.K. Cullum, R.A. Willoughby, *Lanczos Algorithms for Large Symmetric Eigenvalue Computations, Prog. in Scientific Computing* **3** and **4** (Birkhäuser, Boston 1985).
29. G.H. Golub and C.F. van Loan, *Matrix Computations*, 2nd edition (Baltimore: Johns Hopkins University Press, 1989).
30. H.W.J. Blöte and M.P. Nightingale, *Physica* **134A**, 274 (1985).
31. U. Glaus, *Phys. Rev. B* **34**, 3203 (1986).
32. J. Yeomans and B. Derrida, *J. Phys. A* **18**, 2343 (1985).
33. R.J. Baxter, *J. Math. Phys.* **9**, 650 (1968).
34. R.J. Baxter, *J. Stat. Phys.* **19**, 461 (1978).
35. R.J. Baxter, *Exactly Solved Models in Statistical Mechanics* (Academic Press, London 1982).
36. S.K. Tsang, *J. Stat. Phys.* **20**, 95 (1979).
37. R.J. Baxter and S.K. Tsang, *J. Phys. A* **13**, 1023 (1980).
38. R.J. Baxter and I.G. Enting, *J. Stat. Phys.* **21**, 103 (1979).
39. A. Malakis, *J. Stat. Phys.* **27**, 1 (1982).
40. P.A. Pearce and K.A. Seaton, *J. Stat. Phys.* **53**, 1061 (1988).
41. R.J. Baxter and P.J. Forrester, *J. Phys. A* **17**, 2675 (1984).
42. M.E. Fisher, in *Critical Phenomena, Proc. Enrico Fermi School*, vol. 51, edited by M.S. Green (New York, Academic Press); M.E. Fisher and M.N. Barber, *Phys. Rev. Lett.* **28**, 1516 (1972).

43. M.P. Nightingale, *Physica* **83A**, 561 (1976); M.P. Nightingale, *Phys. Lett.* **59A**, 486 (1977).
44. M.N. Barber, *Finite-Size Scaling, Phase Transitions and Critical Phenomena*, vol. 8, edited by C. Domb and J.L. Lebowitz, (Academic Press, New York, 1983).
45. J.L. Cardy, *Finite-Size Scaling*, edited by J.L. Cardy, Current Physics – Sources and Comments, vol. 2 (North-Holland, Amsterdam, 1983).
46. C. Itzykson, H. Saleur, and J.-B. Zuber, eds. *Conformal Invariance and Applications to Statistical Mechanics*, (World Scientific Singapore, 1988).
47. E. Brézin and J. Zinn-Justin, *Nucl. Phys. B* **257**[FS14], 867 (1985); J. Rudnick, H. Guo, D. Jasnow, *J. Stat. Phys.* **41**, 353 (1985); M.P. Dudek, J.M.J. van Leeuwen, and H.W.J. Blöte, *Physica* **147A**, 344 (1987).
48. Th. Niemeijer and J.M.J. van Leeuwen,*Phase Transitions and Critical Phenomena*, vol. 6, edited by C. Domb and M.S. Green, (Academic Press, 1976).
49. F. Wegner, *Phase Transitions and Critical Phenomena*, vol. 6, edited by C. Domb and M.S. Green (Academic Press, 1976).
50. V. Privman and M.E. Fisher, *Phys. Rev. B* **30**, 322 (1984).
51. M.N. Barber, I. Peschel, and P.A. Pearce, *J. Phys. A* **38**, 497 (1984).
52. J.O. Indekeu, W.V. Wang, and M.P. Nightingale, *Phys. Rev. B* **34**, 330 (1986).
53. J.L. Cardy, *Nucl. Phys. B* **270**[FS16], 186 (1986).
54. D. Kim and P.A. Pearce, *J. Phys. A* **20**, L451 (1987).
55. W.L. McMillan, *Phys. Rev. B* **29**, 4026 (1984).
56. K. Binder, it Z. Phys. B **43**, 119 (1981), reprinted in [45].
57. A.L. Stella, X.-C. Xie, T.L. Einstein, and N.C. Bartelt, *Z. Phys. B* **67**, 357 (1987).
58. T.W. Burkhardt and B. Derrida, *Phys. Rev. B* **32**, 7273 (1985).
59. R.R. dos Santos and L. Sneddon, *Phys. Rev. B* **23**, 3541 (1981); H. Takano and Y. Saito, Prog. Theor. Phys. **73**, 1369 (1985).
60. D.W. Wood and A.H. Osbaldestin, *J. Phys. A* **15**, 3579, (1982) and *ibid.* **16**, 1019 (1983).
61. D.A. Lavis and A.G. Quinn, *J. Phys.* **20**, 2129 (1987).
62. N.C. Bartelt, T.L. Einstein, L.D. Roelofs, *Phys. Rev. B* **34**, 1616 (1986).
63. C.N. Yang, Phys. Rev. 85, 808 (1952); K. Uzelac, Thesis, Orsay, 1980; K. Uzelac and R. Julien, *J. Phys. A* **14**, L151 (1981); C.J. Hamer, *ibid.* **15**, L675; *ibid.* **16**, 3085 (1983).
64. W. Kinzel and M. Schick, Phys. Rev. 24, 324 (1981).
65. H.J. Herrmann, *Phys. Lett.* **100A**, 256 (1984).
66. M.N. Barber, *Physica* **130A**, 171, (1985).
67. P.A. Rikvold, W. Kinzel, J.D. Gunton, K. Kaski, *Phys. Rev. B* **28**, 2686 (1983).
68. J.L. Cardy, *J. Phys. A* **17**, L385 (1984), reprinted in [45] and [46].

69. M.P. Nightingale and H.W.J. Blöte, *J. Phys. A* **16**, L657 (1983), reprinted in [45].
70. I. Affleck, *Phys. Rev. Lett.* **56**, 746 (1986), reprinted in [45] and [46].
71. H.W.J. Blöte, J.L. Cardy, and M.P. Nightingale, *Phys. Rev. Lett.* **56**, 343 (1986), reprinted in [45] and [46].
72. J.L. Cardy, *Nucl. Phys. B* **270**[FS16] 186 (1986); *ibid.* **B275**[FS17] 200 (1986), both reprinted in [46].
73. D. Friedan, Z. Qiu, and S. Shenker, *Phys. Rev. Lett.* **52**, 1575 (1985), reprinted in [46].
74. V.G. Kac, *Group Theoretical Methods in Physics*, edited by W. Beiglbock and A. Bohm, *Lecture notes in Physics* Vol. 94 (Springer- Verlag, New York, 1979).
75. T.W. Burkhardt, *J. Phys. A* **18**, L307 (1985); T.W. Burkhardt and I. Guim, *J. Phys. A* **18**, L25, (1985); L33.
76. M.T. Batchelor and H.W.J. Blöte, *Phys. Rev. B* **39**, 2391 (1989).
77. H.W.J. Blöte and B. Nienhuis, *J. Phys. A* **22**, 1415 (1989).
78. H. Saleur, *Phys. Rev. B* **35**, 3657 (1987).
79. H.W.J. Blöte and M.P.M. den Nijs, *Phys. Rev. B* **37**, 1766 (1988).
80. R.J. Baxter, *J. Phys. A* **13**, L61 (1980); *J. Stat. Phys.* **26**, 427 (1981); R.J. Baxter and P.A. Pearce, *J. Phys. A* **15**, 897 (1982); *ibid.* bf 16, 2239 (1983)
81. D.A. Huse, *J. Phys. A* **16**, 4357 (1983).
82. V. Privman and L. Shulman, *J. Stat. Phys.* **29**, 205 (1984); also see M.A. Novotny, W. Klein, and P.A. Rikvold, *Phys. Rev. B* **33**, 6523 (1986).
83. P.A. Pearce and D. Kim, *J. Phys. A* : **20**, 6471 (1988).
84. J.L. Cardy, *J. Phys. A* **20**, L891 (1987), reprinted in [46].
85. J.V. José, L.P. Kadanoff, S. Kirkpatrick,and D.R. Nelson, *Phys. Rev. B* **16**, 1217 (1977); L.P. Kadanoff and A.C. Brown, *Ann. Phys.* **121**, 318 (1979); H.J.F. Knops, *Ann. Phys.* **128**, 448 (1981).
86. G.F. Andrews, R. Baxter, and P.J. Forrester, *J. Stat. Phys.* **35**, 193 (1984).
87. D.A. Huse, *Phys. Rev. B* **30**, 3908 (1984), reprinted in [46].
88. A.B. Zamolodchikov and V.A. Fateev, *Sov. Phys. JETP* **62**, 215 (1985), reprinted in [46].
89. P. Wills, *J. Phys. A* **20**, 5219 (1987).
90. J.L. Cardy, *J. Phys. A* **19**, 364 (1986), reprinted in [45].
91. H. Kitani and T. Oguchi, *J. Phys. Soc. Jpn.* **57**, 1344 (1988).
92. H.H. Roomany and H.W. Wyld, *Phys. Rev. D* **21**, 3341 (1980).
93. M.P. Nightingale and M. Schick, *J. Phys. A* **15**, L39 (1982);
94. Y.Saito and K. Igeta, *J. Phys. Soc. Jpn.* **53**, 3060 (1984).
95. H. Kawamura, *Prog. Theor. Phys.* **70**, 352 (1983); C.L. Henley, *J. Phys. A* **21**, 1649 (1988); M. Widom, D.P. Deng, and C.L. Henley, *Phys. Rev. Lett.* **63**, 310 (1989).
96. P.D. Beale, P.M. Duxbury, and J. Yeomans, *Phys. Rev. B* **31**, 7166 (1985); P. Rujan, G. Uimin, W. Selke, *Phys. Rev. B* **32**, 7453 (1985); W. Pesch and

J. Kroemer, *Z. Phys. B* **59**, 317 (1985); H. Takano and Y. Saito, *Prog. Theor. Phys.* **73**, 1369 (1985); P.D. Beale, *Phys. Rev. B* **33**, 1717 (1986); J. Oitma, M.J. Velgakis, M.T. Batchelor, M.N. Barber, *J. Magnetism and Magnetic Materials* **54- 57**, 665 (1986); M.D. Grynberg and H. Ceva, *Phys. Rev. B* **36**, 7091 (1987); *ibid.* **38**, 9172 (1988).

97. P.M. Duxbury, J. Yeomans, and P.D. Beale, *J. Phys. A* **17**, L179 (1984).
98. W. Kinzel, *Phys. Rev. Lett.* **51**, 996 (1983); B. Schaub and E. Domany, *Phys. Rev. B* **28**, 2897 (1983).
99. V.L. Pokrovsky and A.L. Talapov, *Phys. Rev. Lett.* **42**, 65 (1979); for a review see M.P.M. den Nijs, *Phase Transitions and Critical Phenomena*, vol. 12, edited by C. Domb and J.L. Lebowitz, (Academic Press, London, 1988).
100. W. Kinzel, W. Selke, and K.Binder, *Surface Sci.* **121**, 13 (1982).
101. J. Amar, S. Katz, and J.D. Gunton, *Surface Sci.* **155**, 667 (1985).
102. R.Q. Hwang, E.D. Williams, N.C. Bartelt, and R.L. Park, *Phys. Rev. B* **37**, 5870 (1989).
103. P.Bak, P.Kleban, W.N. Unertl, J. Ochab, G. Ackinci, N.C. Bartelt, *Phys. Rev. Lett.*, **54** 1539 (1985).
104. L.D. Roelofs, T.L. Einstein, N.C. Bartelt, and J.D. Shore, *Surface Sci.* **176**, 295 (1986).
105. N.C. Bartelt, *Phys. Rev. Lett.*, **59**, 244 (1987).
106. P.A. Rikvold, J.B. Collins, G.D. Hansen, and J.D. Gunton, *Surface Sci.* **203**, 500 (1988).
107. J.B. Collins, P.A. Rikvold, and E.T. Gawlinski, *Phys. Rev. B* **23**, 6741 (1988).
108. N.C. Bartelt and T.L. Einstein, *Phys. Rev. B* **30**, 5339 (1984).
109. N.C. Bartelt, T.L. Einstein, and L.T.Wille, *Phys. Rev. B and Physica C*, in print.
110. I. Morgenstern and K. Binder, *Phys. Rev. Lett.* **43**, 1615 (1979); *Phys. Rev. B* **22**, 288 (1980).
111. J.F. Fernández, *J. Phys. C* **13**, L555 (1980).
112. I. Morgenstern and K. Binder, *Z. Phys. B* **39**, 277 (1980).
113. I. Morgenstern, *Z. Phys. B* **42**, 23 (1981); **43**, 33 (1981); I. Morgenstern, *Phys. Rev. B* **25**, 6067; 6071 (1982); J.F. Fernández, C.Z. Andérico, and T.S. J. Streit, *J. Appl. Phys.* **53**, 7991 (1982); J.F. Fernández, *Phys. Rev. B* **38**, 6901 (1988); J.F. Fernández, *J. Appl. Phys.* **63**, 3738 (1988)
114. I. Morgenstern, K. Binder, and R.M. Hornreich, *Phys. Rev. B* **23**, 287 (1981); I. Morgenstern and H. Horner, *Phys. Rev. B* **25**, 504 (1982); J.F. Fernández, *Phys. Rev. B* **25**, 417 (1982); J.F. Fernández and E. Pytte, *Phys. Rev. B* **31**, 2886 (1985); E. Pytte and J.F. Fernández, *J. Appl. Phys.* **57**, 3274 (1985).
115. K. Binder, *Phys. Rev. B* **29**, 5184 (1984).
116. A.J. Bray and M.A. Moore, *J. Phys. C* **17**, L463 (1984).
117. D.A. Huse and I. Morgenstern, *Phys. Rev. B* **32**, 3032 (1985).
118. Y. Ozeki and H. Nishimori, *J. Phys. Soc. Jpn.* **56**, 3265 (1987).
119. H. Nishimori, *Prog. Theor. Phys.* **66**, 1169 (1981).

120. E. Domany, *J. Phys. C* **12**, L119 (1979).
121. D. M. Ceperley and M. H. Kalos, *Monte Carlo Methods in Statistical Physics*, edited by K. Binder (Springer, Berlin, 1979).
122. M.P. Nightingale and H.W.J. Blöte, *Phys. Rev. Lett.* **60**, 1662, (1988).
123. M.P. Nightingale, K.J. Runge, and C.J. Umrigar, to be published.
124. J.H. Hetherington, *Phys. Rev. A* **30**, 2713 (1984); M.P. Nightingale and H.W.J. Blöte, *Phys. Rev. B* **33**, 659 (1986).
125. M.H. Kalos, *J. Comp. Phys.* **1**, 257 (1966); the original idea of "forward walking" predates this paper (M.H. Kalos private communication).
126. C.J. Umrigar, K.G. Wilson, and J.W. Wilkins, *Phys. Rev. Lett.* **60**, 1719 (1988); *Computer Stimulation Studies in Condensed Matter Physics*, edited by D.P. Landau, K.K. Mon, and H.-B. Schüttler, *Springer Proceedings in Physics* 33 (Springer-Verlag, Berlin, 1988) p.185.
127. M.P. Nightingale, to be published.
128. M.J. George and J.J. Rehr, *J. Stat. Phys.*, in print; A.J. Liu and M.E. Fisher, *Physica* **A156**, 35 (1989); B.G. Nickel and J.J. Rehr, *J. Stat. Phys.* in print.
129. G.S. Pawley, R.H. Swendsen, D.J. Wallace, and K.G. Wilson, *Phys. Rev. B* **29**, 4030 (1984).
130. H.W.J. Blöte and M.P. Nightingale, to be published.
131. M.Ferer, M.A. Moore, and M. Wortis, *Phys. Rev. B* **8**, 5205 (1973); J.-C. le Guillou and J. Zinn-Justin, *Phys. Rev. Lett.* **39**, 95 (1977); G.A. Baker, Jr., B.G. Nickel, D.I. Meiron, *Phys. Rev. B* **17**, 1365 (1978); K. Ohno, Y. Okabe, and A. Morita, *Prog. Theor. Phys.* **71**, 714 (1984).

VIII. APPLICATIONS OF THE HAMILTONIAN LIMIT TO CRITICAL PHENOMENA, FINITE-SIZE SCALING AND CONFORMAL INVARIANCE

Malte Henkel

TABLE of CONTENTS

(continued)

APPLICATIONS OF THE HAMILTONIAN LIMIT TO CRITICAL PHENOMENA, FINITE-SIZE SCALING AND CONFORMAL INVARIANCE

Malte Henkel[1)]

Service de Physique Théorique de Saclay

Laboratoire de l'Institut de Recherche Fondamentale

du Commissariat à l'Energie Atomique

F-91191 Gif-sur-Yvette Cedex

[1)] Address after 1st October 1989:

Departement de Physique Théorique

Université de Genève,

CH 1211 Genève 4, Switzerland

1. INTRODUCTION

Finite-lattice techniques are a useful tool in studying critical phenomena. Among the various approaches to the subject, Hamiltonian limit studies are widely used to obtain precise information about the phase diagram and critical exponents. As will be made explicit below, taking the Hamiltonian limit amounts to taking an extremely anisotropic limit of the original lattice model. The problem of calculating the free energy of a $d+1$-dimensional isotropic (classical) system is thereby transformed in finding the ground state energy of an appropriate d-dimensional quantum system. Although the approach may appear a little abstract, the built-in simplifications as compared to isotropic systems are helpful in understanding a given model and also contribute that Hamiltonian methods are among the most precise available for the explicit calculation of critical parameters.

Here, we shall review applications of the Hamiltonian limit technique to critical phenomena in two and three dimensions, or, in the usual jargon, of (1+1)D and (2+1)D quantum systems. Emphasis will be laid on finite-size scaling techniques.

Several reviews on the subject are available. The classical introduction to the Hamiltonian limit is the review by Kogut [1]. Current finite-size scaling techniques with some applications to the Hamiltonian limit are beautifully reviewed by Barber [2] and summarize the general development just before the advent of conformal invariance. A review on the Hamiltonian limit would be incomplete without at least some discussion of conformal invariance in (1+1)D. A nicely readable review on early developments was given by Cardy [3], for an excellent description of recent developments see [4].

The equivalence between a (d+1)-dimensional isotropic classical system and its d-dimensional quantum analogue was first pointed out by Suzuki and developed systematically by Fradkin and Susskind [5]. The (1+1)D Ising model had already been studied previously [6,116,117] in its own right. Although the quantum Hamiltonian can be shown to commute with the transfer matrix precisely at the critical point even for a fully isotropic system [215], it was realized that the extreme anisotropic limit (to be discussed below) is an essential ingredient. In view of this equivalence, the Hamiltonian limit was done for systems more general than the Ising model and a one-to-one correspondence between the physical quantities in the two formulations was established. The real-space renormalization group was formulated (for the $n-$state (1+1)D Potts model) by Solyom and Pfeuty [216]. For practical calculations, however, it has turned out that the phenomenological renormalization of Nightingale [30] is more accurate. This concept as well as finite-size scaling was translated into the Hamiltonian limit by Hamer and Barber [51], to be followed by the introduction of the Lanczos scheme [217].

Let us give an example how the Hamiltonian limit (which is also often referred to as τ-continuum limit or extreme anisotropic limit) is taken in practice [5]. Consider a classical, 2D Z_3−symmetric spin system in the three-states Potts model universality class with the action

$$\begin{aligned} S = -\sum_{\vec{r}} \quad & \{J_x \cos(\theta(\vec{r}) - \theta(\vec{r}+\vec{x})) \\ & + J_\tau \cos(\theta(\vec{r}) - \theta(\vec{r}+\vec{\tau}))\} \end{aligned} \tag{1.1}$$

where $\vec{r}$ runs over all sites of a square lattice and $\vec{x}$ and $\vec{\tau}$ are the nearest-neighbor displacement vectors on this lattice. (S is normally called the classical Hamiltonian $\mathcal{H}$. We do not use this terminology to avoid later confusion with the quantum Hamiltonian H). θ takes the values $2\pi j/3, j = 0,1,2$. Taking an $N \times N_\tau$ lattice, the partition function Z is written in terms of the transfer matrix T

$$Z = \mathrm{tr}\ T^{N_\tau} \tag{1.2}$$

and periodic boundary conditions are taken in the τ−direction (which will be referred to as "time"). The transfer matrix may be regarded as a "time"-evolution operator, thus defining the quantum Hamiltonian H

$$T = \exp(-\tau H) = 1 - \tau H + O(\tau^2) \tag{1.3}$$

where τ is the lattice spacing in "time" direction. T acts between two rows of the lattice, described by configurations $\{\theta_j\}$ and $\{\tilde{\theta}_j\}$, $j = 1, ...N$ and is a $3^N \times 3^N$ matrix. The Hamiltonian limit consists of taking $\tau \longrightarrow 0$ in such a way that H assumes a simple form.

The diagonal element of T is obtained by letting $\theta_j = \tilde{\theta}_j$, $j = 1,2,...N$

$$\begin{aligned} T_{diag} &= \exp\left\{\sum_{j=1}^{N} J_x \cos(\theta_j - \theta_{j+1})\right\} \\ &= 1 + 2\tau\lambda \sum_{j=1}^{N} \cos(\theta_j - \theta_{j+1}) + O\left(\tau^2\right) \end{aligned} \tag{1.4}$$

where we set

$$J_x = 2\tau\lambda \tag{1.5}$$

The off-diagonal element where $\theta_k - \tilde{\theta}_k = \frac{2\pi}{3}$ for some k and $\theta_j = \tilde{\theta}_j$ for $j \neq k$ is

$$T_{off} = \exp(-\frac{1}{2}J_\tau)(1 + O(\tau)) = \tau + O\left(\tau^2\right) \tag{1.6}$$

where we set

$$J_\tau = -2 \ln\tau \tag{1.7}$$

Now, we come to the essential point: let $\tau \longrightarrow 0$ such that λ stays fixed. Then all higher excitations, where at least two "spins" in the configurations $\{\theta_j\}$ and $\{\tilde{\theta}_j\}$ are different, are at least of order $O(\tau^2)$ because of (1.7) and can be neglected. It is this simplification which makes the Hamiltonian limit attractive for practical calculations. Note that taking this limit implies that $J_x \longrightarrow 0$ and $J_\tau \longrightarrow \infty$ in a peculiar way.

Introduce an operator $\hat{\theta}_j$ with eigenvalues 0, $\frac{2\pi}{3}$, $\frac{4\pi}{3}$ and eigenstates $|\theta_j\rangle$ such that $\hat{\theta}_j\,|\theta_j\rangle = \theta_j\,|\theta_j\rangle$. Its conjugate operator $\hat{p}_j$ is defined by

$$\exp(in\hat{p}_j) \mid \theta_j >=\mid \theta_j + n\,\frac{2\pi}{3} > \tag{1.8}$$

which yields, up to additive constants

$$H = -2\sum_{j=1}^{N}\left\{\lambda\,\cos(\hat{\theta}_j - \hat{\theta}_{j+1}) + \cos\,\hat{p}_j\right\} \tag{1.9}$$

Another commonly used formulation defines the matrices σ_j and $\Gamma_j(\omega = \exp\,2\pi i/3)$

$$\sigma_j = \exp(i\hat{\theta}_j) = \begin{pmatrix} 1 & & \\ & \omega & \\ & & \omega^2 \end{pmatrix} \tag{1.10}$$

$$\Gamma_j = \exp(i\hat{p}_j) = \begin{pmatrix} 0 & 0 & 1 \\ 1 & 0 & 0 \\ 0 & 1 & 0 \end{pmatrix} \tag{1.11}$$

and the quantum Hamiltonian for the three-states Potts model becomes

$$H = -\sum_{j=1}^{N}\left\{\lambda(\sigma_j\sigma_{j+1}^{+} + \sigma_j^{+}\sigma_{j+1}) + \Gamma_j + \Gamma_j^{+}\right\} \tag{1.12}$$

Compared to the transfer matrix, H only contains one spin-flip operators which increases the sparsity of the matrices involved. This fact is very useful in numerical calculations and is a consequence of the limit $\tau \longrightarrow 0$ performed. The price to pay for this is that the critical point must be expressed in terms of λ. From (1.5) it follows that λ takes the role of an inverse temperature. There is, however, no relationship to the critical point $(J_{x,c}, J_{\tau,c})$ of the classical system. This is not very serious, since the critical point, being a non-universal quantity, does depend on all the details of a physical system and one should not expect to be able to calculate accurately critical points of substances studied in experiments from simple lattice models as studied here. Note that the prescription (1.5),(1.7) leaves the total normalization of H arbitrary. Fixing the normalization of H corresponds to a certain choice for the energy scale. Since the partition function Z is related by (1.2) to the largest eigenvalue of T, it follows from (1.3) that the free energy is given by the ground-state energy E_0 of H.

It is important for later applications of conformal invariance that taking the Hamiltonian limit geometrically corresponds to studying a system on an infinitely long strip (corresponding to the time direction) of finite width Na, where a is the lattice constant (we usually put $a = 1$).

For completeness, we also write down the quantum Hamiltonian for the Ising model (which can be derived following the lines given above)

$$H = -\sum_{j=1}^{N} \{\sigma^x(j) + \lambda\, \sigma^z(j)\sigma^z(j+1)\} \tag{1.13}$$

where σ^x and σ^z are Pauli matrices [1,5]. The generalization of (1.12, 1.13) to higher dimensions is obvious.

Let us state some simple properties of the Potts-3 quantum chain. H (1.12) commutes with the charge operator (for simplicity take periodic boundary conditions $\sigma_{N+1} = \sigma_1$)

$$\hat{Q} = \prod_{j=1}^{N} \Gamma_j \tag{1.14}$$

which has eigenvalues ω^Q ($Q = 0,1,2$). The corresponding eigenspaces of $\hat{Q}$ are called *charge sectors*. The parity operator P is defined by

$$P\,\sigma_j P^+ = \sigma_{N-j+1}, \qquad P\,\Gamma_j P^+ = \Gamma_{N-j+1} \tag{1.15}$$

and has eigenvalues $P = \pm 1$. One can define the charge conjugation operator

$$C = \prod_{j=1}^{N} c_j, \qquad c_j = \begin{pmatrix} 1 & 0 & 0 \\ 0 & 0 & 1 \\ 0 & 1 & 0 \end{pmatrix} \tag{1.16}$$

with eigenvalues $C = \pm 1$. Since C commutes with H, the spectrum of H in the two sectors $Q = 1$ and $Q = 2$ is identical. Sometimes it is useful to rewrite (1.12) in a "high-temperature" basis. This can be done by the unitary transformation

$$\begin{aligned} \Gamma &\longrightarrow \mathcal{U}\Gamma\mathcal{U}^+ = \sigma \\ \sigma &\longrightarrow \mathcal{U}\sigma\mathcal{U}^+ = \Gamma^+ \end{aligned} \tag{1.17}$$

which keeps intact the commutation relation $\sigma\Gamma = \omega\Gamma\sigma$. The matrices σ_j, Γ_k commute on different sites. The duality transformation is defined by

$$\begin{aligned} \sigma_j \longrightarrow \tilde{\sigma}_j &= \Gamma_{j+1}\Gamma_j^+ \; ; \qquad j < N \\ \sigma_N \longrightarrow \tilde{\sigma}_N &= \Gamma_N^+ \\ \Gamma_j \longrightarrow \tilde{\Gamma}_j &= \prod_{\ell=1}^{j} \sigma_\ell \end{aligned} \tag{1.18}$$

and leave the algebra of σ, Γ invariant. The spectrum of H then has the *self-duality* property

$$H(\lambda) = \lambda H(1/\lambda) \tag{1.19}$$

and it follows that if the system has only one transition, it must be at $\lambda_c = 1$. It is straightforward to restate these properties for the Ising model Hamiltonian (1.13) (see [1,5] and sections 3.1 and 4.2).

We now make explicit the relationship between the spectrum of a d–dimensional quantum system and its classical analogue in $d+1$ dimensions. For a short "dictionary", see table 1.1.

Statistical mechanics	Quantum system
Hamiltonian $-\beta\mathcal{H}$	action S
infinitesimal transfer matrix T	quantum Hamiltonian H
Equilibrium state	ground state
Ensemble averages	ground state expectation values
Correlation functions	Propagators
Free energy	ground state energy E_0 of H
inverse correlation length ξ^{-1}	lowest energy gap of H

Table 1.1 : Some equivalences between statistical mechanics and its corresponding quantum system.

From these correspondences, some further relationships can be obtained. Recall that λ takes the role of an (inverse) temperature. Then the analogy of the specific heat C is

$$C = -\frac{1}{N}\frac{d^2 E_0(\lambda)}{d\lambda^2} \sim (\lambda - \lambda_c)^{-\alpha} \tag{1.20}$$

which defines the critical exponent α. An order parameter can also be defined easily. In the example of the three-states Potts model (1.12), one adds to the Hamiltonian

$$-h\sum_{j=1}^{N}(\sigma_j + \sigma_j^+) \tag{1.21}$$

Calculating the spectrum as a function of h, the analog of the magnetic susceptibility is

$$\chi = -\frac{1}{N}\frac{d^2E_0(h)}{dh^2} \sim (\lambda - \lambda_c)^{-\gamma} \tag{1.22}$$

Finally, if E_1 is the energy of the lowest excited state of H, the correlation length ξ is

$$\xi^{-1} = E_1 - E_0 \sim (\lambda - \lambda_c)^{\nu} \tag{1.23}$$

where γ and ν are critical exponents. In the language of field theory, the gap(s) of H describe the mass(es) present in the theory. It is often very useful that correlation length(s) or mass(es) are directly accessible since many theoretical predictions are phrased in terms of them. This fact has become extremely useful in 2D conformal invariance applications. Further physical quantities will be defined when needed.

Since the lattice spacing is believed to be an irrelevant parameter at any phase transition, the continuum limit (1.5,1.7) involved in defining H should not affect the universal features of the critical behaviour. It is this universality which makes work the analogies listed in table 1.1 and in eqs. (1.20,1.22,1.23). Analytical solutions allow to check this universality explicitly for two cases: the (1+1)D Ising model [6] and the $(r+1)$D spherical model ($r > 1$) [7]. It is found that not only the critical exponents of the original isotropic statistical mechanics model and its Hamiltonian limit are the same, but also that universality holds for the finite-size scaling functions. For a wealth of other systems universality of the critical exponents was also checked numerically, some of which will be mentioned in the later sections. Note, however, that for genuinely anisotropic systems, like for example the three-states asymmetric clock model there exists a continuing discrepancy between the statistical mechanics model and its quantum analogue on the existence of a Lifshitz point in the phase diagram [8,9].

This article is organised into three large sections. In section 2, one of the principal methods of applying the quantum Hamiltonian to critical phenomena is presented. It consists of diagonalising the Hamiltonian H on finite lattices, which is normally done using the Lanczos algorithm. Finite-size scaling is used to extract finite-lattice estimates for the critical point, exponents or amplitudes. This is followed by some extrapolation to the $N \longrightarrow \infty$ limit. We hope to have covered in this section the principal stages of the calculations involved. Section 3 describes applications to systems in (2+1)D, mainly the (2+1)D Ising model. In this example, Hamiltonian limit calculation are among the most precise methods available for the calculation of critical exponents. The existing results are in excellent agreement with universality.

Section 4 deals with (1+1)D systems and is intended to illustrate some of the powerful methods of conformal invariance. These techniques allow to identify the *complete* operator content of a model, rather than a few leading exponents as in (2+1)D systems. Taking the Hamiltonian limit of a classical spin system, in addition to the conventional transitions mentioned above, new types of phase transitions can occur, which manifest themselves by a level crossing in the ground state energy, rather than a conventional non-analyticity of a simple level as corresponds to the types of transitions known in the classical systems. This kind of behaviour is often linked to commensurate-incommensurate transitions and it can happen that the scaling form of the correlation length(s) is *not* in agreement with conformal invariance. Some examples will be described.

It should be noted that this review suffers from many omissions, but inclusion of these topics would have increased the size of this work beyond any bounds [10]. Quantum systems have very intimate relationships with classical statistical mechanics and many of the existing links could not be mentioned. Besides their interpretation as the Hamiltonian limit, they can also be considered as zero-temperature quantum systems in their own right, with many interesting applications. For a review, see [11]. Series expansion techniques are only briefly mentioned, mainly in the context of the (2+1)D Ising model. Also, although many numerical results are to be discussed, only few original data could be included [12]. Relations to experimental data are not discussed [13] which is mainly due to the fact that experiments for (1+1)D, for which many predictions can be made, are extremely hard to perform.

In two dimensions, conformal invariance has revealed deep relationships to integrable systems [14]. A rich mathematical structure has been developed. All the extensions of the conformal (Virasoro) algebra like superconformal, Zamolodchikov-Fateev or Kac-Moody algebras or quantum groups are not treated. Unfortunately, Bethe ansatz techniques could not be covered. For reviews and a description of the links between (1+1)D conformal invariance and integrable systems, see [14,100]. The inclusion of these topics would have transformed this work into a review on conformal invariance, which was not intended. I hope the references provided give a sufficient entry into the rapidly expanding literature. No attempt was made for a complete bibliography [15]. Apologies are offered to all authors whose work could not be mentioned.

Still, I hope to have collected some interesting and sometimes amusing examples of critical quantum systems which might serve to stimulate the reader's imagination and phantasy.

2. NUMERICAL METHODS

In this section, the most common approach using the quantum Hamiltonian in critical phenomena applications is described. The technique proceeds in four steps. First, using known symmetries, the Hamiltonian H is broken into some still relatively large blocks. Second, these blocks are diagonalised numerically by the Lanczos algorithm. Third, finite-size scaling is used to obtain estimates of the critical parameters from the spectrum of H. Special attention is paid to the finite-size scaling of the magnetization and the latent heat, where the Hamiltonian limit opens up new possibilities. Finally, an extrapolation towards the infinite system must be done.

Possible alternatives to this approach are discussed in section 3.4.

2.1. Symmetries

In section 1, some symmetries of the three-states Potts quantum chain were mentioned. Using the quantum numbers of operators like charge, parity or charge conjugation, H is broken into independent blocks which can be studied separately. For example, the $n-$states Potts quantum chain is

$$H = -\sum_{j=1}^{N} \sum_{k=1}^{n-1} \{\sigma_j^k + \lambda\, \Gamma_j^k \Gamma_{j+1}^{n-k}\} \tag{2.1}$$

where σ and Γ are the obvious generalisations of (1.10,1.11) and satisfy $\sigma^n = \Gamma^n = 1$, $\sigma\Gamma = \omega\Gamma\sigma$, $\omega = \exp(2\pi i/n)$. Note that (2.1) can be rewritten in the form (1.12) by a change of basis (1.16). Each of the n charge sectors of H is a $n^{N-1} \times n^{N-1}$ matrix, which may eventually be further decomposed according to parity or charge conjugation eigenstates.

For convenient boundary conditions, the invariance under translations (and lattice rotations in (2+1)D) serves to introduce momentum and angular momentum quantum numbers. Generally, this can be implemented in a straightforward way for the case of the so-called toroidal boundary conditions, which are for the example (2.1)

$$\Gamma_{N+1} = \exp(2\pi i\tilde{Q}/n)\Gamma_1 \tag{2.2}$$

where $\tilde{Q} = 0, 1, \ldots n-1$. For $\tilde{Q} = 0$ which corresponds to periodic boundary conditions, the implementation of the translation operator poses no problem. For $\tilde{Q} \neq 0$, the translation operator T is replaced by some modified operator $T_{\tilde{Q}}$.

For a quantum chain of Z_n symmetry, in the charge sector Q and boundary conditions $\tilde{Q}$ (see (2.2)), eigenstates of H with momentum P are defined on N sites by [16]

$$\begin{aligned} \mu(N)^{1/2} \mid P, Q, \ldots > \; &= \mid \alpha_1 \alpha_2 \ldots \alpha_N > \\ &+ \exp(2\pi i C_1/nN) \mid \alpha_2 \alpha_3 \ldots \alpha_N \alpha_1 > \\ &+ \ldots \\ &+ \exp(2\pi i C_{N-1}/nN) \mid \alpha_N \alpha_1 \ldots \alpha_{N-1} > \end{aligned} \tag{2.3}$$

where

$$C_k = \tilde{Q} N(\alpha_1 + \ldots + \alpha_k) - (Q\tilde{Q} + Pn)k \tag{2.4}$$

$\mu(N)$ is the number of states $|\alpha_1, \ldots \alpha_N\rangle$ appearing in $|P, Q, \ldots\rangle$, and the spin on the k-th site is labelled by $\alpha_k \in \mathbb{Z}_n$. If H commutes with an operator R such that $R^2 = 1$ (R may be the parity or charge conjugation operator, for example), one forms eigenstates of R by $|\psi; \pm\rangle = \frac{1}{\sqrt{2}}(|\psi\rangle \pm R|\psi\rangle)$. Recently, systems with non-toroidal boundary conditions but still invariant under translations have been considered [17]. Here the boundary conditions are implemented by choosing a basis which diagonalizes the charge conjugation operator C (which commutes with H). (2+1)D lattices are discussed in [18,19]. While there is an expression similar to (2.3) defining the angular momentum eigenstates (with the simplification that there is no analog to states with $\tilde{Q} \neq 0$), for a square lattice one can consider invariance under mirror reflection by four axes, which map the lattice onto itself.

2.2. Diagonalization

The calculation of the lowest eigenvalues of H is usually done using the Lanczos [20] algorithm. Although in the mathematical literature this algorithm has a reputation of being numerically unstable [21], the practical application shows that these instabilities are normally not that harmful and can usually be avoided.

If H is a symmetric $n \times n$ matrix, a convenient representation of the Lanczos algorithm is (see [21])

$$\begin{array}{ll} & [v_i, w] \\ & \downarrow \\ w' = w + H v_i & [v_i, w'] \\ \alpha_i = < v_i \mid w' > & \downarrow \\ w'' = w' - \alpha_i . v_i & [v_i, w''] \\ \beta_i = (< w'' \mid w'' >)^{1/2} & \downarrow \\ v_{i+1} = w'' / \beta_i & [v_i, v_{i+1}] \\ & \downarrow \\ swap(v_i, v_{i+1}) & [v_{i+1}, v_i] \\ & \downarrow \\ w = -\beta_i v_i & [v_{i+1}, w] \end{array} \tag{2.5}$$

where the v_i $(i = 1, ...n), w, w', w''$ are vectors of n components. As *initial conditions*, one has $w = 0$ and a starting vector v_1 must be supplied. Practical experience suggests that at least for Z_n spin systems the actual choice of v_1 is not very important provided it has a sizeable overlap with the true ground state. The particular form given in (2.5) makes explicit the fact that only two vectors (rather than three [21]) have to be kept in storage. To illustrate this, the square brackets $[A, B]$ indicating which vectors take the places A and B in storage are included for each step in (2.5). This observation is of relevance since the main limitation in practical applications is storage and not CPU time.

In the basis spanned by the vectors v_i $(i = 1, ...n)$, H takes a tridiagonal form

$$H = \begin{bmatrix} \alpha_1 & \beta_1 & & & \\ \beta_1 & \alpha_2 & \beta_2 & & \\ & \ddots & \ddots & \ddots & \\ & & \ddots & \ddots & \ddots \\ & & & \ddots & \ddots & \beta_{n-1} \\ & & & & \beta_{n-1} & \alpha_n \end{bmatrix} \tag{2.6}$$

However, if one only computes a truncated tridiagonal matrix using only m steps of the Lanczos algorithm, the eigenvalues of the $m \times m$ principal submatrix of H converge extremely rapidly to those of H. [21] It is this observation which makes the Lanczos algorithm useful for practical calculations. For example, the lowest lying eigenvalue is determined, for matrices of size $n \lesssim 10^6$, to about 8 digits in $m \simeq 25$ iterations.

If one is only interested in the leading critical exponents, it is enough to obtain the lowest two eigenvalues of H. In this case, one does not encounter the instability of the Lanczos algorithm due to round-off errors which we shall discuss now.

Choosing a starting vector v_1 with a non-zero overlap to the true ground state, one produces a sequence of eigenvalues and the corresponding eigenvectors. While the second eigenvector is in principle orthogonal to the first, round-off errors will make it to get a significant overlap with the first eigenvector after sufficiently many iterations. Then the estimates for the second eigenvalue will cross over to the first, producing an artificial degeneracy. One can prevent this by re-orthogonalising the set of the v_i sufficiently often (not necessarily after each step).

On the other side, practical experience [22] gives the following. Suppose one is interested in the higher eigenvalues but not in the eigenvectors. Then one observes that the convergence of the higher eigenvalues increases after each cross-over of an eigenvector into the ground state. This trick allows a faster determination of the higher eigenvalues (which are needed when one wishes to use the methods of conformal invariance, see section 4). Some caution is in order, however. The procedure described yields artificial degeneracies not only for the ground state, but also for the higher states. The question whether an eigenvalue obtained is degenerate must be answered by looking at the evaluation of the eigenvalues with an increasing number of iterations. The accuracy obtained (which is usually checked by looking at the variation with m) does depend strongly on the parameters in the Hamiltonian. For example, in a Z_6-system [23], one needs to obtain, say, the fifth eigenvalue, to an accuracy of 10^{-8} about $m \simeq 50-200$ iterations, depending on the parameters.

Obtaining *eigenvectors* with the Lanczos algorithm probably does require to re-orthogonalise the set of the v_i. Since the calculation of eigenvectors offers additional insight into the critical behaviour of a model (see section 3.1) we mention two methods which have been tried in practice.

One method uses the conjugate gradient method [24] and was applied to obtain finite-size estimates for the magnetization [25] and latent heat [26] of Z_n quantum chains. The other method is the BMKD algorithm [27]. This is a two-step Lanczos algorithm where the resulting 2×2 matrix is diagonalised and the lowest eigenvalue and its eigenvector are then used in the next step. The method yields only the lowest eigenvalue of H and its eigenvector but is numerically extremely stable. Due to the larger number of applications of H to some vector v_i and a slower convergence as compared to the Lanczos algorithm [27], the computational effort required is raised considerably. The method was used to calculate the lowest eigenvectors in the (2+1)D Ising model [19].

2.3. Finite-size scaling

Having computed the eigenvalues of H for a finite system with $\mathcal{N} = N^{d-1}$ sites, one has to extrapolate for $N \longrightarrow \infty$. The basis for this comes from the theory of finite-size scaling. Here, we merely collect some results which shall be useful below. For complete descriptions, we refer to the literature [2,28].

Finite-size scaling asserts that:

$$\begin{aligned} f^{(s)} &= N^{-d} Y(C_1 t N^{1/\nu},\ C_2 h N^{\Delta/\nu}) \\ \xi_i^{-1} &= N^{-1} S_i(C_1 t N^{1/\nu},\ C_2 h N^{\Delta/\nu}) \end{aligned} \tag{2.7}$$

where $f^{(s)}$ is the singular part of the free energy density, ξ_i is the correlation length where the index i labels various thermodynamic fields (e.g. spin-spin or energy-energy correlation), $t = (T - T_c)/T_c$ is the reduced temperature and h is a magnetic field. The critical exponents ν and $\Delta = \beta + \gamma$ as well as the scaling functions Y and S_i are the same for all systems in a given universality class. All the details of the model are summarized into the non-universal constants C_1 and C_2. It should be stressed that C_1 and C_2 are the *only* non-universal constants, in particular, there is *no* non-universal prefactor, which means that $Y(0,0)$ and $S_i(0,0)$ are universal numbers but depend on the shape (aspect ratio) of the lattice. This statement will be referred to as Privman-Fisher hypothesis [29].

For two-dimensional systems, $Y(0,0)$ and $S_i(0,0)$ become related to the central charge and the anomalous dimensions of the model (see section 4.1).

From (2.7), one can derive the corresponding finite-size scaling formula for, say, the specific heat and the magnetic susceptibility. When applying (2.7) to the Hamiltonian limit, one should keep in mind that the normalization of H is a priori arbitrary. Unless it can be fixed (as in two dimensions) one has to add a non-universal normalization factor in (2.7). The free energy density f is obtained from the ground state energy per spin $f = E_0/\mathcal{N}$ and d is the number of "space" dimensions plus one (for the "time" direction).

The critical point T_c is now found using the phenomenological renormalization [30]

$$\frac{1}{N}\xi_i(N, T_c) = \frac{1}{N-1}\xi_i(N-1, T_c) \tag{2.8}$$

Solving this equation yields a set of estimates $T_c(N)$ to be extrapolated for $N \longrightarrow \infty$ (see section 2.4). Critical exponents are then obtained as follows. Suppose a quantity X (like specific heat or susceptibility) has for the infinite system a power-law singularity $X \sim (T - T_c)^x$ close to the critical point T_c. One then has the finite-size scaling behaviour at $T = T_c$ [2,28]

$$X_N \sim N^{-x/\nu} \tag{2.9}$$

Estimates for the exponent x/ν are then obtained either from

$$\left(\frac{x}{\nu}\right)_N = -N\left[(X_N/X_{N-1}) - 1\right] \tag{2.10}$$

or else from

$$\left(\frac{x}{\nu}\right)_N = -\frac{\ln X_N - \ln X_{N-1}}{\ln N - \ln (N-1)} \tag{2.11}$$

Practical experience suggests that the correction-to-scaling terms are smaller for (2.11) in many but not all systems (see [31,32]). Note that the values X_N (2.9) can also be taken at $T = T_c(N)$, rather than $T_c(\infty)$. In this sense the estimates for x/ν do not depend on the estimate for T_c. We now discuss the finite-size scaling of the magnetization M and of the latent heat L, following the lines of Uzelac and Hamer [33,25,26]. These two quantities have in common that on a finite lattice they vanish. This comes from the fact that the free energy and its derivatives are analytic on any finite lattice. In order to have a non-vanishing spontaneous magnetization or latent heat, however, the free energy must have a discontinuous derivative (e.g. [34]). Eq.(2.9) therefore cannot be applied in order to estimate the critical exponent of either L or M at $T = T_c$ or their non-zero value away from the critical point. In principle, one could use a double limiting process: that is, by calculating the free energy and its derivatives for $T \neq T_c$ (that is, outside the scaling region), taking first the limit $N \longrightarrow \infty$ and then extrapolating back to the transition point from either side. In practice, however, one should not expect to get accurate results.

A possible way around this problem goes as follows. This method assumes that the critical limit (e.g. $T \longrightarrow T_c$ or $h \longrightarrow 0$) and the thermodynamic limit $N \longrightarrow \infty$ are treated in a particular way. The assumptions involved in this approach have been carefully studied in the literature [34] (for the case of a non-vanishing magnetic field). To be specific, consider the latent heat in the Ising model. Write the Hamiltonian as

$$H = H_0 + tV \tag{2.12}$$

where H_0 is the Hamiltonian at the critical point and $t = T - T_c$ measures the distance from the transition. Let $|0>$ be the ground state eigenvector and $|\,\varepsilon>$ the eigenvector the first excited state in the sector $Q = 0$ (its energy gap is the inverse energy correlation length ξ_ε^{-1}). Suppose that at $t = 0$, both $|\,0>$ and $|\,\varepsilon>$ have the same eigenvalue E_0. Then the ground state energy near $t = 0$ is the lowest eigenvalue of the matrix $E = E_0 + tE_1$ where

$$E_1 = \begin{pmatrix} \langle 0|V|0\rangle & \langle 0|V|\varepsilon\rangle \\ \langle \varepsilon|V|0\rangle & \langle \varepsilon|V|\varepsilon\rangle \end{pmatrix} \tag{2.13}$$

Let e_+ and e_- the higher and lower eigenvalues of E_1. Then the latent heat L per site is [26]

$$\begin{aligned} L &= (e_+ - e_-)/\mathcal{N} \\ &= \left[(\langle 0|V|0\rangle - \langle \varepsilon|V|\varepsilon\rangle)^2 + 4\langle 0|V|\varepsilon\rangle^2\right]^{1/2} /\mathcal{N} \end{aligned} \tag{2.14}$$

The magnetization per spin M can be treated analogously [25]

$$M = \frac{1}{\mathcal{N}}\langle 0|V|\sigma\rangle \tag{2.15}$$

where $\mid \sigma >$ is the lowest eigenvector in the sector $Q = 1$ and V is the magnetization operator.

If one has a system with, say, a global Z_n symmetry, the magnetization is obtained from the eigenvalues of an $n \times n$ matrix, rather than (2.15). Both (2.14) and (2.15) can be used at a first order transition to obtain L and M. Hamer [26] obtained L for the $q-$state Potts model and found good agreement with the known [35] exact results.

On the other hand, for a second order transition one does have a finite-size scaling behaviour [19,25], at the critical point

$$\begin{aligned} L \sim N^{-x_\varepsilon} \quad &, \quad M \sim N^{-x_\sigma} \\ x_\varepsilon = \frac{1-\alpha}{\nu} \quad &\quad x_\sigma = \frac{\beta}{\nu} \end{aligned} \tag{2.16}$$

where α and β are the exponents of the specific heat and the magnetization. While there is no other direct way to obtain β using finite-size scaling, the exponent α could in principle also be obtained from the specific heat. For the (2+1)D Ising model however, the specific heat estimates for α converge very slowly, since α is rather small (see section 3.1.2). Using (2.16) offers a means to avoid this problem and to obtain an accurate estimate [19].

2.4. Extrapolation

Having used finite-size scaling to obtain sequences $\{f_N\}$ of finite-size estimates for quantities like the critical temperature or critical exponents, we now have to carry out the limit $N \longrightarrow \infty$. In doing so, we must necessarily make some assumption on the asymptotic behaviour of the f_N if N becomes large. Further, we have to assume that the asymptotic regime goes well down to the values of N for which finite-size data were obtained. As an example, we give in Fig. 1 finite-lattice estimates for the critical exponent $x_\varepsilon = (1-\alpha)/\nu$ for the quantum Ising chain (as obtained from conformal invariance techniques to be described in section 4). Even for very small lattices, one finds a rapid convergence towards the exact result $x_\varepsilon = 1$.

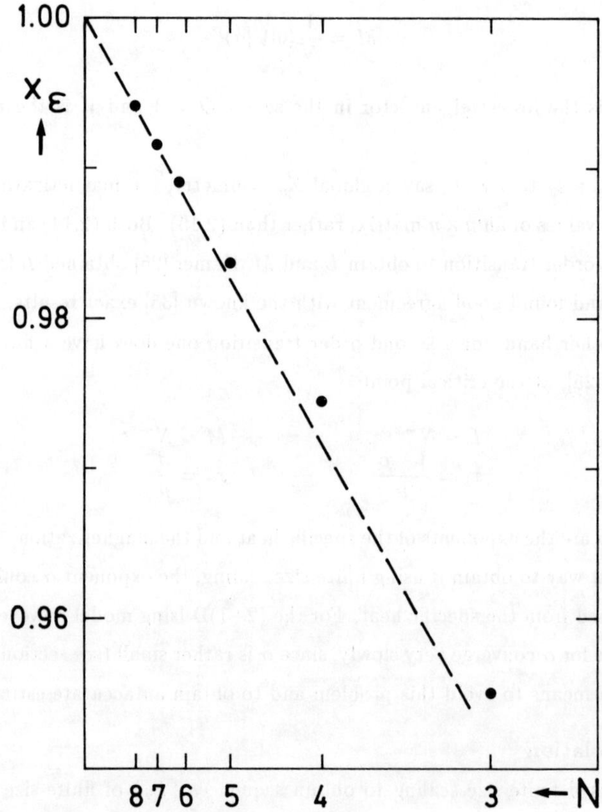

Fig.1: Finite-size estimates for the critical exponent x_ε in the (1+1)D Ising model. The dashed line describes the leading finite-size correction $x_\varepsilon = 1 - \frac{\pi^2}{24}N^{-2} + \ldots$

Although the assumptions mentioned above are reasonably well satisfied for most of the systems studied, they remain dangerous. In fact, it is quite common that finite-size data show a non-monotonic behaviour for small values of N. A striking example has been given by Luck [36] for the 3D spherical model. In Fig.2, we show finite-size estimates $\beta_c(N)$ for the inverse critical temperature β_c as obtained from (2.13). The full line gives the leading asymptotic behaviour [36]

$$\beta_c(N) - \beta_c = a \cdot N^{-2} + ... \tag{2.17}$$

with $a = 0.0278073...$. Considering square lattices with sizes (up to $N = 9$) which are already considerably larger than those which can be treated numerically for other (2+1)D systems (for the Ising model $N_{max} \lesssim 5$), even the sign of the finite-size correction is found to be different with the analytical prediction. In fact, $\beta_c(N)$ shows a maximum at $N = 11$ and only then one finally finds agreement with the prediction (2.17).

This kind of phenomenon is not at all restricted to the spherical model. Similar effects show up in the (1+1)D Ising model in a magnetic field and seem to appear generically [37].

It should be stressed that the process of extrapolation is to be judged to be the most difficult part in the whole calculation. (The numerical diagonalization may be computationally demanding, but is in principle straightforward). Although there exist algorithms to do the limit $N \longrightarrow \infty$, their application to a given system does require care and a considerable amount of judgement. By necessity, this implies that the final estimates for $N \longrightarrow \infty$ are somewhat subjective. In particular, this statement holds for the *accuracy* of the final results since the estimates eventually obtained will almost always contain systematic errors. If at all possible, one should therefore try to obtain a second estimate of the same quantity by a different method. For example, one possibility is to study several models in the same universality class. We shall do so in the sequel, using as examples the (1+1)D and (2+1)D Ising model. Usually the accuracies quoted in the literature only take into account the internal errors of the extrapolations and neglect the systematic error.

We now discuss some commonly used extrapolation algorithms. Techniques for accelerating the convergence of sequences to their limits have been widely developed in applied mathematics (see [38,39,40] and references therein). One distinguishes two situations [39]:

1. If, for $N \longrightarrow \infty$

$$f_N = f + O(\mathrm{e}^{-aN}) \tag{2.18}$$

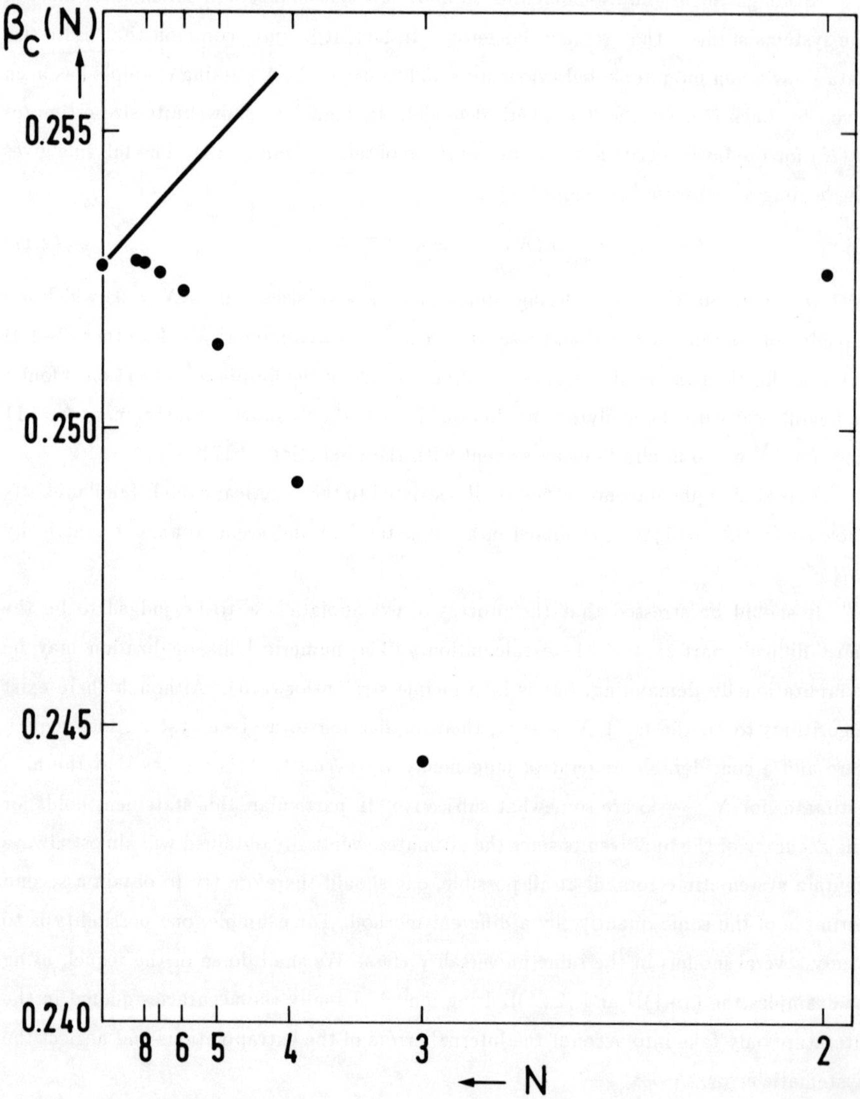

Fig.2: Finite-size estimates $\beta_c(N)$ for the inverse critical temperature in the 3D spherical model (after Luck [36]).

the sequence f_N is said to converge *linearly* towards its limit f. In statistical mechanics, this corresponds to systems away from criticality.

2. On the other hand, if

$$f_N = f + a_1 N^{-\omega_1} + a_2 N^{-\omega_2} + \dots \tag{2.19}$$

the sequence is said to be *logarithmically* convergent. This situation corresponds to critical systems, in particular the finite-size estimates for the critical point and the critical exponents obtained from eqs (2.13, 2.15, 2.16) fall into this class.

We shall concentrate here on the logarithmically convergent sequences and shall discuss two algorithms for their extrapolation. The first one was proposed by van den Broeck and Schwartz (VBS) [41] and first applied to critical phenomena by Hamer and Barber [42]. If in (2.19) the ω_i are *non-integer* numbers, practical experience shows that the VBS algorithm produces more accurate results than other known algorithms [43]. Most of the practical calculations to be discussed in later sections were performed using it.

A second algorithm, originally invented as a variant of Romberg integration by Burlirsch and Stoer (BST) [44], can be adapted to statistical mechanics problems. The first application to critical phenomena was made by Henkel and Patkos [45]. We shall discuss only these two algorithms [46]. For a detailed review of further extrapolation algorithms, see [47].

2.4.1. VBS - algorithm

The VBS algorithm is defined as follows. Take a *consecutive* sequence of lattices, that is, sequences like 0,2,6,3,... or 0,1,3,7,8,11,... should not be taken, but 0,2,4,6,... can be used. Form approximants $[N, M]$, arranged as a triangular table

$$\begin{array}{lll} [0,0] & & \\ [1,0] & [1,1] & \\ [2,0] & [2,1] & [2,2] \\ [3,0] & [3,1] & \\ [4,0] & & \end{array} \tag{2.20}$$

where the $[N, M]$ are defined recursively:

$$\begin{aligned} &[N,-1] = \infty \\ &[N,0] = f_N \\ &([N,M+1]-[N,M])^{-1} + \alpha([N,M-1]-[N,M])^{-1} \\ &= ([N+1,M]-[N,M])^{-1} + ([N-1,M]-[N,M])^{-1} \end{aligned} \tag{2.21}$$

The original sequence $[N,0]$ is transformed into a new one (e.g. $[N,1],...$) which is expected to converge faster towards the desired limit f. Special cases are the Padé table ($\alpha = 1$) or the Aitken-Shanks table ($\alpha = 0$). Barber and Hamer [42,43] studied the convergence of the VBS algorithm in some detail. In particular, if $\alpha = -1$, for a logarithmically convergent sequence (2.19) they find for the second iteration:

$$[N,2] = f + O(N^{-\omega'}) \tag{2.22}$$

where ω' is the minimum of $\omega_1 + 2$ and ω_2. From this result they suggest to use a different α for each step of the iteration, namely

$$\alpha_M = -\frac{1}{2}(1-(-1)^M) \tag{2.23}$$

2.4.2. BST - algorithm

Consider a sequence $h_N(N = 0,1,2,...)$ converging to zero as $N \longrightarrow \infty$. In many applications, $h_N = 1/N$, where N stands for the size of a finite system. There is no need for the h_N to be consecutive and N may even take non-integer values. Form a table of extrapolants

$$\begin{array}{ccccc} T_0^{(0)} & & & & \\ & T_1^{(0)} & & & \\ T_0^{(1)} & & T_2^{(0)} & & \\ & T_1^{(1)} & & & \\ T_0^{(2)} & & & & \end{array} \tag{2.24}$$

where

$$\begin{aligned} T_{-1}^{(N)} &= 0 \\ T_0^{(N)} &= T(h_N) = f_N \\ T_m^{(N)} &= T_{m-1}^{(N+1)} + \left(T_{m-1}^{(N+1)} - T_{m-1}^{(N)}\right)\left[\left(\frac{h_N}{h_{N+m}}\right)^{\omega}\left(1 - \frac{T_{m-1}^{(N+1)} - T_{m-1}^{(N)}}{T_{m-1}^{(N+1)} - T_{m-2}^{(N+1)}}\right) - 1\right]^{-1} \end{aligned} \tag{2.25}$$

and ω is a free parameter. This algorithm arises by approximating the function $T(h)$ by a sequence of rational functions of h^ω where the degree of the polynomial in the denominator is equal to or large by one than the degree of the polynomial in the numerator [44]. The sequence $T_1^{(N)}$ is expected to converge faster than the original sequence $T_0^{(N)}$ and so on.

If $T_0^{(N)} = f_N$ has the form (2.19), it is straightforward to show that [46]

$$T_1^{(N+1)} = f + O(N^{-\omega'}) \tag{2.26}$$

where ω' is the minimum of ω_2 and $2\omega_1$, provided $\omega = \omega_1$ is chosen. In practice, one chooses ω such that the different columns of (2.24) are as close as possible, e.g. one tries to minimize simultaneously the differences $T_m^{(N+1)} - T_m^{(N)}$ [46].

Comparing with (2.22), we see that the leading correction is already "absorbed" in the first BST-iteration, rather than the second for the VBS algorithm. On the other hand, the BST-algorithm also yields an estimate for the exponent of the leading correction term. In principle, this exponent should be equal to the leading irrelevant RG eigenvalue. Although this turns out to be true in many cases, it was pointed out that cancellations between various correction terms, if they occur, do alter the value of the (effective) exponent ω [48].

2.4.3. Discussion

A detailed comparison of the VBS and the BST algorithm was done by Henkel and Schütz [46], leading to the following conclusions:

1. Both algorithms have to be applied iteratively to yield reliable estimates.
2. The BST algorithm (besides being more flexible in applications) allows the computation of higher approximants and consequently more iterations than the VBS algorithm.
3. The BST algorithm is less sensitive to round-off errors than the VBS algorithm.
4. Both algorithms tend to "absorb" the leading corrections to the limit in one or two iterations, for the BST and VBS algorithm, respectively.
5. Both algorithms contain a free parameter (α or ω). For a large range of values of α and ω, the estimates of the limit are independent of this parameter.
6. Even in regions of stability with respect to α, examples show that the VBS algorithm can fail to converge to the theoretical prediction. There is no sign for such an effect for the BST algorithm. An example will be given below.

Consequently, the BST-algorithm should be considered as an alternative to the VBS algorithm. In particular, its usage should be advantageous in situations when only a very small number of finite lattices can be considered. This situation arises generically for (2+1)D systems and is also encountered it (1+1) dimensions in studying the higher energy levels. Note, however that these algorithms do not yield satisfactory results if *logarithmic* correction terms are present in the data (corresponding to "sub-logarithmic" convergence). Since there exist techniques to predict the appearance of logarithmic corrections in finite-size data [135,202], these terms can be subtracted explicitly, if present.

To close this section, we briefly discuss the possibility of a lacking convergence in the VBS algorithm. As an example, we use the determination of the exponent ν of the Z_3 quantum chain. (At the same time, this example serves to illustrate the conclusions listed above. For further examples, see [46].) Estimates $\rho_N(\varepsilon)$ for the exponent $1/\nu$ are obtained from the finite-lattice beta function β_N [42] (see section 3.1)

$$\rho_N(\varepsilon) = (N + \varepsilon)(\beta_{N-1} - \beta_N)/\beta_{N-1} \tag{2.27}$$

On expects $\rho_N(\varepsilon) \longrightarrow 1/\nu$ in the limit $N \longrightarrow \infty$, independent of the free parameter ε. Letting $\varepsilon \neq 0$ is sometimes referred to as "N–shift technique" [42]. In table 2.1, we give the results of the VBS algorithm, for $\varepsilon = 0$ and $\varepsilon = 0.7$, respectively [42]. Comparing with the expected value $1/\nu$=1.2 [49], we see that the VBS algorithm fails to converge to the correct result in the case $\varepsilon = 0$. This is probably due to the fact that the sequence $\rho_N(0)$ is not monotonous, but shows a minimum at $N \simeq 6$. Hamer and Barber [42] proposed to let $\varepsilon \neq 0$, which indeed yields a monotonous sequence. Moreover, the estimated limit is stable with respect to ε for $\varepsilon \gtrsim 0.4$ [42].

The results of the BST-algorithm, with $\varepsilon = 0$ and $\omega = 0.91$, are listed in table 2.2. Although the sequence $\rho_N(0)$ is non-monotonous, the final result for $1/\nu$ is as precise as the one obtained from the VBS with $\varepsilon = 0.7$. Thus, using the BST-algorithm, the N–shift technique [42] can be avoided, if so desired.

Table 2.1

a) $\varepsilon = 0.0$

1.276678				
1.197294	1.17915			
1.182527	1.17809	1.17848		
1.179118	1.17879	1.17896	1.17892	
1.178816	1.17903	1.17892	1.17896	1.17897
1.179506	1.17727	1.17832	1.17897	
1.180504	1.16260	1.16954		
1.181550	1.21223			
1.182581				

b) $\varepsilon = 0.7$

1.723515				
1.476662	1.34185			
1.389470	1.29529	1.20047		
1.344194	1.27184	1.20039	1.20054	
1.316345	1.25765	1.20025	1.19989	1.20002
1.297457	1.24812	1.20015	1.20016	
1.283798	1.24128	1.20016		
1.273459	1.23611			
1.265362				

VBS-extrapolants of the sequence $\rho_N(\varepsilon)$ of eq. (2.27) for $N = 2, 3, ..., 10$. (after Hamer and Barber [42]).

Table 2.2

1.276678							
	0.94873						
1.182527		1.17537					
	1.16421		1.18498				
1.179118		1.17845		1.15696			
	1.17715		1.20027		1.20012		
1.178816		1.17923		1.20002		1.19624	
	1.18411		1.20015		1.20031		1.20008
1.179506		1.17519		1.19704		1.19997	
	1.18286		1.19876		1.20011		
1.180504		1.14543		1.22498			
	1.19088		1.20621				
1.181550		1.65118					
	1.19293						
1.182581							

BST-extrapolants of the sequence $\rho_N(0)$ of eq. (2.27) as obtained with $\omega = 0.91$ for $N = 2, 3, ..., 10$.

3. THREE-DIMENSIONAL SYSTEMS

The techniques presented in section 2 are now applied to study the critical behaviour of (2+1)D systems. The same methods can of course (and have been) be used in (1+1)D as well. Here, very little will be said on these applications, (see [2]), the reason being that conformal invariance offers by far more efficient methods to describe two-dimensional statistical systems in the scaling region. This will be described in section 4.

In this section, the discussion concentrates mainly on the (2+1)D Ising model. The sheer size of the matrices involved has so far prevented studying other spin systems using finite-size scaling methods. In section 3.4, we describe possible alternatives to circumvent to massive computational problems of the conventional approach of section 2. The calculation of critical temperature, amplitudes and exponents of the (2+1)D Ising model illustrates the features of the finite-size scaling approach, the results for the exponents ranking among the most accurate results available (sections 3.1, 3.3). Some results for the $O(n)$ model are reviewed in section 3.2.

3.1 Ising model

Numerical studies of the (2+1)D Ising model give an excellent opportunity to check the power of the finite-size scaling method. Since results of the universal quantities of the 3D Ising universality class have been obtained from various different methods, comparison of these results does yield a valid impression of the reliability of the approaches used.

Consider the Hamiltonian, defined on a square $N \times N$ lattice [50]

$$H = -t \sum_n \sigma^z(n) - \frac{1}{2} \sum_{(n,n')} \{(1+\eta)\sigma^x(n)\sigma^x(n') + (1-\eta)\sigma^y(n)\sigma^y(n')\} \tag{3.1}$$

where nearest neighbour interactions are understood and σ^x, σ^y and σ^z are the Pauli matrices. For $\eta = 1$, H is obtained from the τ-continuum limit of the 3D Ising model. Although the case $\eta = 1$ is the one mainly studied, the model stays in the 3D Ising universality class if $\eta \neq 0$ (the global symmetry group is Z_2). For $\eta = 0$, one has a $U(1)$ symmetry and the critical behaviour is different. t plays the role of a temperature. Varying η, one obtains several systems in the same universality class. This can be used to check explicitly the universality of critical exponents and critical amplitudes. On the other hand, taking universality for granted, one obtains several independent results for the same quantity which should allow a realistic estimate of the accuracy of the results obtained.

H commutes with the charge operator

$$Q = \frac{1}{2}\left(1 - \prod_n \sigma^z(n)\right) \tag{3.2}$$

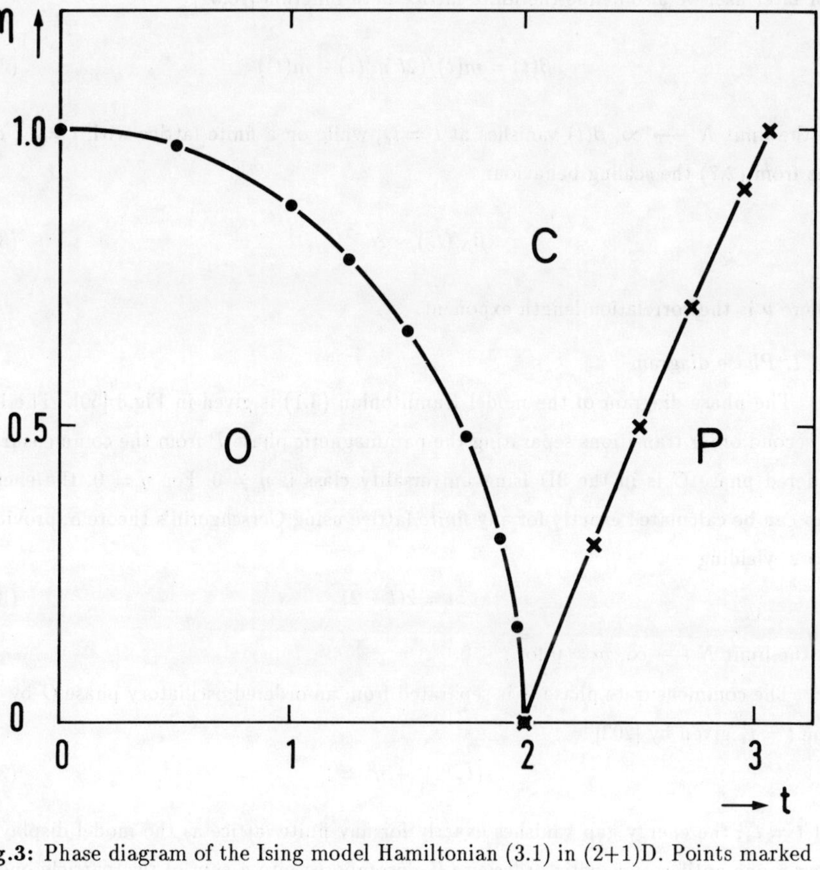

Fig.3: Phase diagram of the Ising model Hamiltonian (3.1) in (2+1)D. Points marked × are obtained from finite-size scaling. The dash-dotted line is given by (3.7).

and is thus decomposed into two blocks of charge 0 and 1, according to the eigenvalues of Q. Let $E_Q(t)$ the lowest eigenvalue of H in the sector Q. The energy gap is

$$m(t) = E_1(t) - E_0(t) \tag{3.3}$$

For later use, we also define the finite-lattice beta function [18,51]

$$\beta(t) = m(t)/(2t\, m'(t) - m(t)) \tag{3.4}$$

In the limit $N \longrightarrow \infty$, $\beta(t)$ vanishes at $t = t_c$, while on a finite lattice with $t = t_c$, one has from (2.7) the scaling behaviour

$$\beta_N(t_c) \sim N^{-1/\nu} \tag{3.5}$$

where ν is the correlation length exponent.

3.1.1. Phase diagram

The phase diagram of the model Hamiltonian (3.1) is given in Fig.3 [50]. The line of second order transitions separating the paramagnetic phase P from the commensurate ordered phase C is in the 3D Ising universality class if $\eta \neq 0$. For $\eta = 0$, the energy gap can be calculated exactly for any *finite* lattice using Gerschgorin's theorem, provided $t \geq 2$, yielding

$$m = 2(t - 2) \tag{3.6}$$

In the limit $N \longrightarrow \infty$, $m = 0$ for $t \leq 2$.

The commensurate phase C is separated from an ordered oscillatory phase O by the line $t = t_z$ given by [203]

$$(t_z/2)^2 + \eta^2 = 1 \tag{3.7}$$

At $t = t_z$, the energy gap vanishes exactly for any finite lattice as the model displays a long-range antiferromagnetic ordering (H decomposes into a sum of two-particle operators). This mechanism is completely analogous to the one found in the spin-s Heisenberg quantum chain [52]. While in the P and C phases the ground state eigenvector of H stays in the $Q = 0$ sector, it changes frequently between the two sectors in the oscillatory phase. In addition, there is a non-vanishing wave-vector K which modulates the correlation function. A finite-size scaling theory for this transition will be given in section 4.5.

It should be noted that the vanishing of the energy gap on a finite lattice does not imply any pathology. Rather, it should be regarded as a signal for an oscillatory or incommensurate phase. While in the transfer matrix formalism the correlation length always stays finite on finite lattices, taking the τ-continuum limit one neglects the terms in the Hamiltonian which would prevent the energy gap to vanish [203].

We now concentrate on the (2+1)D Ising line. The numerical determination of the critical point (and the critical exponents) is particularly difficult since only the square $N \times N$ lattices with $\mathcal{N} = N^2$ sites and $N = 2, 3, 4, 5$ are available. After having used invariance under translations, rotations and reflections to reduce the size of the Hamiltonian matrix by a factor $8N^2$ (see section 2.1), the $N = 5$ lattice still involves matrices of the size 86056 [18,19,31].

We shall describe the determination of the critical "temperature" $t_c(\eta)$ in some detail, since it illustrates how the convergence of the finite-size data can be improved before extrapolation [50,19]. Consider two types of boundary conditions:

1. periodic boundary conditions in both charge sectors (referred to as PP).
2. antiperiodic boundary condition in the $Q = 0$ sector and periodic ones in the $Q = 1$ sector (referred to as AP).

The choice of the AP is motivated by the observation that it does give the *exact* ($N \longrightarrow \infty$) gap on any finite lattice for the (1+1)D Ising model. Even if this should not hold in general, the convergence of the estimates of the critical parameters is improved, as first demonstrated for the Z_3 quantum chain [53].

	$\eta = 1.0$		$\eta = 0.7$	
N	PP	AP	PP	AP
2	3.841 10	4.800 00	3.926 37	4.431 75
3	3.164 56	2.999 79	2.711 80	2.715 68
4	3.084 05	3.022 97	2.732 45	2.709 74
5	3.060 95	3.036 05	2.726 31	2.713 59

Table 3.1. : Finite – size data for $t_c(\eta)$as obtained from solving (2.13) for both PP and AP boundary conditions (from Henkel [50].For $\eta = 1$ PP, the same results were obtained previously by Hamer [18]).

Considering the finite-size data t_N in table 3.1, one notes that t_N^{PP} is decreasing with N while t_N^{AP} is increasing. Thus the mean

$$\bar{t}_N = \frac{1}{2}\left(t_N^{PP} + t_N^{AP}\right) \tag{3.8}$$

shows a better convergence then t_N^{PP} or t_N^{AP} alone. Since there are so few lattices available, it is important to make the convergence of the raw data as good as possible. This is especially valid in this example since the VBS [41] and Lubkin [54] algorithms used to extrapolate [18,50] the data can be only applied once. Lacking the possibility of iterating them, these algorithms are unlikely to yield a large improvement on the raw data (see section 2.4).

Let us compare several results for $t_c(1)$, obtained by different methods. In table 3.2, we quote results from high-temperature series (HT) [55,56], low-temperature series (LT) [57] and finite-size scaling (FSS) using t_N^{PP} [18] or $\bar{t}_N$ [19,50]. Rather than entering here into a discussion on the precision of the different approaches used, we stress that all results are remarkably close together and do confirm each other. It is very satisfying to see that although the possibilities for extrapolating the finite-size data are very limited, the results obtained using finite-size scaling are at least as reliable as those found from series expansion.

Hamer and Irving	HT [55]	3.040 (9)
Hamer and Guttmann	HT [56]	3.044 (3)
Marland	LT [57]	3.044 (5)
Hamer	FSS [18]	3.040 (9)
Henkel	FSS [19]	3.047 (1)

Table 3.2 : Comparison of calculations of $t_c(1)$ on the square lattice in the $(2+1)$D Ising model. The numbers in brackets give the estimated uncertainty in the last digit.

The same impressive agreement of the values of $t_c(1)$ also holds for the triangular lattice [56].

It is well known that the (2+1)D Ising model is dual to the (2+1)D Ising gauge theory [58]. For comparison, we quote an estimate $t_c = 3.00(5)$ as obtained from series analysis [59] on the square lattice.

3.1.2. Critical Exponents

We now turn to the determination of the critical exponents of the (2+1)D Ising model. In particular, we shall be interested in comparing with results obtained on other realisations of the 3D Ising universality class. We shall see that universality is impressively well satisfied.

Let us recall how the critical exponents are determined. Estimates for the correlation length exponent ν and the susceptibility exponent γ are obtained from the beta function β_N (see (3.4) [18,31,50], or else the derivative dm/dt [50]) or the susceptibility χ_N, respectively; using the finite-size behaviour at $t = t_c$:

$$\beta_N \sim N^{-1/\nu}, \qquad \chi_N \sim N^{\gamma/\nu} \tag{3.9}$$

Then, from either (2.10) or (2.11), one obtains sequences converging to $1/\nu$ or γ/ν, respectively. Note that the estimates for the exponents do not depend on the final estimate for t_c. Rather, one can use in (3.9) the values for β_N or χ_N evaluated at the finite-size estimates t_N (i.e. the solutions of (2.8)). Since these values are by construction inside the scaling region, one can expect (3.9) to hold [2,28,18].

In order to estimate the exponent β, one calculates finite-size "magnetizations" M_N from Eq.(2.15) which scale as $M_N \sim N^{-\beta/\nu}$. As discussed in section 2.3, this involves the calculation of eigen*vectors* of the Hamiltonian.

While the determination of ν, γ and β can be done relatively easily, the determination of the specific heat exponent α from finite-size data has proved to be difficult. In principle, one should have a scaling of the specific heat $C_N \sim N^{\alpha/\nu}$, yielding estimates for α/ν. The results obtained, however, are $\alpha/\nu = 0.3$ [18] and 0.34 [31] for the square and triangular lattice respectively, to be compared with an expected value ~ 0.17. Since α/ν is rather small, it is possible that for the small lattices considered ($N \leq 5$), the finite-size behaviour is not dominated by the singular term but rather reflects the influence of some analytic contribution to the specific heat C_N, which are absent in the susceptibility χ or the correlation length ξ. This would mean that the lattices considered are far too small to allow a reliable determination of α/ν. (A similar problem also occurs in the (2+1)D Ising gauge theory [59]).

Henkel suggested to use different finite-size quantities to circumvent this problem [19]. The energy density operator is ($\mathcal{N} = N^2$ is the number of sites)

$$\varepsilon = \frac{1}{\mathcal{N}} \sum_n \sigma^z(n) \tag{3.10}$$

Let $|0\rangle$ and $|\varepsilon\rangle$ be the two lowest states in the $Q = 0$ sector. The matrix element $\langle 0|\varepsilon|\varepsilon\rangle$ should scale at $t = t_c$ as

$$\langle 0|\varepsilon|\varepsilon\rangle \sim N^{-x_\varepsilon} \tag{3.11}$$

with $x_\varepsilon = (1-\alpha)/\nu$. A second set of estimates for x_ε is obtained using finite-size data for the latent heat L (see Eq.(2.12)). It turns out that the two sequences for x_ε are increasing or decreasing with N for $\langle 0|\varepsilon|\varepsilon\rangle$ and L, respectively [19]. Taking the mean of the two sequences (see section 3.1.1.), one finally obtains $x_\varepsilon = 1.42(2)$ and, using the known value of ν (see below)

$$\alpha = 0.11(1) \tag{3.12}$$

in good agreement with other estimates (see table 3.3).

Let us compare the results for the critical exponents of the 3D Ising model obtained in various realisations of this universality class, as listed in table 3.3. For the isotropic 3D (classical) Ising model we quote results from high-temperature series for various lattices [60,61] (HT), a Monte Carlo renormalisation group study (MCRG) [62] and finite size scaling (FSS) using data from a specifically designed Ising processor [63]. Estimates using the field theory renormalisation group (RG)[64] rank among the most accurate results available. Recently, one has tried to improve on these estimates by constraining the analysis to reproduce the exactly known result in two dimensions [65]. Concerning the Hamiltonian limit, critical exponents have been calculated using high-temperature (HT) [55,56,68] and low-temperature (LT) series [57] (which yield the exponents α' and γ' which we take here to be equal to α and γ, respectively). Finite-size scaling (FSS) results have been obtained for the square [18,19,50] and the triangular [31] lattice. Further results include $\nu = 0.628(9)$ from low-temperature expansion [66] and $\gamma/\nu = 1.964(3)$ from Monte Carlo simulations [67], both in the isotropic 3D Ising model. A complementary approach involves the zeroes of the partition function in the complex activity plane. If $u_1(N)$ is the zero closest to the real positive $u-$axis and u_c the critical (thermal) activity, one has a scaling form $|u_1(N) - u_c| \sim N^{-1/\nu}$ [206]. From a numerical estimate of the partition function of the isotropic 3D Ising model, an estimate $\nu = 0.6295(10)$ was obtained [207].

Although the methods used to calculate the critical exponents are very different in character, the results are extremely close together. This means that the different kinds of assumptions implicitly contained in each method are sufficiently well satisfied in the 3D Ising model. In particular one should expect any exact calculation to reproduce precisely these results or else explain why all these numerical approaches should fail at the same time.

α	β	γ	ν	Method
0.12(2)	0.321*	1.238(3)	0.629(2)	HT [60]
0.10*	0.329*	1.239(3)	0.632(3)	HT [61]
0.113*	0.324*	1.239*	0.629(4)	MCRG [62]
0.11*	0.32*	1.25*	0.63(2)	FSS [63]
0.110*	0.3250(15)	1.2405(15)	0.6300(15)	RG [65]
0.098(3)	0.315(2)	1.2505(12)	0.634*	LT [57]
0.08*	0.34*	1.247(5)	0.64(2)	HT [55]
0.08*	0.34*	1.245(4)	0.640(5)	(s) HT [56]
0.08*	0.34*	1.242(2)	0.640(5)	(t) HT [56]
0.11(1)	0.324(9)	1.239*	0.629(2)	(s) FSS [19,50]
0.12*	0.322(6)	1.236(8)	0.627(4)	(t) FSS [31]

Table 3.3 : Critical exponents of the 3D Ising model. Entries marked with a star ($*$) are computed from other exponents using scaling relations (including hyperscaling) (s) and (t) refer to square and triangular lattices, respectively.

There are enough independent estimates of the critical exponents to allow for a test of hyperscaling. Freedman and Baker [69] have claimed that hyperscaling might be violated in the 3D Ising model. A careful study on the possible forms of hyperscaling violations has pointed out some loopholes to be avoided in numerical tests for hyperscaling [70]. The hyperscaling relations

$$\gamma + 2\beta = 3\nu, \qquad x_\varepsilon + 1/\nu = 3 \tag{3.13}$$

were checked for the triangular lattice [31] and the square lattice [19], respectively. From the numbers in table 3.3, one sees that (3.13) is satisfied with a precision of 0.3% , well within the uncertainties in the exponents.

To conclude, the results obtained for the critical exponents using the Hamiltonian limit rank among the most accurate methods available, especially when finite-size scaling techniques are used. What are the prospects to obtain yet more precise results? While in principle the existing HT-series or the ε-expansion (now known to order ε^5 [65]) can be pushed to yet higher orders, it appears at present out of reach to deal with the matrices of size 1.2 10^8 and storage requirements of 800 GBytes (!) necessary for the $N = 6$ lattice, although the reward would be considerable, since the improved possibilities for extrapolation could probably yield another accurate digit in the exponents. In section 3.4 we discuss briefly some suggestions for doing finite-size calculations on larger lattices.

We also mention that a RG study was tried in the (2+1)D three-states Potts model [71] yielding a first-order transiton.

3.2. O(n)-models

Although the (2+1)D Ising model is by far the most studied case, there do exist some results for the $O(n)$-model, namely $n = 2, 3$ and the "spherical" limit $n \longrightarrow \infty$. For the $O(n)$-model, the Hamiltonian is [56]

$$H = \sum_i \vec{J}^2(i) - x \sum_{(i,j)} \vec{n}(i) \cdot \vec{n}(j) - h \sum_i n_1(i) \tag{3.14}$$

where $\vec{n}(i)$ is an n-component spin "vector" at site i and $\vec{J}(i)$ is a "vector" angular momentum operator (with one or three components, respectively). Nearest neighbour interactions are understood, x is a temperature-like variable and h is a magnetic field. For $n = 2, 3$, the models were studied by calculating strong-coupling series for the mass gap or the susceptibility [72] which have been extended recently [56], on both the square and the triangular lattice. The series were analysed by comparing the results of seven different extrapolation algorithms by Hamer and Guttmann [56]. They find for $n = 2$

$$\begin{array}{llll} \text{(sq)} & x_c = 0.3480(16), & \nu = 0.676(10), & \gamma = 1.310(5) \\ \text{(tri)} & x_c = 0.2207(1), & \nu = 0.686(3); & \gamma = 1.334(5) \end{array} \tag{3.15}$$

for square (sq) and triangular (tri) lattice, respectively and for $n = 3$

$$\begin{array}{llll} \text{(sq)} & x_c = 1.080(5), & \nu = 0.724(4), & \gamma = 1.40(1) \\ \text{(tri)} & x_c = 0.6783(4), & \nu = 0.713(2), & \gamma = 1.395(5) \end{array} \tag{3.16}$$

which agree well with those obtained from the ϕ^4 field theory renormalisation group [64], thus again confirming universality.

In the $n \longrightarrow \infty$ limit, the $O(n)$ model goes into the exactly solvable spherical model [73]. The Hamiltonian reads [7] in $d = r + 1$ dimensions

$$H = -\frac{1}{2}\Delta + \frac{1}{2}\left(\chi\lambda r\vec{x}^2 - \frac{1}{2}\lambda\vec{x}M\vec{x}\right) \tag{3.17}$$

where $\vec{x}$ is an r-dimensional vector, Δ is the vector Laplacian, $\lambda = 2/T^2$ where T is the temperature, M is the interaction matrix describing the interactions in the r-dimensional hyperplanes (not necessarily nearest neighbour interactions) and χ is determined from the spherical constraint:

$$\tilde{N} = \langle 0 | \vec{x}^2 | 0 \rangle \tag{3.18}$$

where $\tilde{N}$ is the number of sites in the r-dimensional quantum plane.

The model can be solved exactly in terms of free bosonic oscillators and its critical behaviour is identical to the one of the isotropic version [7]. This even holds for more complicated cases like competing next-nearest neighbour interactions. In addition, the finite-size scaling behaviour was calculated [74] and is in complete agreement with the results obtained for the isotropic version [75,36]. This is one of the very few *analytical* confirmations of universality between the isotropic spin models and their Hamiltonian limits. The spherical model Hamiltonian becomes at $T = T_c$ on a square $(N)^r$ lattice [74], to leading order in $1/N$

$$H = \frac{1}{T_c}\frac{2\pi}{N}\sum_{\vec{k}}\left(\frac{y^2}{\pi^2}+\sum_{j=1}^{r}k_j^2\right)^{1/2} a_{\vec{k}}^{+}a_{\vec{k}} \tag{3.19}$$

where $\vec{k}$ is a r-dimensional vector whose components are integers or half-integer for periodic or antiperiodic boundary conditions, respectively. y is determined from the constraint (3.18) via the relation $y = N[(\chi-1)r/2]^{1/2}$. Note that the scaled one-particle energies are those of a massive particle. At the critical point $T = T_c$, one has for $r = 1+\varepsilon$, $y \sim \varepsilon$, for $r = 3-\varepsilon$, $y \sim \varepsilon^{1/3}$ [36] and for $r = 2$, $y \sim 0.1$. It is a universal number.

We close by noting that the problem of directed percolation can be mapped onto an eigenvalue problem of a non-hermitian Hamiltonian [76] (see section 4.5). Estimates for critical exponents were obtained in (2+1)D from high-temperature series [77].

3.3. Critical amplitudes

Having studied the critical exponents in some detail, we now turn to the second set of universal quantities, the critical amplitudes. In particular, one is interested in the scaling functions $Y(z_1, z_2)$ and $S_i(z_1, z_2)$ (see (2.7)) at the point $z_1 = z_2 = 0$. Since taking the τ-continuum limit yields an arbitrary, non-universal factor multiplying H (see section 1), one can, in the context examined here, only consider *ratios* of critical amplitudes.

As an example, we give the ratio A_ε/A_σ of the amplitudes A_σ, A_ε of the spin-spin and energy-energy correlation length ($i = \sigma, \varepsilon$) on the square lattice

$$\xi_i^{-1} = A_i N^{-1} \tag{3.20}$$

for the variant of the (2+1)D Ising model defined in Eq.(3.1), for both periodic and antiperiodic boundary conditions. (Recall that the finite-size scaling functions do depend on the shape and on the boundary conditions [2,29]). Varying η allows to check explicitly for universality [78]

η	1.0	0.9	0.7	0.5	0.3
Periodic	3.64	3.64	3.66	3.67	3.5
Antiperiodic	2.76	2.78	2.70	2.80	2.77

Table 3.4 : Ratios of amplitudes A_ε/A_σ (from Henkel [78]).

Indeed, the ratios A_ε/A_σ are seen to be independent of η (up to a few per cent, which is of the same order as the numerical errors), thus confirming universality. In fact, one can use one of the gaps to fix the non-universal normalisation factor and then check the universality of all other physical quantities [19,79,80].

At this point, another question arises. For (1+1)D systems, one knows that conformal invariance implies [81]

$$A_i = 2\pi\ x_i \tag{3.21}$$

where x_i is the exponent of the two-point correlation function

$$\langle \phi_i(r)\phi_i(0)\rangle \sim r^{-2x_i} \tag{3.22}$$

where ϕ_i is the $i-$th scaling field (e.g. spin or energy density). Before its explanation from 2D conformal invariance, Eq.(3.21) had been conjectured from both analytical and numerical results for a large variety of models (isotropic 2D as well as (1+1)D quantum chains) [110]. Does (3.21) also hold in higher dimensions?

Cardy [82] has shown the following. Define a quantum Hamiltonian on a sphere S^2 of radius R, rather than on a torus $S^1 \otimes S^1$ as usual. Then one has for the correlation lengths $\xi_i = R/x_i$, very much like (3.21). Alcaraz and Herrmann [83] tried to verify this for the Ising model by approximating the sphere S^2 by the five regular Platonic solids. They did not arrive at conclusive results, however, since each of the platonic solids used has a different number of links per site and therefore in some sense a different t_c.

On the other hand, one has $x_\varepsilon/x_\sigma = 2.76(2)$ (see section 3.1). Comparing with the numbers in table 3.4, Henkel observed that a linear relationship $A_i \sim x_i$ seems to hold, provided *antiperiodic* boundary conditions are taken [78] (as opposed to the periodic ones required in (1+1)D [81]). This surprising finding is further supported by an exact result in the (2+1)D spherical model with antiperiodic boundary conditions, where

$A_\varepsilon/A_\sigma = x_\varepsilon/x_\sigma = 2$ [74] (which by itself is not surprising, because of the free-field nature of the model) [84].

At present, there does not seem to be an argument to explain the above observation. There are some studies on the correlation functions in higher dimensions, using conformal invariance, [85], but these have not yet yielded some useful approach for numerical applications. Further studies on different models are needed to clarify the issue.

We close this section by describing possible checks on the universality of the finite-size scaling function Y and S_i (see Eq.(2.7)).

$$\begin{aligned} f^{(s)} &= N^{-d} Y\left(C_1 t\, N^{1/\nu},\, C_2 h\, N^{\Delta/\nu}\right) \\ \xi_i^{-1} &= N^{-1} S_i\left(C_1 t\, N^{1/\nu},\, C_2 h\, N^{\Delta/\nu}\right) \end{aligned} \tag{3.23}$$

(with the same notation as in section 2.3).

In analogy to the checks done for $S_i(0,0)$, one can also show that $Y(0,0)$ is universal by using free energy data. Antiperiodic boundary conditions are treated by calculating the surface tension $\tau = f^{(s)}_{\text{anti}} - f^{(s)}_{\text{peri}}$ [86]. All data available are in agreement with the expected universality [19]. A more stringent test involves calculating the magnetization M and the susceptibility χ

$$\begin{aligned} M &= N^{-\beta/\nu} C_2 \cdot Y'(0,0) &= N^{-\beta/\nu} a_M \\ \chi &= N^{\gamma/\nu}\, C_2^2 \cdot Y''(0,0) &= N^{\gamma/\nu} a_\chi \end{aligned} \tag{3.24}$$

where a prime denotes differentiation with respect to h. Now, a_M^2/a_χ should be universal, as was checked numerically [19]. Related calculations were performed in the 3D Ising model on the SC and BCC lattice for the free energy and the surface tension [87]. (Note, however, that Eqs.(3.23) and (3.24) assume that the normalisation of H is fixed in a unique way).

There is yet another method to check the universality of the scaling functions. Let $\phi(x)$ a scaling field, $|\phi\rangle = \phi(0)|0\rangle$ and define

$$\langle\psi|\phi(0)|\theta\rangle = N^{-x_\phi} a_{\psi\phi\theta} \tag{3.25}$$

where ψ, θ are further fields and x_ϕ is given by (3.21). Consider the quantities

$$C_{\varepsilon\sigma\sigma} = \frac{a_{\varepsilon\sigma\sigma}}{a_{0\sigma\sigma}}, \qquad C_{\sigma\varepsilon\sigma} = \frac{a_{\sigma\varepsilon\sigma}}{a_{0\varepsilon\varepsilon}} \tag{3.26}$$

where σ, ε are the magnetisation density and energy density, respectively [88]. For a quantum chain, these quantities are (universal) operator product expansion coefficients and one has $C_{\varepsilon\sigma\sigma} = C_{\sigma\varepsilon\sigma}$ [89]. In table 3.5, we give results obtained for periodic boundary conditions. Although universal, the C's do not appear to be related to some kind of operator product expansion since they are found to be not equal.

η	1.0	0.7	0.5
$C_{\varepsilon\sigma\sigma}$	0.850	0.855	0.848
$C_{\sigma\varepsilon\sigma}$	1.08	1.10	1.13

Table 3.5 : Finite – size scaling amplitudes as a function of η for the model (3.1) (from Henkel [19]).

How is this observation of universality related to the Privman-Fisher hypothesis? Decompose the Hamiltonian, e.g.

$$H = H_c + t\,\varepsilon \tag{3.27}$$

where H_c corresponds to the critical point. Then, writing $z = t\,N^{1/\nu}$ and $E_{0c} = \langle 0\,| H_c |\, 0\rangle$ one has from perturbation theory (see e.g. [80])

$$E_0 = E_{0,c} + \frac{1}{N}\left(a_{0\varepsilon 0}\cdot z + \sum_{\psi\neq 0} a_{0\varepsilon\psi} a_{\psi\varepsilon 0} g_{\psi}^{-1} z^2 + \cdots \right) \tag{3.28}$$

where g_ψ is the (universal) finite-size scaling amplitude of the energy gap of the scaling field ψ calculated at $t = 0$. In principle, one can check this way the universality of the entire series (3.28). For two-dimensional systems, this expression serves as a starting point for a systematic calculation of the finite-size scaling functions. Using the spectral representation, the $n-$th order term can be rewritten in terms of $(n-2)-$point correlation functions. At the 2D critical point, these correlation functions can be calculated from conformal invariance (see section 4.3).

3.4. Alternative Methods for the Calculation of Eigenvalues

Only the (2+1)D Ising model has so far been studied using finite-size scaling techniqes. The sheer size of the matrices involved has prevented the evaluation of eigenvalues on the $N = 6$ lattice for the Ising model and also the study of other systems. It would be desirable to have alternatives to the conventional Lanczos method. Indeed, several proposals have been made, but unfortunately, none of them has already been applied succesfully to a system in (2+1)D.

Irving and Hamer [90] suggested two methods of truncating the Hilbert space involved. The first ("N-cut-off") breaks up the Hamiltonian into two pices $H = H_0 + \lambda V$

and uses the eigenstates of H_0 as a basis. Calculating, say, the ground state energy, one retains only those states which are reached by applying V N times to the ground state of H_0. It is unlikely, "that the usual finite-size scaling techniques for obtaining critical parameters can be applied" [90].

The second scheme ("E_0-cut-off") only retains those states where eigenvalues of H_0 is below a given cut-off E_0 [90]. Patkos and Rujan [91] proposed an improvement, by using a variational estimate for the ground-state, rather than an eigenstate of H_0. They further proposed to avoid choosing the starting vector in the strong coupling regime (i.e. $\lambda \ll 1$) by the following scheme ("basis vector importance sampling") [91]: consider the series of Hilbert subspaces arising from subsequent applications of V to the $\lambda = 0$ ground state vector

$$\begin{gathered} \mathcal{H}_1 \subset \mathcal{H}_2 \subset \cdots \subset \mathcal{H}_m \subset \cdots \\ \mathcal{H}_m = V(\mathcal{H}_{m-1}) \cup \mathcal{H}_{m-1} \end{gathered} \tag{3.29}$$

Use the BMKD algorithm [27] (see section 2.2) to find the ground state vector ψ_m in $\mathcal{H}_m$. Then, retain a state $b_m \in \mathcal{H}_m$, only if $\langle \psi_m \mid b_m \rangle \geq r$, where r is a fixed constant (e.g. $r \simeq 10^{-3} - 10^{-5}$ [91]). Only the states b_m retained form the new $\mathcal{H}_m$.

This method was applied successfully to the (1+1)D $O(2)$ quantum chain [91,92]. It should also be feasible for calculating the $N = 6$ lattice in the (2+1)D Ising model [93], but results are not yet available.

Variational estimates for the ground state in combination with the Lanczos algorithm have also been studied [94] and are frequently used for applications in gauge theories. The accuracy of the method for spectrum calculations was tested for the Z_2 and Z_3 quantum chains by comparing with the known results of conformal invariance [95].

Recently, a new stochastic truncation method for quantum Hamiltonians was proposed and tested successfully for the (2+1)D Z_2 and U(1) gauge theories [205].

In an interesting attempt, Hamer [96] investigated the anisotropic $N \times N$ square lattice in the (2+1)D Ising model. For large anisotropies, eigenvalues correct to 10^{-7} were obtained on the $N = 7$ lattice. However, corrections to scaling and crossover terms due to the nearby (1+1)D critical point become so large that the accuracy of the estimates of the critical parameters is poor.

Finally, it was proposed to use rectangular $N_1 \times N_2$ lattices [97,18]. The effective linear dimension of the system could be obtained from $N = \sqrt{N_1 N_2}$ or else [97,98]

$$N = \sqrt{2} N_1 N_2 / \left(N_1^2 + N_2^2\right)^{1/2} \tag{3.30}$$

Considering $N_1 \neq N_2$ will probably be necessary in studying systems other than the Ising model. In any case, the BST-algorithm (see section 2.4) should be well suited for non-integer values of N.

3.5. Further Development

Finite-size scaling techniques applied to the Hamiltonian limit have yielded accurate estimates for critical parameters. The main limitation to the accuracy of the estimates is that the lattices available are small ($N \leq 5$ for the (2+1)D Ising model) and extrapolation techniques do not yet reach their full power on the short sequences available. While the conventional approach using the Lanczos algorithm certainly reached its limit, there are sufficiently promising alternatives, which could be used in performing successful calculations.

Besides improving the existing estimates of the (2+1)D Ising model, other systems like the Ashkin-Teller model should be studied. This could help to clarify whether there exists indeed a linear amplitude exponent relation (see (3.21)) for the correlation length in 3D. Another topic could be the study of systems with oscillating energy gaps.

4. TWO-DIMENSIONAL SYSTEMS

The critical behaviour of two-dimensional *isotropic* systems can be described in great detail by using the powerful techniques of conformal invariance. In section 4.1, we restate some basic results of 2D conformal invariance which are needed in practical applications. Some applications of conformal invariance to a few simple systems are presented in section 4.2. This cannot, however, be a complete account on all applications of conformal invariance in the Hamiltonian limit, rather, these examples should illustrate the principal ideas. For a full account on conformal invariance, one should consult the many excellent reviews available [e.g. 3,4,99,100,101,102,218].

In section 4.3, the neighbourhood of the critical point is studied. Conformal invariance allows the systematic calculation of finite-size scaling functions and corrections to scaling. A recent development goes even further and attempts to predict the masses of particles occurring in the off-critical theories having a factorizable S-matrix. Hamiltonian studies are ideal to check these predictions.

Rather than elaborating on more sophisticated applications of conformal invariance, in section 4.4 it is asked how much a perfect system may be distorted without loosing conformal invariance completely, using as an example the 2D Ising model with defect lines. Section 4.5 deals with systems with different correlation length exponents $(\nu_{/\!/}, \nu_{\perp})$

in different space-time directions. These systems are manifestly not conformal invariant albeit critical. They often show level crossings in the ground state energy and require extended finite-size scaling techniques. The type of transition found here is characteristic of the quantum nature of the model under consideration.

4.1 Conformal Invariance

We now briefly restate some basic results following from the hypothesis of conformal invariance at the critical point. This should therefore not be taken as a complete review on conformal invariance, which is a topic on its own. Excellent reviews are available in the literature [3,4,99,100,102,218].

Conformal invariance can only be applied to systems with short-ranged interactions which are (in the continuum limit) invariant under translations and rotations. Then scale invariance also implies conformal invariance. It is therefore a necessary condition for conformal invariance to hold that the system has a linear dispersion relation [103]

$$E = \left|\vec{k}\right| \tag{4.1}$$

(where the speed of light is normalized to unity) [104]. We stress that systems with anisotropic interactions (in particular quantum Hamiltonians!) can be conformal invariant provided their critical point is described by an isotropic theory. This is the case in most practical applications. (For other situations, see section 4.5.) The two-point correlation function of a scaling field $\phi(x)$ is at the critical point:

$$\langle\phi(x)\phi(y)\rangle = |x-y|^{-2x_\phi} \tag{4.2}$$

For example, if ϕ is the order parameter σ, then $x_\sigma = \beta/\nu$ and if ϕ is the energy ε, then $x_\varepsilon = (1-\alpha)/\nu$, where α, β, ν are the usual bulk critical exponents.

The particular relevance of 2D conformal invariance to critical phenomena was pointed out in the seminal paper of Belavin, Polyakov and Zamolodchikov [101]. The conformal algebra is given by the Virasoro algebra

$$[L_n, L_m] = (n-m)L_{n+m} + \frac{c}{12}\left(n^3 - n\right)\delta_{n+m,0} \tag{4.3}$$

The real parameter c is called the *central charge*. The representations of the Virasoro algebra are characterized by so-called *primary fields* $\phi(x)$, $|\phi\rangle = \phi(0)|0\rangle$ such that

$$\begin{aligned} L_n|\phi\rangle &= 0 \qquad \text{if } n > 0 \\ L_0|\phi\rangle &= \Delta_\phi|\phi\rangle \end{aligned} \tag{4.4}$$

where Δ_ϕ is the scaling dimension of ϕ. L_0 is related to the Hamiltonian H as obtained from the transfer matrix by the τ-continuum limit. For a quantum chain of N sites one has for free boundary conditions (see [81,111], the ground state energy is discussed below)

$$H = \frac{\pi}{N} L_0 \tag{4.5}$$

and for periodic boundary conditions

$$H = \frac{2\pi}{N} \left(L_0 + \bar{L}_0\right) \tag{4.6}$$

The $\bar{L}_n$ satisfy a Virasoro algebra with the same central charge c as the L_n and commute with all L_n. The complete excitation spectrum of the critical point Hamiltonian can therefore be computed from representation theory. In order to have a positive metric state, one requires *unitarity*

$$L_n^\dagger = L_{-n} \tag{4.7}$$

which guarantees the hermiticity of H. Excited states are built from the primary states, e.g.

$$L_{-r_1} L_{-r_2} \cdots |\phi\rangle, \qquad 0 < r_1 \leq r_2 \leq \cdots \tag{4.8}$$

(r_1, r_2 are integers). The critical exponents x_ϕ in (4.2) are obtained from [105]

$$\begin{aligned} x_\phi &= \Delta_\phi + r_1 + r_2 + \cdots = \Delta_\phi + R && \text{(free)} \\ x_\phi &= \Delta_\phi + \bar{\Delta}_\phi + r_1 + r_2 + \cdots + \bar{r}_1 + \bar{r}_2 + \cdots && \text{(periodic)} \\ &= \Delta_\phi + \bar{\Delta}_\phi + R + \bar{R} \end{aligned} \tag{4.9}$$

For periodic boundary conditions, the spin s_ϕ is

$$\begin{aligned} s_\phi &= \Delta_\phi - \bar{\Delta}_\phi + r_1 + r_2 + \cdots - \bar{r}_1 - \bar{r}_2 - \cdots \\ &= \Delta_\phi - \bar{\Delta}_\phi + R - \bar{R} \end{aligned} \tag{4.10}$$

Note that not all of the states (4.8) have to be independent. The degeneracies of the excited states can be calculated from the character functions of the Virasoro algebra [119]. (For explicit tables for $c = \frac{1}{2}, \frac{4}{5}$, see [118,124]). For unitary systems, the primary fields (4.4) are subject to many constraints which can be used to derive linear partial differential equations for their n-point correlation functions [101,105].

The requirement of unitarity imposes strong restrictions on the possible values of c and Δ_ϕ. These restrictions are described by the Kac-table. Unitarity is only possible in the two cases [107]

$$
\begin{aligned}
&1. \qquad c \geq 1 \quad \text{and} \quad \Delta_\phi \geq 0 \\
&2. \qquad c = 1 - \frac{6}{m(m+1)}, \qquad m = 2,3,4,5,\cdots \\
&\text{and} \qquad \Delta_\phi = \Delta_{pq} := \frac{[(m+1)p - mq]^2 - 1}{4m(m+1)} \\
&\qquad 1 \leq p \leq m-1, \quad 1 \leq q \leq p
\end{aligned}
\tag{4.11}
$$

Conversely, using the coset construction, one can also show that each of the cases listed in (4.11) is indeed unitary [141]. In table 4.1, we give the allowed values of Δ_ϕ for the cases $m = 3,4$ and 5, corresponding to the central charges $c = \frac{1}{2}, \frac{7}{10}$ and $\frac{4}{5}$. We also give one example of a quantum chain for each value of c, but stress that there normally do exist several distinct universality classes with the same value of c.

c	model	Δ_ϕ
$\frac{1}{2}$	Ising	$0, \frac{1}{16}, \frac{1}{2}$
$\frac{7}{10}$	tricritical Ising	$0, \frac{3}{80}, \frac{1}{10}, \frac{7}{16}, \frac{3}{5}, \frac{3}{2}$
$\frac{4}{5}$	Potts − 3	$0, \frac{1}{40}, \frac{1}{15}, \frac{1}{8}, \frac{2}{5}, \frac{21}{40}, \frac{2}{3}, \frac{7}{5}, \frac{13}{8}, 3$

Table 4.1 : Allowed values of Δ_ϕ from (4.11).

The set of the primary field with dimensions Δ_{pq}, $\bar{\Delta}_{pq}$ and all its excited states (4.8) is called the *conformal tower* and denoted by $(\Delta_{pq}, \bar{\Delta}_{pq})$.

In particular, for each given $c < 1$, the complete list of possible critical exponents is known! Let us see how to select the exponents realized in a given model from the (finite) list of possible values. Practical calculations are done, as first suggested by Cardy [81], by mapping the infinite 2D plane, described by a complex coordinate z, onto an infinitely long strip of finite width N by the mapping [81] (see also Fig.6)

$$
w = \frac{N}{2\pi} \ln z \tag{4.12}
$$

($w = u + iv$, $-\infty < u < \infty$, $0 < v < N$) with periodic boundary conditions. Note that taking the Hamiltonian limit automatically yields a system in the strip geometry.

At the critical point, the free energy per site is [112]

$$
\frac{E_0}{N} = a_0 + a_1 N^{-1} - \frac{\pi\zeta}{6} N^{-2} + \cdots \tag{4.13a}
$$

where ζ is related to the central charge:

$$\begin{aligned}\zeta &= c \quad \text{(periodic)}\\ \zeta &= c/4 \quad \text{(free)}\end{aligned} \tag{4.13b}$$

and a_0 and a_1 are (non-universal) constants ($a_1 = 0$ for periodic boundary conditions). The critical exponents are obtained from the spectrum (or the inverse correlation length ξ_i^{-1}), as shown by Cardy [81]

$$\begin{aligned}\xi_i^{-1} &= E_i - E_0 = 2\pi\ x_i\ N^{-1} \quad \text{(periodic)}\\ \xi_i^{-1} &= E_i - E_0 = \pi\ x_i^{(s)} N^{-1} \quad \text{(free)}\end{aligned} \tag{4.14}$$

(compare with (4.9)). For periodic boundary conditions, x_i is related to bulk critical exponents, while for free boundary conditions, $x_i^{(s)}$ is related to surface exponents [108]. Eqs.(4.13) and (4.14) define a powerful method to determine the *complete* operator content of a given statistical system, that is, one obtains the leading as well as *all* non-leading exponents. Practical examples are given in the next section [109]. Eq.(4.14) had been conjectured previously from analytical or numerical results of many systems [110], both classical spin systems and in their Hamiltonian limit.

There is a caveat, however. Usually one takes the dispersion relation to be given by (4.1) with the speed of light set equal to one. This implies a certain choice of the normalization of the quantum Hamiltonian which was implicitly assumed throughout this section. A possible way to fix the normalization is to use the result $x_\varepsilon^{(s)} = 2$ in Eq.(4.14) [111]. A useful alternative consists in calculating the proper normalization of H from *several* different conformal tower spacings [111].

The possibilities of combining fields with dimensions $(\Delta_{pq}, \bar{\Delta}_{pq})$ are further restricted by the requirement that the critical point partition function Z (defined on the torus) is modular-invariant [113,114]. For example, for $c = \frac{1}{2}$, only the combinations $(0,0)+(\frac{1}{2},\frac{1}{2})$, $(\frac{1}{16},\frac{1}{16})$ and $(0,\frac{1}{2})+(\frac{1}{2},0)$ are allowed. A classification of the modular invariant partition functions was obtained by Capelli, Itzykson and Zuber [114]. However, already for $m = 4$ (see (4.11)) it was pointed out that there exist several different models with the same modular invariant partition function [115]. Although these models contain scaling fields with the same critical exponents, at least some of these fields appear in different sectors of the model. That means that modular invariance of Z on the torus alone does not yet yield a complete classification of 2D critical models.

4.2. Application to Simple Systems

In order to illustrate the techniques of conformal invariance, we shall now apply them to some simple systems. It should be noted that conformal invariance alone does not "solve" a model. Rather, it provides a scheme to analyze, say, finite-size data which may have been obtained by the methods described in section 2 or, alternatively, from Bethe ansatz techniques. The systematic application of these conformal techniques was started by Gehlen, Rittenberg and Ruegg [111] and further developed by many authors.

4.2.1. Ising model

Consider the quantum Ising chain defined by [116]

$$H = -\frac{t}{2\eta}\sum_{n=1}^{N}\sigma^z(n) - \frac{1}{4\eta}\sum_{n=1}^{N}\{(1+\eta)\sigma^x(n)\sigma^x(n+1) + (1-\eta)\sigma^y(n)\sigma^y(n+1)\} \quad (4.15)$$

where N is the number of sites, t takes the role of temperature,η is a free parameter and σ^x, σ^y, σ^z are Pauli matrices. If $\eta \neq 0$, there is at $t_c = 1$ a transition in the Ising universality class [117]. Note that H is already normalized in order to have the correct dispersion relation [111,118]. The boundary conditions are given by (σ is either σ^x or σ^y)

$$\sigma(N+1) = (-1)^{\tilde{Q}}\sigma(1) \quad (4.16)$$

H commutes with the charge operator

$$Q = \frac{1}{2}\left(1 - \prod_{n=1}^{N}\sigma^z(n)\right) \quad (4.17)$$

and can thus be decomposed into the blocks $H_Q^{(\tilde{Q})}$. Let $E_0^{(0)}$ the lowest energy of $H_0^{(0)}$ and $E_Q^{(\tilde{Q})}(P,r)$ the $r-$th eigenvalue of $H_Q^{(\tilde{Q})}$ with momentum P. Then compute the amplitudes

$$\varepsilon_Q^{(\tilde{Q})}(P,r) = \frac{N}{2\pi}\left(E_Q^{(\tilde{Q})}(P,r) - E_0^{(0)}\right) \quad (4.18)$$

The $\varepsilon_Q^{(\tilde{Q})}$ are related to the Virasoro algebra as follows [118]. Let $r = r_1 + r_2 + \cdots$ in Eq.(4.8) and let $d(\Delta_\phi, r)$ denote the degeneracy of this state (see table 4.2). These numbers can be obtained from the known character functions of the Virasoro algebra [119].

Δ_ϕ	r=0	1	2	3	4	5	6	7	8
0	1	0	1	1	2	2	3	3	5
$\frac{1}{2}$	1	1	1	1	2	2	3	4	5
$\frac{1}{16}$	1	1	1	2	2	3	4	5	6

Table 4.2 : Some values of $d(\Delta_\phi, r)$, for $c = \frac{1}{2}$.

A primary field $(\Delta_\phi, \bar{\Delta}_\phi)$ generates the spectrum

$$\begin{aligned} \varepsilon(P) &= (\Delta_\phi + r) + (\bar{\Delta}_\phi + \bar{r}) \\ P &= (\Delta_\phi + r) - (\bar{\Delta}_\phi + \bar{r}) + \frac{1}{2}\tilde{Q} \end{aligned} \tag{4.19}$$

with degeneracy $d(\Delta_\phi, r) \cdot d(\bar{\Delta}_\phi, \bar{r})$.

As is well known, H can be diagonalized in terms of fermionic operators [116]

$$\begin{aligned} H &= \sum_{m=0}^{N} \Lambda(\varphi(m)) \left(\eta_m^+ \eta_m - \frac{1}{2} \right) \\ \Lambda(\varphi) &= \left[(t - \cos\varphi)^2 + \eta^2 \sin^2\varphi \right]^{1/2} \Big/ \eta \end{aligned} \tag{4.20}$$

and the "momentum" $\varphi(m)$ is discretized by

$$\begin{aligned} \varphi(m) &= (2m+1)\pi/N && \text{if} \quad Q + \tilde{Q} = 0, 2 \\ \varphi(m) &= 2m\pi/N && \text{if} \quad Q + \tilde{Q} = 1 \end{aligned}$$

At $t = t_c = 1$, the ground-state energy per site is [118]

$$\frac{E_0^{(0)}}{N} = A_0(\eta) - \frac{\pi}{12N^2} + \frac{7\pi^3}{960N^4} \left(\frac{1}{\eta^2} - \frac{4}{3} \right) + O\left(N^{-6}\right) \tag{4.21}$$

where $A_0(\eta)$ is a calculable non-universal function. From (4.13) we read off $c = 1/2$, as expected [101]. We see explicitly that the $O\left(N^{-2}\right)$ term is $\eta-$independent and thus universal. Further, the spectra are given by (compare with table 4.1)

$$\begin{aligned} H_0^{(0)} &= (0,0) + \left(\frac{1}{2}, \frac{1}{2} \right) \\ H_1^{(0)} &= H_0^{(1)} = \left(\frac{1}{16}, \frac{1}{16} \right) \\ H_1^{(1)} &= \left(0, \frac{1}{2} \right) + \left(\frac{1}{2}, 0 \right) \end{aligned} \tag{4.22}$$

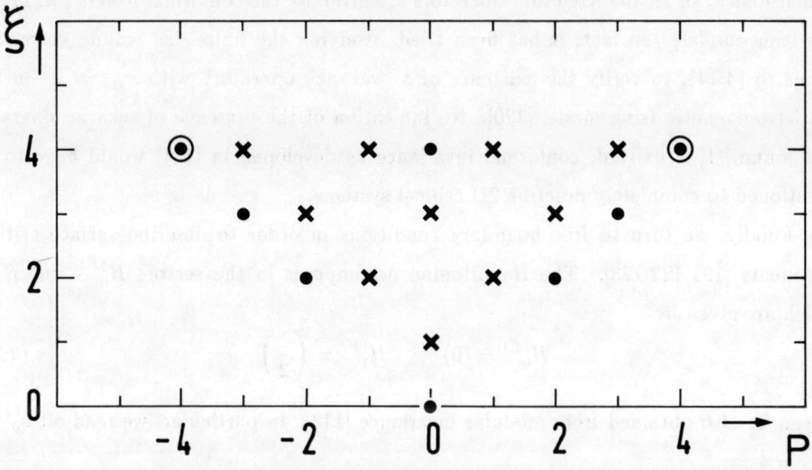

Fig.4: Low-lying scaled spectrum of the Ising quantum chain. The states marked by $\cdot$ and $\times$ are in the conformal towers of (0,0) and $\left(\frac{1}{2}, \frac{1}{2}\right)$. $\odot$ is a double degenerate state in the tower (0,0).

which are just the combinations allowed by modular invariance for $c = \frac{1}{2}$ (see section 4.1). In Fig.4, the low-lying spectrum of $H_0^{(0)}$ is shown. Obviously, the linear dispersion relation (4.1) is satisfied. This also follows immediately from (4.20), since $\Lambda(\varphi) = \varphi + \cdots$. Physically, $H_0^{(0)}$ contains the energy operator ε, $H_1^{(0)}$ and $H_0^{(1)}$ contain the order (σ) and disorder (μ) operator, respectively, and $H_1^{(1)}$ contains the spinor ψ and the antispinor $\bar{\psi}$. In particular, from (4.14) one has $x_\varepsilon = 1$, $x_\sigma = x_\mu = 1/8$ and $x_\psi = x_{\bar{\psi}} = 1/2$, as it should be (recall that $x_\varepsilon = \varepsilon_0^{(0)}(0,1)$, $x_\sigma = \varepsilon_1^{(0)}(0,1)$, $x_\psi = \varepsilon_1^{(0)}(\frac{1}{2},1)$).

While the bulk exponents are thus obtained, conformal invariance states that the dimensions x_i of *all* the irrelevant operators are given by the conformal towers (4.22) (for the Ising model). In fact, it has been tried, studying the finite-size scaling correction terms to (4.14), to verify the existence of a "vacancy operator" with $x_{\text{vac}} = \frac{10}{3}$ in the antiferromagnetic Ising model [120]. No indication of the presence of such an operator was found. If it existed, conformal invariance as developed in [101] would have to be questioned to completely describe 2D critical systems.

Finally, we turn to free boundary conditions in order to describe surface critical exponents [121,122,123]. The Hamiltonian decomposes in the sectors $H_0^{(F)}$ and $H_1^{(F)}$ which are given by

$$H_0^{(F)} = (0) \ , \quad H_1^{(F)} = \left(\frac{1}{2}\right) \tag{4.23}$$

as can be also obtained from modular invariance [113]. In particular, we read off $x_\sigma^{(s)} = \frac{1}{2}\eta_{/\!/} = \frac{1}{2}$.

4.2.2. Three-states Potts model

The correctly normalised Hamiltonian is [124]

$$H = -\frac{2}{3\sqrt{3}} \sum_{i=1}^{N} \{\sigma_i + \sigma_i^+ + \lambda(\Gamma_i \Gamma_{i+1}^+ + \Gamma_i^+ \Gamma_{i+1})\} \tag{4.23}$$

where the 3×3 matrices σ_j, Γ_k satisfy (see section 1)

$$\begin{aligned} \sigma_j \Gamma_k &= \delta_{jk}\omega\ \Gamma_k \sigma_j + (1 - \delta_{jk})\Gamma_k \sigma_j \\ \Gamma_k^3 &= \sigma_k^3 = 1, \qquad \omega = \exp(2\pi i/3) \end{aligned} \tag{4.24}$$

The critical point is at $\lambda_c = 1$. The central charge of the three-states Potts model is expected to be $c = 4/5$ [106], which is confirmed from numerical finite-size data for both periodic and free boundary conditions using (4.13) [124].

For free boundary conditions, one has the sectors $H_{0,\pm}^{(F)}$, $H_{1,\pm}^{(F)} = H_{2,\pm}^{(F)}$ where "±" labels states of positive or negative parity. One has [122]

$$H_{0,+}^{(F)} = (0),\ H_{0,-}^{(F)} = (3),\ H_{1}^{(F)} = H_{1,+}^{(F)} + H_{1,-}^{(F)} = \left(\frac{2}{3}\right) \tag{4.25}$$

Table 4.3 shows some numerical values to illstrate the precision by which (4.25) is confirmed. Eq. (4.25) was also obtained by modular invariance [113], and is consistent with previous estimates for the surface exponent $\eta_{\|} = 4/3$ [111], [123].

$\Delta + r$	$d\left(\frac{2}{3}, r\right)$	$\varepsilon_{1,+}^{(F)}$	$\varepsilon_{1,-}^{(F)}$
0.666	1	0.6662(4)	–
1.666	1	–	1.668(2)
2.666	2	2.66(1), 2.68(4)	–
3.666	2	–	3.64(4), 3.66(2)
4.666	4	4.65(4), 4.66(3), 4.68(3), 4.67(2)	–
5.666	5	–	5.58(8), 5.65(7), 5.65(5), 5.66(6), 5.66(4)
6.666	8	6.6(2), 6.55(10)	–
7.666	10	–	> 7.5(?)

Table 4.3: Numerical estimates for the amplitudes $\varepsilon_{1,\pm}^{(F)}$. The first column gives the expected exponent, the second column the expected degeneracy of the state. The notation is the same as in section 4.2.1. (after Gehlen and Rittenberg [122])

Now, consider the boundary conditions preserving the Z_3−symmetry (for $\tilde{Q} = 0$ the symmetry group is S_3) and compatible with translation invariance

$$\Gamma_{N+1} = \omega^{\tilde{Q}}\ \Gamma_1, \qquad \tilde{Q} = 0, 1, 2 \tag{4.26}$$

which yields the sectors $H_Q^{(\tilde{Q})} = H_{\tilde{Q}}^{(Q)}$ (and also $H_1^{(\tilde{Q})} = H_2^{(\tilde{Q})}$); $\tilde{Q} = 0, 1, 2$. The states of $H_0^{(0)}$ can be further decomposed by charge conjugation. Eq.(4.19) can still be used, provided $\frac{1}{2}\tilde{Q}$ is replaced by $\frac{1}{3}\tilde{Q}$. The five independent sectors are [124]

$$\begin{aligned}
H_{0,+}^{(0)} &= (0,0) + \left(\frac{2}{5},\frac{2}{5}\right) + \left(\frac{7}{5},\frac{7}{5}\right) + (3,3) \\
H_{0,-}^{(0)} &= \left(\frac{2}{5},\frac{7}{5}\right) + \left(\frac{7}{5},\frac{2}{5}\right) + (3,0) + (0,3) \\
H_{1}^{(0)} &= \left(\frac{1}{15},\frac{1}{15}\right) + \left(\frac{2}{3},\frac{2}{3}\right) \\
H_{1}^{(1)} &= \left(\frac{2}{5},\frac{1}{15}\right) + \left(0,\frac{2}{3}\right) + \left(\frac{7}{5},\frac{1}{15}\right) + \left(3,\frac{2}{3}\right) \\
H_{1}^{(2)} &= \left(\frac{1}{15},\frac{2}{5}\right) + \left(\frac{2}{3},0\right) + \left(\frac{1}{15},\frac{7}{5}\right) + \left(\frac{2}{3},3\right)
\end{aligned} \tag{4.27}$$

One recognizes from $H_{0,+}^{(0)}$ $x_\varepsilon = \frac{4}{5}$ and from $H_1^{(0)}$ $x_\sigma = \frac{2}{15}$, as expected [49,111,123]. These results agree with those from modular invariance [113].

Finally, one can also consider boundary conditions which preserve only a Z_2 global symmetry [115].

$$\Gamma_{N+1} = \omega^{\tilde{Q}}\Gamma_1^+ \; ; \qquad \tilde{Q} = 0,1,2 \tag{4.28}$$

This gives blocks $K_Q^{(\tilde{Q})}$ (with $Q = 0,1$ only) and $K_Q^{(0)} = K_Q^{(1)} = K_Q^{(2)}$. One has [115], where the ground state energy is still the lowest level of $H_{0,+}^{(0)}$

$$\begin{aligned}
K_0^{(0)} &= \left(\frac{1}{40},\frac{1}{40}\right) + \left(\frac{1}{8},\frac{1}{8}\right) + \left(\frac{21}{40},\frac{21}{40}\right) + \left(\frac{13}{8},\frac{13}{8}\right) \\
K_1^{(0)} &= \left(\frac{21}{40},\frac{1}{40}\right) + \left(\frac{1}{40},\frac{21}{40}\right) + \left(\frac{1}{8},\frac{13}{8}\right) + \left(\frac{13}{8},\frac{1}{8}\right)
\end{aligned} \tag{4.29}$$

It is interesting to note that by choosing the two sets of boundary conditions (4.26), (4.28), one gets two mutually exclusive sets of critical exponents (see also table 4.1). For a more systematic study, see [17].

4.2.3. Tricritical Ising model

One of the possible realizations of the 2D tricritical Ising universality class [125] is given by the Blume-Capel quantum chain [126,127]

$$H = -\sum_{i=1}^{N} \{S^z(i)S^z(i+1) - \alpha(S^z(i))^2 - \beta S^x(i)\} \tag{4.30}$$

where S^x, S^z are spin-one matrices

$$S^x = \frac{1}{\sqrt{2}}\begin{pmatrix} 0 & 1 & 0 \\ 1 & 0 & 1 \\ 0 & 1 & 0 \end{pmatrix}, \qquad S^z = \begin{pmatrix} 1 & 0 & 0 \\ 0 & 0 & 0 \\ 0 & 0 & -1 \end{pmatrix} \tag{4.31}$$

and the charge operator is

$$Q = \frac{1}{2}\left(1 - \prod_{i=1}^{N}\left[(S^x(i))^2 - 1\right]\right) \tag{4.32}$$

and one has the sectors $H_Q^{(\tilde{Q})}$ $(Q, \tilde{Q} = 0, 1)$ and $H_Q^{(F)}$ (as for the Ising model, see section 4.2.1).

The tricritical point (α_t, β_t) is found as follows, adapting a well known method used for isotropic tricritical systems [128]. The line of first order transitions close to the tricritical point is characterized by the coexistence of two ordered and one disordered phase which become indistinguishable at the tricritical point. One should therefore define two correlation lengths (using the notation of section 4.2.1)

$$\begin{aligned} \xi^{-1} &= E_1^{(0)}(0,1) - E_0^{(0)} \\ \tilde{\xi}^{-1} &= E_0^{(0)}(0,2) - E_0^{(0)} \end{aligned} \tag{4.33}$$

corresponding to the two lowest gaps of H and the tricritical point is found by solving simultaneously the two equations (see (2.13))

$$\begin{aligned} N\xi^{-1}(N, \alpha_t, \beta_t) &= (N+1)\xi^{-1}(N+1, \alpha_t, \beta_t) \\ N\tilde{\xi}^{-1}(N, \alpha_t, \beta_t) &= (N+1)\tilde{\xi}^{-1}(N+1, \alpha_t, \beta_t) \end{aligned} \tag{4.34}$$

with the result $\alpha_t = 0.910207(4)$, $\beta_t = 0.415685(6)$[219].

Note that the Hamiltonian (4.30) is not yet properly normalised so that the speed of light must be determined numerically [127,219]. Then, using (4.13) (with periodic boundary conditions), the central charge was estimated $c = 0.70(2)$ [127] in good agreement with the expectation $c = 7/10$. The operator content is (compare with table 4.1)

$$\begin{aligned} H_0^{(F)} &= (0), \quad H_1^{(F)} = \left(\frac{3}{2}\right) \\ H_0^{(0)} &= (0,0) + \left(\frac{1}{10}, \frac{1}{10}\right) + \left(\frac{3}{5}, \frac{3}{5}\right) + \left(\frac{3}{2}, \frac{3}{2}\right) \\ H_1^{(0)} &= \left(\frac{3}{80}, \frac{3}{80}\right) + \left(\frac{7}{16}, \frac{7}{16}\right) \\ H_1^{(1)} &= \left(0, \frac{3}{2}\right) + \left(\frac{3}{2}, 0\right) + \left(\frac{1}{10}, \frac{3}{5}\right) + \left(\frac{3}{5}, \frac{1}{10}\right) \end{aligned} \tag{4.35}$$

The values of the critical exponents to be read from (4.35) agree nicely with previous results [125]. It should be noted that the spectrum (4.35) is also a representation of the $N = 1$ superconformal algebra [129].

4.2.4. Other systems

Rather than giving a detailed exposition of all possible applications of conformal invariance, we shall limit ourselves to briefly mention just some further examples. The selection is somewhat arbitrary and certainly not complete. The techniques of conformal invariance, in particular eqs. (4.13,4.14) have become common as an additional technique besides conventional finite-size scaling to determine the universality class. Still, finding the central charge c is not completely conclusive, since there may be several distinct universality classes having the same c (and even the same modular invariant partition function [115]).

Several variants of the quantum Ising chain have been studied [130], in particular the Ising chain defined on the Fibonacci lattice (a 1D prototype of a quasicrystal) was solved [131], having the same operator content as the normal Ising chain. Another possibility is to study quantum hard rods, which are Ising chains with up to $m-$th neighbour exclusion. For $m = 2$ one has again Ising critical behaviour [132]. These systems are closely related the Ising chains with multi-spin interactions, which may be more realistic models for applications either in solid state or liquid physics. Hamiltonians of the type [133]

$$H = -\sum_{n=1}^{N} \{\sigma^z(n)\sigma^z(n+1)...\sigma^z(n+m-1) - t\sigma^x(n) - h\sigma^z(r)\} \tag{4.36}$$

have been studied. For $h = 0$, these systems are self-dual [134]. For $m = 2$, one recovers the Ising model, while for $m = 3$ ($h = 0$) the system is in the four-states Potts universality class. It is not clear yet whether these results depend on the spin of the lattice variables. The numerical analysis is very difficult, since logarithmic corrections arise in the finite-size data, which must be explicitly subtracted from the data, using predictions from conformal invariance [135]. For $h < 0$ and $m = 3$, the transition is in the three-states Potts universality class [136]. These findings have let to hypothesize that the critical behaviour depends solely on the ground state degeneracy.

Another intensively studied system is the Ashkin-Teller quantum chain [16,137]. Since it has $c = 1$, the critical exponents may change continuously with the parameters of the model and in fact they do, as it is well known. The phase diagram of the Ashkin-Teller quantum chain was studied in considerable detail. It contains two critical lines (dual to each other) in the 2D Ising universality class which meet at a four-state Potts critical point. Along the self-dual critical line joining the Potts point and the Kosterlitz-Thouless transition point critical exponents vary continuously. Numerical finite-size data have led to the identification of the complete operator content, while Bethe-ansatz techniques have allowed to calculate analytically finite-size scaling amplitudes and corrections [138] (on the self-dual line).

The operator content was also obtained for the self-dual transitions of the D_n−symmetric n−states quantum chains [139]. As for the Ashkin-Teller quantum chain, the numerical results are confirmed by being compatible with modular invariance [140].

It is a common phenomenon for 2D critical systems with $c \geq 1$ that the conformal algebra is larger than the Virasoro algebra. As additional symmetries Kac-Moody algebras, $N = 1$ and $N = 2$ superconformal algebras [141] and Zamolodchikov-Fateev [142] invariance were found in some points of the phase diagram. A puzzling example was given by Schütz [143] using a D_6−symmetric 6-states quantum chain. From numerical data it is claimed that the central charge and critical exponents might change without the presence of a marginal operator !

The study of these quantum chains yields that "a universality class is not characterized by some critical exponents, but rather by specific sets of exponents, namely sets which lead to modular invariant partition functions of the system" [143]. One needs thus a more precise notion of the universality classes of 2D critical systems. The non-self-dual transitions in the Z_n quantum chain was also studied [92], one finds a conventional second order transition for $n = 5$ (with an exponent $\nu = 2.19$), while one has Kosterlitz-Thouless type transitions for $n = 7, 9$, with an exponent $\sigma \simeq 0.1$, where σ is defined by

$$\beta(t) \sim (t - t_c)^{1+\sigma} \tag{4.37}$$

where t is the temperature and β the beta function (see section 3) which implies for the energy gap the form $m \sim \exp(-c/(t - t_c)^\sigma)$, rather than a conventional power law.

A very detailed study was presented for the O(2) quantum chain [32]. Finite-size scaling data were used to determine $\sigma = 0.501(5)$ in excellent agreement with the Kosterlitz-Thouless prediction $\sigma = \frac{1}{2}$ [144]. The order parameter correlation exponent η was determined as a function of temperature and found to agree well with low-temperature series results. The data are consistent with the expected value for the central charge $c = 1$.

As an example for quantum chains with higher spin S, the Takhtajan-Babujian model was studied [145]. Its continuum limit was found to be described by the $k = 2S$ Wess-Zumino-Witten model. The numerical analysis is complicated due to the presence of logarithmic finite-size corrections, which can be calculated by Bethe-ansatz techniques.

Finally, we mention that antiferromagnetic SU(n) quantum chains were shown to be equivalent to n^2−state Potts quantum chains [146].

A generic model is the XXZ quantum chain

$$H = -\frac{\gamma}{2\pi\sin\gamma}\sum_{n=1}^{N}\{\sigma^x(n)\sigma^x(n+1)+\sigma^y(n)\sigma^y(n+1)-\cos\gamma\ \sigma^z(n)\sigma^z(n+1)\} \quad (4.38)$$

with $0 \leq \gamma \leq \pi$. For a detailed review, see [100]. This model contains all the spectra of all (unitary) $c < 1$ systems. The projection rules were given explicitly by Alcaraz, Grimm and Rittenberg [115]. Physically, this corresponds to the fact that the 2D Coulomb gas acts as a general frame in 2D critical phenomena [147] as was made explicit by di Francesco, Saleur and Zuber by rewriting the modular invariant partition functions of [114] in terms of coulombic partition functions [148]. This can be also restated by linking the XXZ chain to the $RSOS$ model of Andrews, Baxter and Forrester [149]. Pasquier suggested an explicit lattice derivation of modular invariant partition functions [150]. The spectrum of (4.38) can be studied in great detail. Analytic results are available using Bethe ansatz techniques as developed by de Vega, Eckle, Hamer, Karowski, Woynarovich and others. We are not going to describe this in detail, but refer to the literature [100]. Recently, the construction of RSOS critical integrable models was extended from systems based a SU(2) (like the XXZ chain) to SU(N) [209].

Geometric phase transitions have also been studied. Critical exponents were calculated and observed to fit into the Kac-table, although the models are not unitary. For a detailed description, see [210].

4.3. Systems close to the critical point

We now consider systems close to the critical point. First, we discuss finite-size corrections and the explicit calculation of finite-size scaling functions from conformal invariance for small values of the scaling variables. Then, we report results for the other extreme case of large scaling variables, which are quite surprising.

In the Ising model, finite-size scaling calculations go back to Ferdinand and Fisher [98]. The translation to the τ−continuum limit was made by Hamer and Barber [51], who studied the lowest excitation in the spectrum. Their techniques were extended to the full Hamiltonian (4.15) and the entire spectrum [118] and are then sufficiently general to allow the calculation of the finite-size scaling functions of the correlation lengths also for the *isotropic* 2D Ising model [151]. The results are in complete agreement with universality. Introducing the parameter η (see (4.15)) allows a full test of the Privman-Fisher hypothesis (2.7) and a clear identification of the various operators contributing. Using the notation of section 4.2.1, the finite-size corrections (at $t = t_c = 1$) are for the two lowest levels [118]

$$\begin{aligned} x_\varepsilon &= 1 + \frac{\pi^2}{8N^2}\left(\frac{1}{\eta^2} - \frac{4}{3}\right) \\ &+ \frac{\pi^4}{16N^4}\left(\frac{1}{8}\left(\frac{1}{\eta^2} - \frac{4}{3}\right)^2 + \frac{1}{3}\left(\frac{1}{\eta^2} - \frac{16}{15}\right)\right) + O(N^{-6}) \\ x_\sigma &= \frac{1}{8} - \frac{\pi^2}{128N^2}\left(\frac{1}{\eta^2} - \frac{4}{3}\right) + O(N^{-4}) \end{aligned} \tag{4.39}$$

($x_\psi = \frac{1}{2}x_\varepsilon$) and similar expressions for the higher levels. Since the critical exponents and all correlation functions of all possible irrelevant operators can be calculated at the critical point from conformal invariance [106,152], these results can be reproduced from conformal perturbation theory. We stress that these methods do not require the analytic solution of the system at the critical point. Writing (see [3,113])

$$H = H_c + \sum_j a_j(\eta)\phi_j \tag{4.40}$$

where H_c is the critical point Hamiltonian described by conformal invariance, ϕ_j some perturbing operator and $a_j(\eta)$ a non-universal coupling. Since the leading corrections to the universal terms in (4.40) and also to the ground-state energy (4.21) have the same $\eta-$dependence, these terms are generated by a single field ϕ_1. The next order has contributions from ϕ_1 and at least one more field ϕ_2. Reinicke [153] has identified $\phi_1 = L_{-1}\bar{L}_{-1}\varepsilon + \alpha(L_{-2}^2 + \bar{L}_{-2}^2)1$ in the Ising model (α is a free parameter). Note that the fields from the conformal tower of the identity operator are always present, leading to the analytical correction terms or order N^{-2}, N^{-4} etc... The contribution of n'th order perturbation in the $a_j(\eta)$ can be explicitly calculated from the $(n-2)$-point (critical) correlation functions of the ϕ_j, which are known [101,106,152] from conformal invariance. For the three-state Potts chain, the leading finite-size correction for the gaps and for the order parameter is given by the primary field (7/5, 7/5) [153,154] in good agreement with numerical results [89]. However, the next term of order $N^{-1.6}$ was not found. It is possible that the precision of the numerical data available is not sufficient to separate this term from the contribution of the operator $L_{-2}^2 + \bar{L}_{-2}^2$, of order N^{-2} [153]. The (primary) field describing the leading finite-size corrections was also identified in the spin-s XXZ chain [145].

Perturbation with a relevant operator in (4.40) can be used to obtain the finite-size scaling functions. For the Ising model one has, introducing the scaling variable

$$Z = N(t-1) \tag{4.41}$$

and taking the limit $N \longrightarrow \infty$, $t \longrightarrow 1$, Z fixed [51,118,80]

$$\begin{aligned} x_\varepsilon &= \left[1+\frac{Z^2}{\pi^2\eta^2}\right]^{1/2} + \frac{Z}{2\eta^2 N}\left[1+\frac{Z^2}{\pi^2\eta^2}\right]^{-1/2} + O(N^{-2}) \\ x_\sigma &= \frac{1}{8} + \frac{1}{4\pi}\frac{Z}{\eta} + \frac{\ln 2}{4\pi^2}\frac{Z^2}{\eta^2} \\ &+\frac{1}{2}R_{1\frac{1}{2},0}\left(\frac{Z^2}{4\pi^2\eta^2}\right) - \frac{1}{8}R_{1\frac{1}{2},0}\left(\frac{Z^2}{\pi^2\eta^2}\right) + O(N^{-1}) \end{aligned} \tag{4.42}$$

where $R_{1\frac{1}{2},0}(X)$ is a remnant function [155]. Similar results are obtained for the higher levels [118]. Scaling functions for free, fixed and mixed boundary conditions were also obtained [151]. Since to leading order in $1/N$ only the combination Z/η appears, this result is clearly in agreement with the expectation (2.7) and allows to relate the non-universal constant C_1 with $1/\eta$.

Perturbing with $\phi = \varepsilon$ in (4.40), eq. (4.42) was reproduced up to $O(Z^6)$ by Reinicke [156]. Since the remnant functions are analytic at the origin [155] the resulting perturbation series is convergent. Generally, the Privman-Fisher hypothesis (2.7) is satisfied iff only one relevant field enters in (4.40). This condition is for example satisfied for thermal and magnetic perturbations in the critical and tricritical Ising and three-states Potts models [156]. Similar calculations were done by Saleur and Itzykson [157]. In particular, a fourth-order calculation for the Ising model in a magnetic field reproduces a numerical estimate [158] for the universal ratio of the second and fourth cumulant of the magnetization. Another type of conformal perturbation theory was developed by Dotsenko [159] proposing a new regularisation scheme for relevant operators. Ludwig and Cardy [211] explicitly calculated the effect of a perturbing relevant operator on the central charge.

While the results presented are concerned with the limit of small scaling variables $Z \longrightarrow 0$, we now turn to the other extreme case $Z \longrightarrow \infty$. To be specific, consider a thermal perturbation, with $Z = tN^y$, but generalisations are immediate. In the limit $Z \longrightarrow \infty$ one has from (2.7)

$$\xi_i^{-1} = N^{-1}S_i(CZ) = C^{1/y}t^{1/y}\tilde{S}_i(1/CZ) \tag{4.43}$$

where $\tilde{S}_i$ is a new scaling function. On the other hand, one expects for the infinite system $\xi_i^{-1} \simeq B_i t^{\nu}$ which allows to identify

$$B_i = C^{1/y}\tilde{S}_i(0), \qquad A = C^{d/y}\tilde{Y}(0) \tag{4.44}$$

where the singular free energy density $f^{(s)}$ defines A by $f^{(s)} \simeq At^{2-\alpha}$ and $\tilde{Y}$ is the corresponding scaling function. Consequently the ratios

$$R_{ij} = \frac{B_i}{B_j} = \frac{\tilde{S}_i(0)}{\tilde{S}_j(0)} \quad , \qquad Q_i = \frac{A}{B_i^d} = \frac{\tilde{Y}(0)}{\tilde{S}_i^d(0)} \tag{4.45}$$

are both universal and independent of the boundary conditions, for arbitrary dimensions (provided the global symmetry is not changed by the boundary conditions).

For 2D systems, R_{ij} and Q_i can be calculated exactly. For a thermal perturbation one finds from the c–theorem [160,161]

$$Q_\varepsilon = -\frac{c}{12\pi}\frac{2-\alpha}{1-\alpha} \tag{4.46}$$

where the correlation length ξ_ε is defined [161] through the second moment of the energy-energy correlation function. For magnetic perturbation, Q_σ is obtained by replacing in (4.46) α by the exponent γ [161]. Concerning the R_{ij}, Zamolodchikov [162] has related them to the ratios of masses occuring in factorized S–matrices of integrable 2D systems. Hamiltonian limit studies are ideally suited for mass spectrum calculations of this kind.

As an example, consider the quantum Ising chain in a magnetic field at the critical point. It can be shown that even for a non-zero field the system contains at least six nontrivial conserved quantities of spin 1,7,11,13,17,19 [162]. This is then used as input for a bootstrap calculation of the S-matrix. It is *assumed* that the theory contains a finite number of particles describing the poles of the S-matrix. Trying to satisfy the constraints resulting from the existence of conserved quantities with a given spin in a "minimal" way, the complete S–matrix can be derived, containing eight stable particles. Considering the lightest particle of mass $m_1 = m$ as fundamental, two other states fall below the continuum starting at $2m$, with [162]

$$R_{21} = \frac{m_2}{m_1} = 2\cos\frac{\pi}{5} \; ; \qquad R_{31} = \frac{m_3}{m_1} = 2\cos\frac{\pi}{30} \tag{4.47}$$

Although some assumptions are needed to obtain this result, eq. (4.47) in fact describes the spectrum of the 2D Ising model at $T = T_c$ in a magnetic field as was checked numerically [37], confirming (4.47) up to less than 1 % . The ratios $R_{41}, \ldots R_{81}$ which are larger

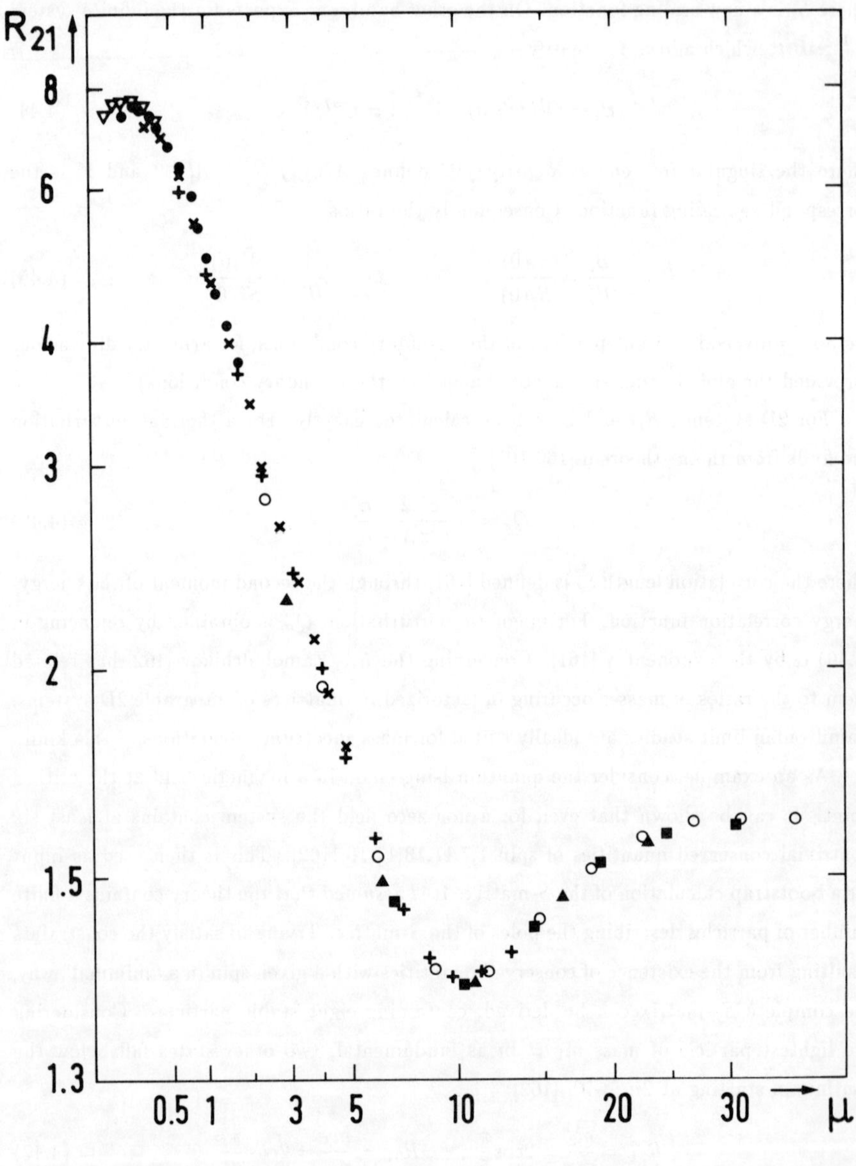

Fig.5: R_{21} (4.47) as a function of $\mu = hN^{15/8}$. The symbols refer to the value of h ($\nabla : h = 0.004$, $\cdot : h = 0.01$, $\times : h = 0.03$, $+ : h = 0.08$, $o : h = 0.3$, $\blacksquare : h = 0.5$)

than 2 could not be distinguished from the states in the continuum. The reason for the stability of these states is not yet understood.

In Fig. 5, R_{21} is shown as a function of the scaling variable $\mu = hN^{15/8}$. For μ small, that is $\xi \gg N$, $R_{21} \simeq 8$ as expected from conformal invariance (see section 4.2.1). There is a minimum at $R_{21} \simeq 1.35$ at $\mu \simeq 10$ and only for $\mu \longrightarrow \infty$ (i.e. $\xi \ll N$) one has $R_{21} \simeq 1.62$, in agreement with (4.47). The precise determination of the $\mu \longrightarrow \infty$ limit is rendered difficult by the non-monotonic behaviour of $R_{21}(\mu)$, which is generic. (For further details on the scaling functions, see [37]).

It is surprising to see that a system away from it critical point should have several discrete correlation lengths below the continuum. A good physical description of these states is still missing.

4.4. Defect lines

In section 4.2, the operator content was given for systems with boundary conditions which respect translational and rotational invariance. Since conformal invariance *assumes* that these invariances are satisfied, it is of interest to ask to what extent translational and rotational invariance can be broken without loosing conformal invariance. Here this question will be studied using as an example the exactly solvable Ising quantum chain with (several) defect lines, the interaction energy being changed from its bulk value J_b to κJ_b, where κ is the *defect strength*.

The critical behaviour of thermodynamic quantities has already received some attention for the case of a single line defect [163]. The critical exponents obtained generalize the known surface exponents [108] to which they reduce for $\kappa = 0$. For the 2D (isotropic) Ising model with one infinitely long defect line, the local magnetization was computed exactly and its critical exponent $\beta(\kappa)$ was found to depend continuously on κ [164]. Assuming conformal invariance, the map (4.12) can be used to relate a star-shaped configuration of half-infinite defect lines in the 2D plane and a set of parallel infinite defect lines on the infinite strip of finite width N (see Fig. 6). Turban showed that the conformal techniques for $\kappa = 0$ [165] can be generalized to yield the amplitude-exponent relation [166]

$$\xi_j^{-1} = E_j - E_0 = 2\pi \; x_j(\kappa) N^{-1} \tag{4.48}$$

Since the complete set $x_j(\kappa)$ of critical exponents is known for the Ising model with one defect line [167], conformal invariance (that is, eq. (4.48)) can be checked from Hamiltonian spectrum calculations [168,169,45]. These checks were extended recently to more complex defect configurations [170].

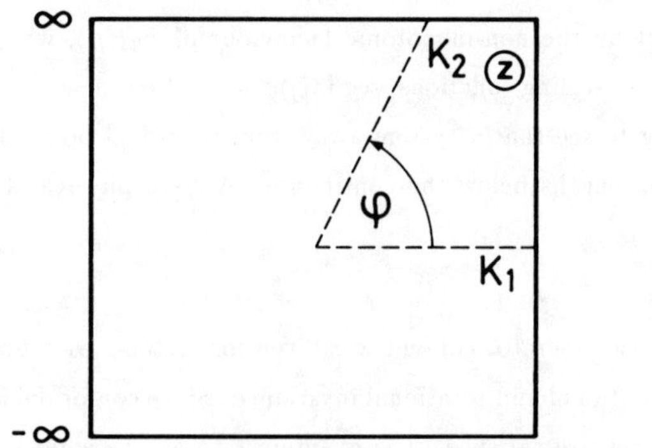

Fig.6: Conformal mapping (4.12) between the 2D plane and the strip, showing the effect on defect lines.

Henkel and Patkos [45,171,172] started a detailed investigation of the quantum Ising chain with n defect lines. The Hamiltonian is (at $T = T_{c,bulk}$)

$$\begin{aligned} H &= -\frac{1}{2}\sum_{k=1}^{N}[\sigma^z(k) + \sigma^x(k)\sigma^x(k+1)] \\ &+\frac{1}{2}\sum_{i=1}^{n}(1-\kappa_i)\sigma^x(N_i)\sigma^x(N_i+1) \end{aligned} \tag{4.49}$$

where the defects of strength κ_i sit at the chosen sites N_i. Note that the distances of the defects scale with the number of sites $N_i - N_{i'} \sim N$.

The exact solution [173] follows the well-established technique of Lieb, Schultz and Mattis [174] and is described in some detail since it has been used time and again in explicit Ising model calculations. By a Jordan-Wigner transformation, H is written as a bilinear expression in fermionic operators $c(n)$

$$H = \sum_{n,m=1}^{N}\left[c^+(n)A_{nm}c(m) + \frac{1}{2}(c^+(n)B_{nm}c^+(m) + h.c.)\right] \tag{4.50}$$

which can be diagonalized

$$H = \sum_k \Lambda(k)(\eta_k^+\eta_k - \frac{1}{2}) \tag{4.51}$$

where the $\Lambda(k)$ are determined from the eigenvalue equations

$$M\psi_k = (A-B)(A+B)\psi_k = \Lambda(k)^2\psi_k \tag{4.52}$$

The resulting N equations fall into two classes. $N - 2n$ of them are the same and yield the dispersion relation

$$\Lambda(k) = 2\mid\sin\frac{k}{2}\mid \tag{4.53}$$

The remaining $2n$ equations describe the physical behaviour since they yield the way k is discretized. How this is done determines crucially whether conformal invariance is maintained or not.

For simplicity, consider the case $n = 2$. Let the angle φ take the value (see Fig. 6)

$$\varphi = 2\pi\delta = 2\pi\frac{p}{q} \tag{4.54}$$

where p, q are coprime integers. In this case, the one-particle energies $\Lambda(k)$ become, up to order $O(1/N^2)$

$$\Lambda^{\pm}(k) = \frac{2\pi\gamma(q)}{N}(k + \frac{1}{2} \pm \Delta_i); \quad k = 0, 1, ...; \quad i = 1, ..., \gamma(q) \tag{4.55}$$

where

$$\gamma(q) = \begin{cases} q & \text{if} \quad q \text{ is odd} \\ q/2 & \text{if} \quad q \text{ is even} \end{cases} \tag{4.56}$$

and $\Delta_i = \frac{1}{2} - \alpha_i/2\pi\gamma(q)$ where the $\alpha_i \in [0, 2\pi]$ are the $\gamma(q)$ solutions of the equation [173]

$$\begin{gathered}(1-\kappa_1^2)(1-\kappa_2^2)\cos[\alpha(1-2\delta)] + (1+\kappa_1^2)(1+\kappa_2^2)\cos\alpha \\ +4\kappa_1\kappa_2(1-2Q) = 0\end{gathered} \tag{4.57}$$

where Q is the eigenvalue of the operator (4.17).

It is crucial that the differences of the one-fermion energies (4.55) are integers. It is this fact which allows the construction of conformal towers and thus guarantees conformal invariance. Since the Hamiltonian contains $2\gamma(q)$ independent Majorana fermions, the central charge is [172,173]

$$c = \gamma(q) \tag{4.58}$$

Note that c depends discontinuously on φ while the energy levels are continuous functions of φ. The conformal towers of the $c \geq 1$ Virasoro algebra (4.58) yield a smooth interpolation between the known cases $\kappa_i = 1$ (no defects) and $\kappa_i = 0$ (free boundaries). The critical exponents $x_j(\kappa)$ as obtained from (4.48,4.55) generalize the known corner exponents (surface exponents for $n = 2$ and $\varphi = \pi$). The interpolation between different conformal structures is illustrated in Fig.7 for $n = 2$, $\varphi = \pi$ and $\kappa_1 = \kappa_2 = \kappa$, in the sector $Q = 1$. For $\kappa = 1$, the spectrum is given by the conformal tower $(\frac{1}{16}, \frac{1}{16})$ of the $c = \frac{1}{2}$ Virasoro algebra, while for $\kappa = 0$, one has $(0, \frac{1}{2})$+ $(\frac{1}{2}, 0)$. In addition, for $\kappa = \sqrt{2} - 1$, one has a single representation $(\frac{1}{2})$, as can be checked by comparing the degeneracies with table 4.2. For the other values of κ, one has several primary fields of a single $c = 1$ Virasoro algebra, the degeneracies $d(\Delta_\phi, r)$ being equal to the number of partitions of r (r =0,1,2,...).

Studying the ground state energy, the effective central charge ζ (see (4.13)) also gets modified, one has $\zeta = \sum_i(\frac{1}{2} - 6\Delta_i^2)$ if the lowest state in the $Q = 0$ sector is taken as the physical ground state [171, 173].

On the other hand, if eq. (4.54) is *not* satisfied, the defects lines break conformal invariance completely. Eq. (4.57) then describes the zeroes of a quasiperiodic function and this renders the construction of conformal towers impossible. (Formally, $\gamma(q) = \infty$).

Similar results hold for any (finite) number n of defects [173]. In fact, it is sufficient to consider a configuration of equidistant defects since any arbitrary defect configuration can be extended to an equidistant one by adding some defect lines with $\kappa_i = 1$. Then the

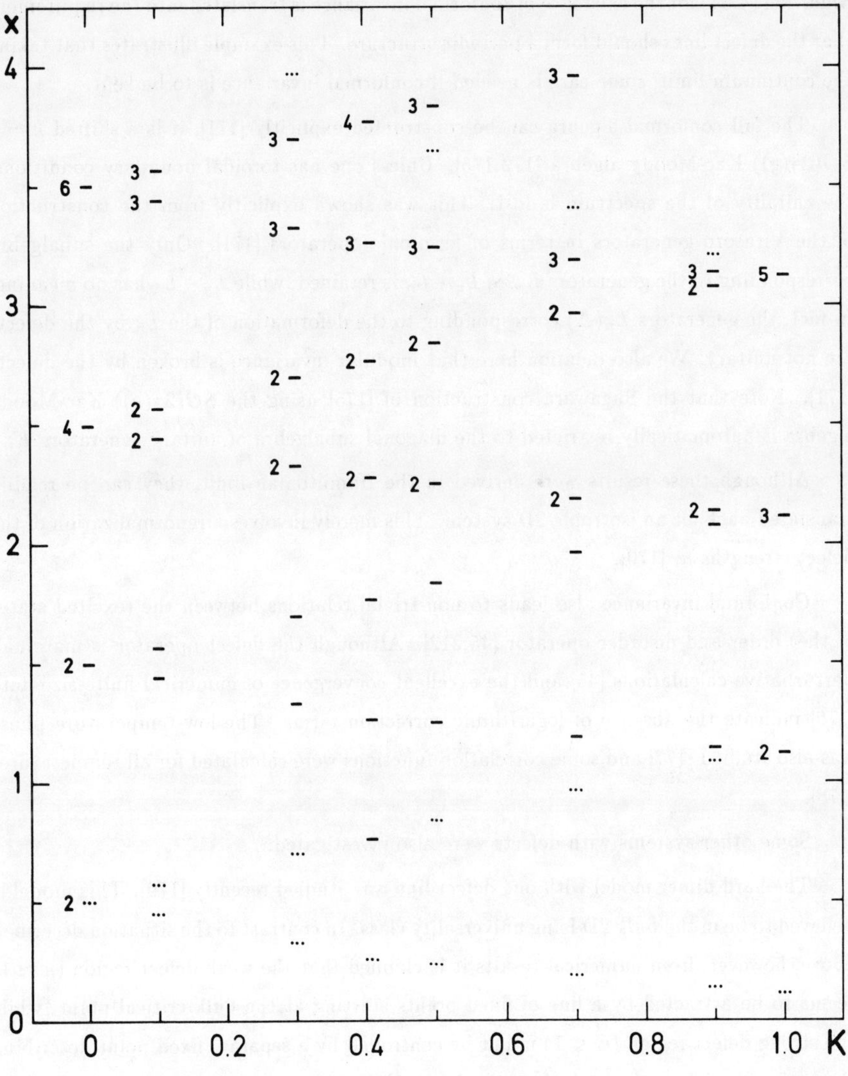

Fig.7: Critical exponents $x_i(\kappa)$ for $n = 2$, $\varphi = \pi$ and $\kappa_1 = \kappa_2 = \kappa$ in the Ising quantum chain with defects. The numbers denote the degeneracy of the levels. Dotted levels correspond to primary fields (after Henkel and Patkos [45]).

condition (4.54) for the existence of conformal invariance is translated into the requirement that the defect lines should form a periodic structure. This example illustrates that taking the continuum limit, some care is needed if conformal invariance is to be kept.

The full conformal algebra can be constructed explicitly [171], it is a shifted $k = 1$ $SO(2\gamma(q))$ Kac-Moody algebra [172,175]. Unless one has toroidal boundary conditions, the chirality of the spectrum is lost. This was shown explicitly from the construction of the Virasoro generators in terms of fermionic operators [171]. Only the subalgebra corresponding to the generators $K_n \sim L_n + \bar{L}_n$ is retained, while $L_n - \bar{L}_n$ has no meaning. In fact, the generators $L_n(\Delta)$ corresponding to the deformation of the L_n by the defects are not unitary. We also mention here that modular invariance is broken by the defects [171]. Note that the Sugawara construction of [175] using the $SO(2\gamma(q))$ Kac-Moody algebra is automatically restricted to the diagonal subalgebra of unitary generators K_n.

Although these results were derived in the Hamiltonian limit, they can be readily translated back for an isotropic 2D system. This merely involves a renormalization of the defect strengths κ_i [170].

Conformal invariance also leads to non-trivial relations between the (excited states of the) order and disorder operator [45,212]. Although the defect operator is marginal, perturbative calculations [45] and the excellent convergence of numerical finite-size data [176] indicate the absence of logarithmic correction terms. The low-temperature phase was also studied [177] and some correlation functions were calculated for all temperatures [178].

Some other systems with defects were also investigated.

The hard dimer model with one defect line was studied recently [179]. This model is believed to be in the *bulk* 2D Ising universality class. In contrast to the situation described above, however, from numerical results it is claimed that the weak defect region ($\kappa \simeq 1$) seems to be attracted by a line of fixed points starting at the bulk critical point, while the strong defect region ($\kappa \ll 1$) might be controlled by a separate fixed point describing the surface transition ($\kappa = 0$) of the model [179].

Conformal invariance was also checked for the three states Potts quantum chain with a defect [180]. The central charge is changed from $c = 4/5$ to $c > 1$ in analogy with the Ising model. From the spectrum, using a standard argument from representation theory (see e.g. [171]), it could be concluded that the model does not contain a $U(1)$ Kac-Moody algebra [180], in contrast to the Ising model result.

Finally, we consider briefly the situation where the mean distance between two defects scales like $d \sim N^\theta$, where θ is a chosen constant. The correlation length should scale as $\xi \sim N^\theta$, which is typical for systems with different correlation length exponents $\nu_\parallel, \nu_\perp$ in different space directions and $\theta = \nu_\parallel/\nu_\perp$, but not compatible with conformal invariance. In the limit $\theta \longrightarrow 0$, one goes into the McCoy-Wu model [181]. Systems with $\theta \neq 1$ will be considered in the next section. In this context, we mention that introducing large-scale inhomogeneities caused by a line defect can lead to a breakdown of the usual scaling behaviour, as was made explicit in the 2D Ising model [208].

4.5 Systems with Oscillatory Energy Gaps

It is a peculiarity of the Hamiltonian limit that in the ordered phase(s) of a quantum system the energy gap may vanish even on a finite lattice. An example was already given in section 3.1.1 and it was also noticed that these level crossings do not imply any pathology, since taking the Hamiltonian limit the terms which would prevent these crossings are neglected. Although level crossings do not occur for the most conventional type of ferromagnetic transition, they are typically present in modulated phases and close to Lifshitz points.

In order to see how finite-size scaling works in these cases, consider as a toy model the Ising quantum chain (4.15). The Ising transition is at $t_c = 1$, the transition to the oscillatory phases is at $t = t_z$ with [117] (compare with the (2+1)D case, Fig.2)

$$t_z^2 + \eta^2 = 1 \tag{4.59}$$

where the lowest gap vanishes exactly for any finite lattice (with periodic boundary conditions). For $\eta = 0$, the lowest gap is $m = |t-1|$, while for $t < 1$ one has ((4.15) should be multiplied with η before taking $\eta = 0$) [182]

$$C = \frac{1}{\pi}\left(t_z^2 - t^2\right)^{-1/2} \tag{4.60}$$

$$K \simeq \frac{1}{2}\left(t_z - t\right)^{1/2} \tag{4.61}$$

where K is the wave-vector since the system is in an incommensurate phase and C is the specific heat. For $\eta \neq 0$, (4.61) still holds, while C is regular at $t = t_z$ [117].

Hoeger, Gehlen and Rittenberg [183] proposed an extension to finite-size scaling when the system contains a wave-vector K (4.61). The scaling variable is $z = N^{1/\nu}(t_z - t)$ and the specific heat becomes

$$C(t, N) = N^{\alpha/\nu} F(z) \tag{4.62}$$

($N \longrightarrow \infty$, z fixed) and one assumes for $z \longrightarrow \infty$ that $F(z) = A\, z^{-\alpha}$. The exponent ν is defined from the wave-vector

$$K \sim (t - t_z)^{\nu} \tag{4.63}$$

while the gap scales as $m \sim (t - t_z)^s$.

Now, the incommensurate phase is characterized by specifying $F(z)$ as a sum of delta functions [183]

$$F(z) = \sum_{k=0}^{\infty} \tilde{C}_k \delta(z - z_k) \tag{4.64}$$

Consistency with (4.62) demands that, if $k \longrightarrow \infty$

$$\tilde{C}_k \sim k^{\lambda}, \quad z_k \sim k^{\mu}, \quad \frac{1}{2}(\mu - \lambda + 1) = \alpha$$

and the specific heat becomes

$$C(t, N) = \sum_{k=0} C_k(N) \delta(t - t_k(N)) \tag{4.65}$$

where

$$\begin{aligned} t_k(N) &= t_z - z_k N^{-1/\nu} \\ C_k(N) &= \tilde{C}_k N^{(\alpha-1)/\nu} \end{aligned} \tag{4.66}$$

where the level crossings in the ground state occur at $t = t_k(N)$.

In writing (4.64), one assumes that the smooth part of the specific heat does not dominate in the thermodynamic limit [183]. Eq.(4.66) allows to determine the transition point t_z and the exponents α and ν in analogy to the methods of section 2.3. Finally, comparing with the scaling of m, one obtains [183]

$$\alpha = 2 - s - d\nu \tag{4.67}$$

for a $d-$dimensional quantum hyperplane. Generalized hyperscaling relations of this type are familiar from systems with different correlation length exponents in different directions of space (e.g. [184]).

Eq.(4.66) can now be applied to the Hamiltonian (4.15). The $t_k(N)$ and $C_k(N)$ can be found from the exact solution [183]. For $\eta = 0$, one obtains $\alpha = \frac{1}{2}$, $\nu = \frac{1}{2}$, $s = 1$ in agreement with (4.60, 4.61). For $\eta \neq 0$, $\nu = \frac{1}{2}$ is found, while (4.67) is no longer applicable, since t_c and t_z do not concide.

These techniques were applied to the three-states asymmetric clock model [185]. There has been a considerable debate on the phase diagram in this model. In particular, a Lifshitz point was found in the isotropic version of the model [8] while its existence was questioned in the Hamiltonian limit [9,186]. This illustrates that universality need not hold for all anisotropic systems.

Taking the $x-$continuum limit, $J_x \longrightarrow 0$, $J_\tau \longrightarrow \infty$ and $\lambda = J_x \exp \frac{3}{2} J_\tau$ fixed (as opposed to the $\tau-$continuum limit of section 1), one obtains a non-hermitian Hamiltonian [9] from the asymmetric clock model

$$H = -\sum_{k=1}^{N} \sigma_k - \lambda \sum_{r=1}^{N} \left(\Gamma_k \Gamma_{k+1}^{+} + \Gamma_k^{+} \Gamma_{k+1} \right) \tag{4.68}$$

where the matrices σ and Γ are defined in section 1. Note that the non-hermiticity is not pathological but merely indicates the presence of a non-vanishing wave-vector.

The energy gaps and the wave-vector are obtained as follows. The eigenvalues of (4.68) can be ordered according to their real part. Then the "lowest" eigenvalue E_0 is real (and is from the $Q = 0$ sector). The first excited state is in the sector $Q = 1$ (the sectors $Q = 1$ and $Q = 2$ give complex conjugate eigenvalues) with real part E_1 and imaginary part F_1. Then the gap is $m = E_1 - E_0$ and the wave-vector $K = F_1$ [9]. One has a scaling behaviour

$$m \sim N^{-\theta}, \qquad K \sim N^{-\theta} \tag{4.69}$$

with $\theta = \nu_\tau / \nu_x$ at $\lambda = \lambda_c$. Techniques for finding the critical point for this more general form of scaling are available and amount to comparing data from the three lattices N, $N+1$, $N+2$ [187,188]. The results are [9]

$$\lambda_c = 0.667(4), \quad \nu_x = 0.95(4), \quad \nu_\tau = 0.67(4), \quad \alpha = 0.48(5) \tag{4.70}$$

and are compatible with the scaling relation $\alpha + \nu_x + \nu_\tau = 2$ (see (4.67)). For an extension of this model, see [213].

Another example of an anisotropic system is the directed percolation problem [191]. It can be reformulated in terms of Reggeon field theory [189]. Taking the Hamiltonian limit, one obtains [76]

$$\begin{aligned} H = -\frac{1}{2} \sum_{n=1}^{N} \{ & t\, \sigma^z(n) + \sigma^x(n)\sigma^x(n+1) + \sigma^y(n)\sigma^y(n+1) - \sigma^z(n)\sigma^z(n+1) \\ & + i\,(1 - \sigma^z(n))\,\sigma^y(n+1) + i\,\sigma^y(n)\,(1 - \sigma^z(n+1)) \} \end{aligned} \tag{4.71}$$

The non-hermiticity of H reflects the directed nature of the problem.

Consequently, there are two correlation length exponents $\nu_{/\!/}$, $\nu_{\perp}$. A finite-size scaling analysis yields [190]

$$t_c = 2.6064, \quad \theta = 1.5810(2), \quad \nu_{/\!/} = 1.752(4) \tag{4.72}$$

in excellent agreement with series expansion [77] and RG results [189] for the Reggeon theory and results in the original percolation formulation [191], thus nicely confirming universality.

We close this section describing a special $n-$states quantum chain which has been of considerable interest. The Hamiltonian is [192,193]

$$H = -\sum_{p=1}^{N}\sum_{k=1}^{n}\{a_k\sigma^k(p) + \lambda\, \bar{a}_k\Gamma^k(p)\Gamma^{n-k}(p+1)\} \tag{4.73}$$

where

$$\begin{aligned} a_k &= \exp[i(2k-n)\phi/n]/2\,\sin(\pi k/n) \\ \bar{a}_k &= \exp\left[i(2k-n)\bar{\phi}/n\right]/2\,\sin(\pi k/n) \end{aligned} \tag{4.74}$$

and $\cos\phi = \lambda\cos\bar{\phi}$. If $\phi = \bar{\phi}$, the model is self-dual. For $n = 2$, it reduces to the Ising quantum chain. We concentrate here on the case $\phi = \bar{\phi} = \frac{\pi}{2}$, where $a_k = \bar{a}_k = \left(1-\omega^{-k}\right)^{-1}$ ($\omega = \exp 2\pi i/n$). The model describes an incommensurate transition.

Early studies of the model yielded a single transition at $\lambda_c = 1$. For $n = 3$, the gap is

$$m = 1 - \lambda \tag{4.75}$$

and the order parameter M takes thé form

$$M = \left(1-\lambda^2\right)^{1/9} \tag{4.76}$$

as confirmed by tenth (or thirteenth) order perturbation theory [192]. The linear form (4.75) was extended to arbitrary values of n using finite-size scaling (up to $n = 6$) [193] and fifth-order perturbation theory [194] for n arbitrary.

Defining $n-1$ order parameters $M_k = \sum_p \sigma^k(p)$, perturbation theory is consistent with the conjecture [195,194] ($M = \sum_{k=1}^{n-1} M_k$)

$$M_k = \left(1-\lambda^2\right)^{k(n-k)/2n^2} \tag{4.77}$$

In fact, the Hamiltonian (4.73, 4.74) is integrable [214]. This was first shown by Gehlen and Rittenberg [193] using the Dolan-Grady theorem [196] to construct an infinite set of commuting conserved charges. The exact solution was constructed by Albertini, Au-Yang, Baxter, Mc Coy, Perk, Sah, Tang and Yan [197,195] by finding a novel solution to the star-triangle equations. The ground state energy is

$$E_0 = -\frac{N}{2}(1+\lambda)\sum_{k=1}^{n-1} F\left(-\frac{1}{2}, \frac{k}{n}, 1; \frac{4\lambda}{(1-\lambda)^2}\right) \tag{4.78}$$

where $F(a,b,c;z)$ is the hypergeometric function. Critical exponents were derived [198]

$$\alpha = 1 - \frac{2}{n}, \quad \nu_x = 1, \quad \nu_\tau = 2/n \tag{4.79}$$

which satisfy (4.67) and are in agreement with the earlier conjectures (4.75), but not compatible with conformal invariance unless $n = 2$.

It was pointed out recently [199], however, that the phase structure is more complicated and that one can have more than one transition [204]. These transitions arise from level crossings in the ground state and invalidate previous results obtained from perturbation theory. In table 4.4, the four phases for $n = 3$, $\phi = \bar{\phi} = \frac{\pi}{2}$ are listed. For example, if the I-II transition point is approached from below, the "one-particle" excitation over the perturbed weak-coupling ground state becomes negative [199]. This means that conjectures like (4.75) or (4.77) only apply to the phase I (and IV by duality). The transition from phase I to II will therefore not show any singularity in thermodynamic functions like the specific heat. This is characteristic of the quantum nature of the transition. Calculations of the excitation spectrum for the gapless phases II and III show a breaking of the left-right symmetry [199] which is incompatible with conformal invariance.

I	*II*	*III*	*IV*
$M \neq 0$ $D = 0$ $m \neq 0$ no oscill.	$M \neq 0$ $D = 0$ $m = 0$ oscill.	$M = 0$ $D \neq 0$ $m = 0$ oscill.	$M = 0$ $D \neq 0$ $m \neq 0$ no oscill.

0	0.9013	1	1/0.9013	∞

Table 4.4 : Schematic list of the four phases for the $n = 3, \phi = \bar{\phi} = \frac{\pi}{2}$ model. M stands for the order parameter, D for the disorder parameter and m for the lowest gap. The lowest line gives approximate values of the transition points (after Albertini et al. [199]).

The study of this model is in full progress and "plays the role with respect to level crossing transitions that the $n = 2$ Ising model has played for 45 years with respect to second order transitions" [199].

4.6. Further Developments

Conformal invariance has brought great progress into the study of 2D critical phenomena. The field is currently expanding very rapidly and has led to an enormous increase in understanding these transitions. While the picture is relatively complete by now for the plane and the torus (genus 0 and 1), theories on higher genera remain to be studied.

Another important development concerns off-critical systems, which is just developing. Present studies aim at finding the masses of the factorizable S-matrix involved for an increasing number of systems [37,200]. In particular, the nature of the stable "particles" above the continuum threshold should be clarified.

The investigation of non-isotropic systems should further clarify whether the conformal approach working for the Ising model with defect lines is essentially independent of the peculiarities of the Ising model. For the different but related subject of diluted systems, the concept of critical exponent needs some generalization [201].

Finally, the Hamiltonian limit offers the opportunity to study a new type of (quantum) phase transitions which is more general than the conventional ones and requires techniques beyond conformal invariance.

5. CONCLUSIONS

Hamiltonian limit studies have proved to be a useful tool in studying critical phenomena. The direct accessibility of the various correlation lengths of the scaling fields present in a given model, makes them ideally suited both for the clear statement of a theoretical prediction as well as the numerical check of existing theoretical ideas or conjectures. This property has been intensively elaborated on for (1+1)D systems in connection with 2D conformal invariance.

The additional features offered by the Hamiltonian limit are computational as well as conceptual. While the calculation of eigenvalues is in a one-to-one correspondence with ensemble averages in isotropic classical systems, obtaining eigenvectors allows the definition of new quantities like a "finite-size magnetization". How this can be used to compute critical exponents and amplitudes was illustrated in the example of the (2+1)D Ising model. Comparing with the results of other techniques (series expansion, MCRG, etc.), Hamiltonian limit methods are among the most precise approaches available, in particular when combined with finite-size scaling. Although the possibilities of the standard Lanczos approach seem to be exhausted in (2+1)D, there are sufficiently promising alternatives, which might allow to study systems on somewhat larger lattices. This should lead to a substantial increase of the precision of the critical parameters obtained.

For quantum chains, calculations of the low-lying spectra has allowed, with the help of 2D conformal invariance, to identify the complete operator content of a still rapidly increasing number of critical systems. Current work tries to include the entire finite-size scaling region into the analysis. At present, the relationship of non-critical theories to factorizable S−matrices seems most promising.

On the other hand, the ordered phases of a quantum system can show structures which escape from conventional transfer matrix studies and require extended techniques for their study. These systems can be so genuinely anisotropic that conformal invariance techniques are no longer applicable. At present, some simple systems have been studied in some detail and the leading exponents are obtained. It would be interesting if some of the ideas of conformal invariance could be borrowed to lead to a better understanding of these models.

Acknowledgement

I am greatly indebted to the following for helping clarify my views on finite-size scaling and conformal invariance. It is a pleasure to thank V. Rittenberg, A. Patkos, E. Domany, B. Duplantier, R. Flume, G.v. Gehlen, H.J. Herrmann, C. Hoeger, J.M. Luck, A. Ludwig, V. Pasquier, H. Saleur and J.-B. Zuber for useful discussions and A. Guttmann, C. Hamer and B. McCoy for useful correspondence.

This work was supported by a grant of the Wissenschaftsausschuß of NATO via DAAD.

REFERENCES

[1] J.B. Kogut, Rev. Mod. Phys. **51**, 659 (1979)

[2] M.N. Barber, in Domb and Lebowitz (Eds) *Phase Transitions and Critical Phenomena*, Vol.8, Academic Press, (1983).

[3] J.L. Cardy, in Domb and Lebowitz (Eds) *Phase Transitions and Critical Phenomena*, Vol.11, Academic Press, (1987).

[4] J.L. Cardy, in Brézin and Zinn-Justin (Eds) *Fields, Strings and Critical Phenomena*, Elsevier (1989).

[5] M. Suzuki, Prog. Theor. Phys. **46**, 1337 (1971)
E. Fradkin and L. Susskind, Phys. Rev. **D17**, 2637 (1978).

[6] P. Pfeuty, Ann. of Phys. **57**, 79 (1970).

[7] M. Henkel and C. Hoeger, Z. Phys. **B55**, 67 (1984).

[8] P.M. Duxbury, J.M. Yeomans and P.D. Beale, J. Phys. **A17**, L179 (1984).

[9] T. Vescan, G.v. Gehlen and V. Rittenberg, J. Phys. **A19**, 1957 (1986).

[10] See the other chapters of this volume for additional information.

[11] I. Affleck, in Dietz and Rittenberg (Eds) *Infinite-Dimensional Lie algebras and Conformal invariance...*, World Scientific (1987) and same volume as Ref.4.

[12] For extensive lists of finite-size data, see the results of the groups in Bonn, Canberra and São Carlos, usually published in J. Phys. A.

[13] See T. Einstein, same as Ref.11.

[14] E.g. R.J. Baxter, *Exactly Solved Models in Statistical Mechanics*, Academic Press (1982).

[15] For a bibliography on conformal invariance see Ref.102.

[16] G.v. Gehlen and V. Rittenberg, J. Phys. **A20**, 227 (1987).

[17] U. Grimm, J. Phys. **A21**, 3013 (1988).

[18] C.J. Hamer, J. Phys. **A16**, 1257 (1983).

[19] M. Henkel, J. Phys. **A20**, 3969 (1987).

[20] C. Lanczos, J. NBS **45**, 255 (1950).

[21] J.H. Wilkinson, *The Algebraic Eigenvalue Problem*, Clarendon Press (1956).
A. Jennings, *Matrix Computation for Engineers and Scientists*, Wiley (1977).

[22] G. Schütz and G.v. Gehlen, Priv. Comm.

[23] G. Schütz, J. Phys. **A20**, L906 (1987).

[24] M.R. Hestemes and E. Stiefel, J. NBS **49**, 409 (1982).

[25] C.J. Hamer, J. Phys. **A15**, L675 (1982).

[26] C.J. Hamer, J. Phys. **A16**, 3085 (1983).

[27] W.A. Berger, H.G. Müller, K.G. Kreuzer and R.M. Dreizler, J. Phys. **A10**, 1089 (1977).

[28] M.E. Fisher, in Green (Ed) *Critical Phenomena*, Proc. Enrico Fermi Vol.51, Academic Press (1971)
V. Privman, this volume

[29] V. Privman and M.E. Fisher, Phys. Rev. **B30**, 322 (1984).

[30] M.P. Nightingale, Physica **83A**, 561 (1976).

[31] C.J. Hamer and C.H.J. Johnson, J. Phys. **A19**, 423 (1986).

[32] C.R. Allton and C.J. Hamer, J. Phys. **A21**, 2417 (1988).

[33] K. Uzelac, Thesis Orsay (1980).
K. Uzelac and R. Jullien, J. Phys. **A14**, L151 (1981).

[34] T.D. Schultz, D.C. Mattis and E.H. Lieb, Rev. Mod. Phys. **36**, 856 (1964).

[35] R.J. Baxter, J. Phys. **C6**, L445 (1973).

[36] J.M. Luck, Phys. Rev. **B31**, 3069 (1985).

[37] M. Henkel and H. Saleur, J. Phys. **A22**, L513 (1989).

[38] D.C. Joyce, SIAM Rev. **13**, 435 (1971).

[39] D.A. Smith and W.F. Ford, SIAM J. Num. Anal. **16**, 223 (1979).

[40] S. Bhowmick, R. Bhattacharya and D. Roy, Comp. Phys. Comm. **54**, 31 (1988).
F. Beleznay, J. Phys. **A19**, 551 (1986).

[41] J.M. van den Broeck and L.W. Schwartz, SIAM J. Math. Anal. **10**, 658 (1979).

[42] C.J. Hamer and M.N. Barber, J. Phys. **A14**, 2009 (1981).

[43] M.N. Barber and C.J. Hamer, J. Austr. Math. Soc. **B23**, 229 (1982).

[44] R. Bulirsch and J. Stoer, Num. Math. **6**, 413 (1964).

[45] M. Henkel and A. Patkos, J. Phys. **A20**, 2199 (1987).

[46] M. Henkel and G. Schütz, J. Phys. **A21**, 2617 (1988).

[47] A.J. Guttmann, in Domb and Lebowitz (Eds) *Phase Transitions and Critical Phenomena*, Vol.13, Academic Press.

[48] V. Privman and M.E. Fisher, J. Phys. **A16**, L295 (1983).

[49] F.Y. Wu, Rev. Mod. Phys. **54**, 235 (1982).

[50] M. Henkel, J. Phys. **A17**, L795 (1984).

[51] C.J. Hamer and M.N. Barber, J. Phys. **A14**, 241 (1981).

[52] J. Kurmann, H. Thomas and G. Müller, Physica **112A**, 235 (1982).

[53] P. Centen, V. Rittenberg and M. Marcu, Nucl. Phys. **B205**, 585 (1982).

[54] S. Lubkin, J. NBS **48**, 228 (1952).

[55] C.J. Hamer and A.C. Irving, J. Phys. **A17**, 1649 (1984).

[56] C.J. Hamer and A.J. Guttmann, J. Phys. A to be published.

[57] L. Marland, J. Phys. **A14**, 2047 (1981).

[58] G. Bhanot, this volume.

[59] A.C. Irving and C.J. Hamer, Nucl. Phys. **B230**, 361 (1984).

[60] J. Adler, J. Phys. **A16**, 3585 (1983).

[61] A.J. Guttman, J. Phys. **A20**, 1855 (1987).

[62] G.S. Pawley, R.H. Swendsen, D.J. Wallace and K.G. Wilson, Phys. Rev. **B29**, 4030 (1984).

[63] M.N. Barber, R.B. Pearson, D. Toussaint, J.L. Richardson, Phys. Rev. **B32**, 1720 (1985).

[64] J.C. Le Guillou and J. Zinn-Justin, Phys. Rev. **B21**, 3976 (1980).

[65] J.C. Le Guillou and J. Zinn-Justin, J. Physique, **50**, 1365 (1989).

[66] K. Decker, Nucl. Phys. **B257**, 419 (1985).

[67] G. Bhanot, D. Duke and R. Salvador, Phys. Rev. **B33**, 7841 (1988).

[68] The HT results for ν [55,56] are somewhat high. This could be corrected by including the confluent correction terms in the analysis (B.G. Nickel Physica **106A**, 48 (1981), A.J. Guttmann, Priv. Comm.).

[69] B.A. Freedman and G.A. Baker Jr., J. Phys. **A15**, L715 (1982).

[70] K. Binder, M. Nauenberg, V. Privman and A.P. Young, Phys. Rev. **B31**, 1498 (1985).

[71] F. Igloi and C. Vanderzande, Physica **135A**, 347 (1986)

[72] C.J. Hamer and J.B. Kogut, Phys. Rev. **B20**, 3859 (1979)
C.J. Hamer, J.B. Kogut and L. Susskind, Phys. Rev. **D19**, 3091 (1979).
G.E. Sobelman, Phys. Rev. **B29**, 1493 (1981).
P.G. Hornby and M.N. Barber, J. Phys. **A18**, 827 (1985).

[73] H.E. Stanley, Phys. Rev. **176**, 718 (1968)
T.H. Berlin and M. Kac, Phys. Rev. **86**, 821 (1952)
H.W. Lewis and G.H. Wannier, Phys. Rev. **88**, 682 (1952).

[74] M. Henkel, J. Phys. **A21**, L227 (1988).

[75] S. Singh and R.K. Pathria, Phys. Rev. **B31**, 4483 (1985), **B32**, 4618 (1985), **B38**, 3769 (1987)
M.N. Barber and M.E. Fisher, Ann. of Phys. **77**, 1 (1973)
E. Brézin, J. Physique **43**, 15 (1982).

[76] R.C. Brower, M.A. Furman and K. Subbarao, Phys. Rev. **D15**, 1756 (1977).

[77] R.C. Brower, M.A. Furman, M. Moshe, Phys. Lett. **76B**, 213 (1978).

[78] M. Henkel, J. Phys. **A20**, L769 (1987).

[79] M. Henkel, J. Phys. **A19**, L247 (1986).

[80] M. Henkel, in Horvath, Palla and Patkos (Eds) *Nonperturbative Methods in Quantum Field Theory*, World Scientific (1987).

[81] J.L. Cardy, J. Phys. **A17** L385 (1984).

[82] J.L. Cardy, J. Phys. **A18**, L757 (1985).

[83] F.C Alcaraz and H.J. Herrmann, J. Phys. **A20**, 5735 (1987).

[84] The form (3.19) of the exactly known spectrum of the spherical model makes it unlikely that a simple spectrum generating algebra can be found in (2+1)D.

[85] J.L. Cardy, Nucl. Phys. **B290**, 355 (1987)
A. Cappelli and A. Coste, Nucl. Phys. **B314**, 707 (1989).

[86] Hamer and Johnson calculated (in Ref.31) τ from the lowest gap in the ordered phase.

[87] K.K. Mon, Phys. Rev. Lett. **54**, 2671 (1985)
K.K. Mon and D. Jasnow, J. Stat. Phys. **41**, 273 (1985).

[88] In calculating $a_{0\varepsilon\varepsilon}$ one should note that the usual energy operator (3.10) contains contributions from both the unit operator and the scaling field ε. The contribution of the unit operator is by definition *not* included in $a_{0\varepsilon\varepsilon}$.

[89] G.v. Gehlen, V. Rittenberg and T. Vescan, J. Phys. **A20**, 2577 (1987). This identity arises since the C's appear in the three-point correlation function.

[90] A.C. Irving and C.J. Hamer, J. Phys. **A16**, 829 (1983).

[91] A. Patkos and P. Rujan, J. Phys. **A18**, 1765 (1985).

[92] A. Patkos and P. Rujan, Phys. Lett. **152B**, 346 (1985).

[93] P. Rujan, Private Communication.

[94] T. Almassy and A. Patkos, J. Phys. **A22**, 1577 (1989)
A. Duncan and R. Roskies, Phys. Rev. **D31**, 364 (1985).

[95] A. Duncan and A. Patkos, Phys. Lett. **128A**, 146 (1988).

[96] C.J. Hamer, Canberra preprint.

[97] H.H. Roomany and H.W. Wyld, Phys. Rev. **D21**, 3341 (1980).

[98] A.E. Ferdinand and M.E. Fisher, Phys. Rev. **185**, 832 (1969).

[99] J.M. Drouffe and C. Itzykson, *Statistical Field Theory,* Cambridge University Press.

[100] F.C. Alcaraz, M.N. Barber and M.T. Batchelor, Ann. of Phys. **182**, 280 (1988).

[101] A.A. Belavin, A.M. Polyakov and A.B. Zamolodchikov, Nucl. Phys. **B241**, 333 (1984).

[102] C. Itzykson, H. Saleur,J.-B. Zuber, *Conformal Invariance and Applications to Statistical Mechanics*, World Scientific (1988).

[103] This condition is not sufficient as shown in section 4.4.

[104] Sometimes it is also referred to as speed of sound in the literature.

[105] Δ_ϕ and $\bar{\Delta}_\phi$ are both real and not the complex conjugates of each other.

[106] Vl.S. Dotsenko and V.A. Fateev, Nucl. Phys. **B240**, 312 (1984)
Vl.S. Dotsenko, Nucl. Phys. **B235**, 54 (1984)

[107] D. Friedan, Z. Qiu and S. Shenker, Phys. Rev. Lett. **52**, 1575 (1984); Comm. Math. Phys. **107**, 535 (1986).

[108] For a review H.W. Diehl in Domb and Lebowitz (Eds) *Phase Transitions and Critical Phenomena*, Vol.10, Academic Press (1988).

[109] Eq.(4.14a) is valid for any boundary condition compatible with the global symmetry of the model. x_i is then the exponent of a "parafermionic" operator.

[110] J.L. Pichard and G. Sarma, J. Phys. **C14**, L617 (1981)
B. Derrida and L. de Seze, J. Physique **43**, 475 (1982)
J.M. Luck, J. Phys. **A15**, L169 (1982)
M.P. Nightingale and H.W. Blöte, J. Phys. **A16**, L657 (1983)
K. Penson and M. Kolb, Phys. Rev. **B29**, 2854 (1983)
G.v. Gehlen, C. Hoeger and V. Rittenberg, J. Phys. **A17**, L469 (1984); **A18**, 187 (1985).

[111] G.v. Gehlen, V. Rittenberg and H. Ruegg, J. Phys. **A19**, 107 (1986).

[112] I. Affleck, Phys. Rev. Lett. **56**, 746 (1986)
H.W. Blöte, J.L. Cardy and M.P. Nightingale, Phys. Rev. Lett. **56**, 742 (1986).

[113] J.L. Cardy, Nucl. Phys. **B270**, 186 (1986), **B275**, 200 (1986).

[114] A. Cappelli, C. Itzykson and J.B. Zuber, Nucl. Phys. **B280**, 445 (1987); Comm. Math. Phys. **113**, 1 (1987).

[115] F.C. Alcaraz, U. Grimm and V. Rittenberg, Nucl. Phys. **B316**, 735 (1989).

[116] S. Katsura, Phys. Rev. **127**, 1508 (1962).

[117] E. Barouch and B.M. Mc Coy, Phys. Rev. **A3**, 786 (1971).

[118] M. Henkel, J. Phys. **A20**, 995 (1987).

[119] A. Rocha-Caridi, in Lepowski, Singer and Mandelstam (Eds) *Vertex Operators in Mathematics and Physics*, Springer (1984).

[120] H.W. Blöte and M.P.M. den Nijs, Phys. Rev. **B37**, 1766 (1988). The models studied cannot be solved exactly using the Onsager solution.

[121] T.W. Burkhardt and I. Guim, J. Phys. **A18**, L33 (1985).

[122] G.v. Gehlen and V. Rittenberg, J. Phys. **A19**, L631 (1986).

[123] M. Droz, A. Malaspinas and A.L. Stella, J. Phys. **C18**, L245 (1985).

[124] G.v. Gehlen and V. Rittenberg, J. Phys. **A19**, L625 (1986).

[125] For a review I.D. Lawrie and S. Sarbach in Domb and Lebowitz (Eds) *Phase Transitions and Critical Phenomena*, Vol.9, Academic Press (1984).

[126] F.C. Alcaraz, J.R. Drugowich de Felicio, R. Köberle and J.F. Stilck, Phys. Rev. **B32**, 7469 (1985).

[127] D.B. Balbão and J.R. Drugowich de Felicio, J. Phys. **A20**, L207 (1987).

[128] P.A. Rikvold, W. Kinzel, J.D. Gunton and K. Kaski, Phys. Rev. **B28**, 2686 (1983)
H.J. Herrmann, Phys. Lett. **100A**, 256 (1984).

[129] D. Friedan, Z. Qiu and S. Shenker, Phys. Lett. **151B**, 37 (1985).

[130] F. Igloi and J. Zittartz, Z. Phys. **B70**, 387 (1988).

[131] F. Igloi, J. Phys. **A21**, L911 (1988).
V.G. Benza, Europhys. Lett. **8**, 321 (1989).

[132] F. Igloi, Phys. Rev. B, to be published.

[133] F.C. Alcaraz and M.N. Barber, J. Phys. **A20**, 179 (1987)
M. Kolb and K.A. Penson, J. Phys. **A19**, L779 (1986)
F. Igloi, J. Phys. **A20**, 5319 (1987)
L. Turban, J. Phys. **A21**, 3035 (1988)
and references therein.

[134] L. Turban, J. Phys. **C15**, L65 (1982)
K.A. Penson, R. Jullien and P. Pfeuty, Phys. Rev. **B26**, 6334 (1982).

[135] J.L. Cardy, J. Phys. **A19**, L1093 (1986).

[136] K.A. Penson, J.M. Debierre and L. Turban, Phys. Rev. **B37**, 7884 (1988).

[137] M. Kohmoto, M. den Nijs and L.P. Kadanoff, Phys. Rev. **B24**, 5229 (1981)

M. Baake, G.v. Gehlen and V. Rittenberg, J. Phys. **A20**, L479 and L487 (1987)

S.K. Yang, Nucl. Phys. **B285**, 183 and 639 (1987)

S.K. Yang and H.B.Zheng, Nucl. Phys. **B285**, 410 (1987)

M. Baake, P. Christe and V. Rittenberg, Nucl. Phys. **B300**, 637 (1988)

F. Igloi, J. Stat. Phys. **55**, 311 (1989)

and references therein.

[138] e.g. H.J. de Vega and F. Woynarovich, Nucl. Phys. **B251**, 439 (1985)

F. Woynarovich and H.P. Eckle, J. Phys. **A20**, L97 (1987)

C.J. Hamer, G.R.W. Quispel and M.T. Batchelor, J. Phys. **A20**, 5677 (1987)

and references therein.

[139] G. v. Gehlen, V. Rittenberg and G. Schütz, J. Phys. **A21**, 2805 (1988).

[140] P. Suranyi, J. Phys. **A21**, L401 (1988).

[141] for a review P. Goddard and D. Olive, Int. J. Mod. Phys. **A1**, 303 (1986); F. Ravinni, NORDITA preprint 89/21P.

[142] A.B. Zamolodchikov and V.A. Fateev, JETP **62**, 215 (1985).

[143] G. Schütz, J. Phys. **A20**, L905 (1987), **A22**, 731 (1989)

G. v. Gehlen and V. Rittenberg, J. Phys. **A19**, 2439 (1986), **A20**, 1309 (1987).

[144] J.M. Kosterlitz and D.J. Thouless, J. Phys. **C6**, 118 (1973).

[145] F.C. Alcaraz and M.J. Martins, J. Phys. **A21**, L381 and 4397 (1988), **A22**, L99 and 1829 (1989).

[146] I. Affleck, Vancouver preprint (1989).

[147] B. Nienhuis in Domb and Lebowitz (Eds): *Phase transitions and critical phenomena*, vol. 11, Academic press (1987).

[148] P. di Francesco, H. Saleur and J.B. Zuber, J. Stat. Phys. **49**, 57 (1987), Nucl. Phys. **B285**, 454 (1987).

[149] G.E. Andrews, R.J. Baxter and J.P. Forrester, J. Stat. Phys. **35**, 193 (1984).

[150] V. Pasquier, J. Phys. **A20**, L1229 and 5707 (1987), Comm. Math. Phys. **118**, 355 (1988)

and references therein.

[151] T.W. Burkhardt and I. Guim, Phys. Rev. **B35**, 1799 (1987).

[152] P. di Francesco, H. Saleur and J.B. Zuber, Nucl. Phys. **B290**, 527 (1987).

[153] P. Reinicke, J. Phys. **A20**, 5325 (1987).

[154] P. Reinicke and T. Vescan, J. Phys. **A20**, L653 (1987).

[155] M.E. Fisher and M.N. Barber, Arch. Rat. Mech. Anal. **47**, 205 (1972). These remnant functions appear generically in finite-size calculations.

[156] P. Reinicke, J. Phys. **A20**, 4501 (1987).

[157] H. Saleur and C. Itzykson, J. Stat. Phys. **48**, 449 (1987).

[158] T.W. Burkhardt and B. Derrida, Phys. Rev. **B32**, 7273 (1985).

[159] Vl.S. Dotsenko, Nucl. Phys. **B314**, 687 (1989).

[160] A.B. Zamolodchikov, JETP Lett. **43**, 730 (1986).

[161] J.L. Cardy, Phys. Rev. Lett. **60**, 2709 (1988).

[162] A.B. Zamolodchikov to appear in: *Advanced Studies in Pure Mathematics*.

[163] T.W. Burkhardt and E. Eisenriegler, Phys. Rev. **B24**, 1236 (1981), **B25**, 3283 (1982).

[164] R. Bariev, JETP **50**, 613 (1979)
B.M. McCoy and J.H.H. Perk, Phys. Rev. Lett. **44**, 840 (1980).

[165] J.L. Cardy, Nucl. Phys. **B240**, 514 (1984).

[166] L. Turban, J. Phys. **A18**, L325 (1985).

[167] L.P. Kadanoff, Phys. Rev. **B24**, 5382 (1981).

[168] I. Peschel and K.D. Schotte, Z. Phys. **B54**, 305 (1984).

[169] L.G. Guimarães and J.R. Drugowich de Felicio, J. Phys. **A19**, L341 (1986).

[170] A.C. Irving, G. Odor and A. Patkos, J. Phys. AL, to be published.

[171] M. Henkel and A. Patkos, Nucl. Phys. **B285**, 29 (1987).

[172] M. Henkel and A. Patkos, J. Phys. **A21**, L231 (1988).

[173] M. Henkel, A. Patkos and M. Schlottmann, Nucl. Phys. **B314**, 609 (1989).

[174] E. Lieb, T. Schultz and D. Mattis, Ann. of Phys. **16**, 407 (1961).

[175] M. Baake, P. Chaselon and M. Schlottmann, Nucl. Phys. **B314**, 625 (1989).

[176] M. Henkel in Dita and Georgescu (Eds) *Conformal invariance and string theory*, Academic Press.

[177] G.G. Cabrera and R. Jullien, Phys. Rev. Lett. **57**, 393 (1986).
J. Zinn-Justin, Phys. Rev. Lett. **57**, 3296 (1986)
M.N. Barber and M.E. Cates, Phys. Rev. **B36**, 2024 (1987).

[178] L.F. Ko, H. Au-Yang and J.H.H. Perk, Phys. Rev. Lett. **54**, 1091 (1985)
D.B. Abraham, L.F. Ko and N.M. Svrakic, Phys. Rev. Lett. **61**, 2393 (1988).

[179] F. Igloi, Leuven preprint (1989).

[180] M. Schlottmann, Diplomarbeit Bonn IR-88-41

G.v. Gehlen and M. Schlottmann, to be published.

[181] B.M. McCoy and T.T. Wu, *The two-dimensional Ising model*, Harvard University Press (1973)

H. Au-Yang and B.M. McCoy, Phys. Rev. **B10**, 886 and 3885 (1974)

C.A. Tracy, J. Stat. Phys. **51**, 481 (1988)

H.A. Cecattoro, Z. Phys. **B75**, 253 (1989)

and references therein.

[182] R. Jullien and P. Pfeuty, Phys. Rev. **B19**, 4646 (1979). Note that for $\eta = 0$, one has a non-linear dispersion relation $E \sim k^2$.

[183] C. Hoeger, G.v. Gehlen and V. Rittenberg, J. Phys. **A18**, 1813 (1985).

[184] R.M. Hornreich, M. Luban and S. Shtrikman, Phys. Rev. Lett. **35**, 1678 (1975).

[185] S. Ostlund, Phys. Rev. **B24**, 398 (1981)

D.A. Huse, Phys. Rev. **B24**, 5180 (1981)

This model is also referred to as chiral Potts model.

[186] G.v. Gehlen and V. Rittenberg, Nucl. Phys. **B230**, 455 (1984).

[187] E. Domany and B. Schaub, Phys. Rev. **B29**, 4095 (1984).

[188] for a review on anisotropic scaling, see

K. Binder and J.S. Wang, J. Stat. Phys. **55**, 87 (1989).

[189] J.L. Cardy and R.L. Sugar, J. Phys. **A13**, L423 (1980).

[190] M. Henkel and H.J. Herrmann, to be published.

[191] W. Kinzel and J.M. Yeomans, J. Phys. **A14**, L163 (1981)

see also W. Kinzel, this volume.

[192] S. Howes, L.P. Kadanoff and M. den Nijs, Nucl. Phys. **B215**, 169 (1983).

[193] G.v. Gehlen and V. Rittenberg, Nucl. Phys. **B257**, 351 (1985).

[194] M. Henkel and J. Lacki, Phys. Lett. **138A**, 105 (1989).

[195] G. Albertini, M.M. McCoy, J.H.H. Perk and S. Tang, Nucl. Phys. **B314**, 741 (1989).

[196] L. Dolan and M. Grady, Phys. Rev. **D25**, 1587 (1982).

[197] H. Au-Yang, B.M.M. McCoy, J.H.H. Perk, S. Tang and M.L. Yan, Phys. Lett. **123A**, 219 (1987)

B.M.McCoy, J.H.H. Perk, S. Tang and C.H. Sah, Phys. Lett. **125A**, 9 (1987)

H. Au-Yang, B.M. McCoy, J.H.H. Perk and S. Tang in Kashiwara and Kawai (Eds) *Algebraic Analysis* Academic Press (1988)

R.J. Baxter, J.H.H. Perk and H. Au-Yang, Phys. Lett. **128A**, 138 (1988).

[198] R.J. Baxter, J. Stat. Phys. **52**, 639 (1988), Phys. Lett. **133A**, 185 (1988).

[199] G. Albertini, B.M. McCoy and J.H.H. Perk, Phys. Lett. **135A**, 159 (1989) and preprints ITP-SB 89-1, 89-24.

[200] for a few recent papers, see
J.L. Cardy and G. Mussardo, Phys. Lett. **225B**, 275 (1989).
T. Eguchi and S.K. Yang, Phys. Lett. **224B**, 373 (1989)
M. Henkel and H. Saleur, Saclay preprint (1989)
P. Christe and G. Mussardo, Santa Barbara preprint (1989)
G. Sotkov and C.J. Zhu, Trieste preprint (1989)
A.B. Zamolodchikov and V.A. Fateev, Landau preprint (1989)
G.v. Gehlen, Bonn preprint (1989).

[201] A.W.W. Ludwig, Saclay preprint 89/041.

[202] J.L. Cardy and I. Peschel, Nucl. Phys. **B300**, 377 (1988)
V. Privman, Phys. Rev. **B38**, 9261 (1988).

[203] In the phases P and C the Hamiltonian (3.1) commutes with the transfer matrix of the (classical) Ising model in (1+1)D (M. Suzuki, ref.5).

[204] The chains used in [193] where probably too short to detect the intermediate phases.

[205] C.R. Allton, C.M. Yung and C.J. Hamer, Phys. Rev. **D39**, 3772 and 3778 (1989).

[206] C. Itzykson, R.B. Pearson and J.B. Zuber, Nucl. Phys. **B220**, 415 (1983).

[207] G. Bhanot, R. Salvador, S. Black, P. Carter and R. Toral, Phys. Rev. Lett. **59**, 803 (1987).

[208] R.Z. Bariev, J. Phys. **A22**, L397 (1989).

[209] P. di Francesco and J.B. Zuber, Saclay preprint SPhT/89-92

[210] B. Duplantier and F. David, J. Stat. Phys. **51**, 327 (1988).

[211] A. Ludwig and J.L. Cardy, Nucl. Phys. **B285**, 687 (1987).

[212] A.C. Brown, Phys. Rev. **B25**, 331 (1982).

[213] H.U. Everts and H. Röder, J. Phys. **A22**, 2475 (1989).

[214] For this reason, it is also referred to as "integrable chiral Potts model".

[215] M.J. Stephen and L. Mittag, Phys. Lett. **41A**, 357 (1972).

[216] J. Solyom and P. Pfeuty, Phys. Rev. **B24**, 218 (1981).

[217] H.H. Roomany, H.W. Wyld and L.E. Holloway, Phys. Rev. **D21**, 1557 (1980).

[218] P. Furlan, G.M. Sotkov and I. Todorov, Riv. del Nuovo Cim. **12**, N°6 (1989).

[219] G.v. Gehlen, ref. 200. See also ref. 126.

IX. FINITE SIZE SCALING IN LATTICE GAUGE THEORY

Gyan V. Bhanot

TABLE of CONTENTS

FINITE SIZE SCALING IN LATTICE GAUGE THEORY

Gyan V. Bhanot

Supercomputer Computations Research Institute
Florida State University
Tallahassee, FL 32306

1. INTRODUCTION

Quantum Chromodynamics (QCD) is a theory that is believed to describe the strong force in nature. It is a theory of the 'color' interaction that binds quarks (q) and anti-quarks ($\bar{q}$) into hadrons (protons, neutrons, etc.) and mesons (pions, kaons, etc.). QCD has an certain symmetry called gauge invariance that is similar to the invariance of the gravitational field equations under coordinate transformations. Wilson [1] invented the lattice version of QCD to be able to study the theory beyond perturbation theory while preserving its gauge invariance. A fruitful way to extract information from the lattice formulation of QCD has been via direct numerical simulation [2,3]. As we shall see, in addition to the thermodynamic limit (infinite volume), contact with the underlying field theory (QCD) is made only in the limit where the lattice correlation length ξ diverges. On a finite lattice, the linear size of the lattice must be several times ξ for finite size effects to be relatively unimportant. In reality, since the problem is four dimensional, this requirement turns out to be very restrictive [3]. Given the limitations of present day computers, most of the numerical simulations done so far have suffered from having lattices that were too small and/or from not making use of finite size scaling to extract useful results about the infinite volume, infinite correlation length limit.

In this paper I will review some recent results on finite size scaling in gauge theories. This is a review for non experts so it is both self contained and sketchy. Depending entirely on personal taste, some topics are treated in detail while others are only mentioned. However, I have tried to give sufficient references so that the motivated reader can find out more on his own. Section 1 is a brief review of the basics of lattice gauge theory. It explains the connection between the lattice theory and the underlying continuum field theory. In addition, it also explains how lattice results answer questions about QCD. Section 2 describes the spectral density method [4] (henceforth called the SDM) which is a new numerical procedure to compute the partition function and order parameters to good accuracy. The method is illustrated using the $Z(2)$ gauge theory in four dimensions [5]. Section 3 describes the finite size scaling theory of the partition function and its zeros [6] and applies these ideas to the SDM calculations of Section 2. In Section 4 the generalization of these ideas to continuous models is discussed with an emphasis on finite size scaling aspects. Section 5 is a brief discussion of the finite size dependence of the mass gap in QCD. Here I mainly discuss the ideas of Lüscher [8] for asymptotic finite size dependences of masses. Section 6 is a discussion of the more realistic situation of small box physics. Here I again describe mainly Lüscher's work on the spectrum of QCD in a small box [9]. Finally in Section 7 I describe briefly some other ongoing finite size scaling studies and ideas.

QCD [10] is the theory of the interaction of quarks and gluons in four dimensions. In a discussion about finite size scaling effects, it is sufficient to discuss QCD without any fermions (quarks). The Lagrangian density of QCD with n colors but no quarks is,

$$L(x) = \frac{1}{4g^2}\mathrm{Tr}[F_{\mu\nu}(x)F_{\mu\nu}(x)] \tag{1}$$

where, g is the bare coupling constant,

$$F_{\mu\nu} = T^a F^a_{\mu\nu}, \tag{2}$$

and

$$F^a_{\mu\nu} = \partial_\mu A^a_\nu - \partial_\nu A^a_\mu + i f^{abc} A^b_\mu A^c_\nu. \tag{3}$$

The T^a's are n^2-1 matrices in the Adjoint representation of the $SU(n)$ algebra and the f^{abc}s are the structure constants of $SU(n)$. The Lagrangian density of Eq. 1 is a generalization to non-abelian $SU(n)$ of the $U(1)$ symmetric theory of Quantum Electrodynamics (QED). The non-abelian nature of QCD is buried in the third term (absent in QED) on the right hand side of Eq. 3. This term is present because the gluon fields A_μ^a themselves carry the color charge (indexed by a). This means that the gluons have self interactions. This is in contrast to QED where the photons carry no electromagnetic charge and hence, have no self interactions.

The field theory dynamics in the Lagrangian formulation is introduced in the usual way through the Feynman sum over histories (also called the path integral or the partition function) in Euclidean space,

$$Z_C = [\Pi_{x\mu} \int dA_\mu^a(x)] \exp[- \int L(x) d^4x] \tag{4}$$

On a naive level, Eq. 4 is meaningless. The integral diverges. The reason is that the Lagrangian of Eq. 1 has a huge symmetry called local gauge invariance or invariance under a choice of coordinates at each space time point. If $\omega^a(x)$ is an arbitrary infinitesimal function (so that ω^2 can be neglected), then L is invariant under,

$$A_\mu^a \to A_\mu^a + \partial_\mu \omega^a(x) + i f^{abc} \omega^b A_\mu^c \tag{5}$$

This means that not all the A_μ^a's are independent (physical) degrees of freedom and the integral over A_μ^a in Eq. 4 can (at least in principle) be broken up into a product of integrations, one over the gauge invariant subspace of A_μ^a and another over the gauge parameters ω^a. The integration over ω^a is undamped because the Lagrangian does not depend on these variables and this causes the result to be infinite.

These problems can be fixed up in a systematic perturbation expansion in g^2 (called weak coupling perturbation theory) [10]. The perturbation expansion seems to predict its own region of validity since it asserts that the color charge vanishes at small distances. This property is called asymptotic freedom. However, the expansion itself has problems. First, it is an asymptotic expansion and so has zero radius of convergence. In fact, for the non-abelian (QCD) case, it is not even Borel summable. The more serious problem is that it is an expansion in powers of the A_μ^a field and so it cannot describe the most important feature of QCD, namely that the color charge is confined and the A_μ^a's never appear as asymptotic states in any physical process. Another problem is that certain amplitudes turn out to be infinite in perturbation theory even after gauge fixing. However, this last problem can be fixed up in a systematic way for certain theories (which are then called renormalizable) by introducing cutoffs in momentum space. The procedure of introducing the cutoffs is called regularization. Fortunately, QCD belongs to the class of renormalizable theories [11].

Wilson's [1,12] elegant solution to all these problem was to define the theory on a lattice in such a way that the gauge integration gives a finite answer *and* also in a way so that he could discuss the theory beyond weak coupling perturbation theory so that color confinement could be studied. In addition, the lattice formulation also highlights a deep connection between field theories and statistical physics so that several techniques (such as high temperature expansions, ϵ-expansions, numerical simulations, etc.) can now be applied to QCD.

Consider a 4-dimensional lattice of points. This can be any lattice structure but for simplicity and concreteness, consider a hypercubic lattice with periodic boundary conditions. Let i label the lattice sites and μ the lattice directions. Instead of using the gluon fields A_μ^a, Wilson introduced the link variables U_μ given by,

$$U_\mu(i) = e^{iaA_\mu^a T^a} \tag{6}$$

The action of the lattice theory (the lattice analogue of the integral of the Lagrangian density in Eq. 4) is given by,

$$S(U) = \frac{1}{2n} \sum_{\mu\nu i} \mathrm{Tr}[U_\mu(i)U_\nu(i+\hat{\mu})U_\mu^\dagger(i+\hat{\nu})U_\nu^\dagger(i) + \mathrm{h.c.}] \tag{7}$$

The $U_\mu(i)$'s are nxn unitary matrices, elements of the $SU(n)$ group and are defined on the links of the hypercubic lattice. The sum over $\mu\nu$ runs over all unit squares (plaquettes) of the lattice, each counted once in the sum. The quantum theory is again defined through the lattice partition function,

$$Z_l = \int dU e^{\beta S(U)} \tag{8}$$

with $dU = \Pi_{\mu i} dU_\mu(i)$ and where $dU_\mu(i)$ is the group invariant Haar measure of $SU(n)$ for the link variable. The lattice partition function Z_l is much better defined than the continuum function of Eq. 4. This is because the group integrations are now over a compact domain.

It is easy to show that when Eq 7 is expanded around $aA_\mu^a = 0$, the leading term is identical to the Lagrangian density of Eq. 1 if one makes the identification

$$\frac{\beta}{2n} = \frac{1}{g^2}. \tag{9}$$

The action of Eq. 7 is certainly not the unique choice. Any action with the correct limit as $a \to 0$ (i.e., any action in the same universality class), would suffice.

There are many good review articles about lattice QCD through which the interested reader can pursue the subject further [2,3,13]. Here, only those issues that are relevant to the discussion at hand will be described. One such issue is that of the continuum limit. It is clear from the definition of the link variables U_μ (Eq. 6) that a provides a cutoff in momentum space since the field A^a_μ, crudely speaking, can only vary over the range $[-\frac{\pi}{a}, \frac{\pi}{a}]$. This is the lattice version of the regularization alluded to before which makes weak coupling perturbation theory sensible. The limit $a \to 0$ corresponds to the limit where the momentum space regulator goes to infinity. This is the limit in which contact is made with the underlying continuum theory. To do this, one must define order parameters in the lattice formulation and connect them with Green's functions (n-point functions) of the underlying field theory.

A useful set of order parameters on the lattice are the so called Wilson loop expectation values. These are the ensemble averages (in the sense of statistical mechanics with Z_l as the partition function) of directed products of the U variables along closed contours on the lattice. Expectation values of U's along open paths vanishes because of gauge invariance. Gauge invariance on the lattice is the invariance of the lattice action (Eq. 7) under,

$$U_\mu(i) \to g^{-1}(i) U_\mu(i) g(i + \hat{\mu}) \tag{10}$$

where $g(i) \subset SU(n)$. This is the lattice analogue of Eq. 5.

One particularly useful quantity is the expectation value of the product of U's along a TxR rectangular contour. It can be shown that in the limit $T \to \infty$ for fixed R, the negative of the logarithm of this quantity gives the potential energy $V(R)$ of two static (infinitely massive) color charges in a color singlet configuration. A strong coupling (large g^2 or small β) expansion then gives the result that the potential energy between the static sources is linear in the separation R with a coefficient (called the string tension σ) proportional to $\ln(g^2)$.

On the lattice one always computes dimensionless quantities. The string tension discussed above is therefore related to a 'physical' string tension (one which has dimensions) through the regulator a. Thus, with $R = r/a$ where r is measured in physical units (fermi),

$$V(R) = V_{phys}(r)a = \sigma R = (\sigma_{phys} r)a \tag{11a}$$

which implies that

$$\sigma = \sigma_{phys} a^2 \sim \ln(g^2) \quad \text{for } g^2 \text{large} \tag{11b}$$

It is clear from Eq. 11 that the limit $a \to \infty$ corresponds to $g^2 \to \infty$. This is the opposite of the continuum limit where contact with QCD is made. However, in this limit, static objects are permanently bound because their interaction potential increases with separation. The limit one would like to explore of course is $a \to 0$ which happens when $g \to 0$. To see this, one has to compute the so called Beta function of QCD in weak coupling perturbation theory. One finds that the relationship between a and g^2 in the limit $g^2 \to 0$ is [3,14],

$$\frac{1}{\xi} \sim \Lambda a = (\beta_0 g^2)^{-\beta_1/2\beta_0} e^{-1/[2\beta_0 g^2]} \tag{12}$$

where

$$\beta_0 = \frac{11}{3} \frac{n}{16\pi^2} \tag{13}$$

and

$$\beta_1 = \frac{34}{3} (\frac{n}{16\pi^2})^2 \tag{14}$$

are the first two coefficients of the theory's perturbative Beta function. All physical quantities on the lattice are measured in units of a. Thus the mass gap m_0 (or inverse correlation length), obtained for example by computing a correlation function, is related to a physical mass m_{phys} by

$$m_0 = a m_{phys} = \frac{1}{\xi} \tag{15}$$

Asymptotic freedom is the observation (see Eq. 12) that for a fixed Λ, $a \to 0$ is the same as $g^2 \to 0$. The color charge vanishes at zero distance.

The limit $m \to 0$ where the correlation length diverges, is therefore identical to the continuum limit where one would like to study the theory. But this is the limit where the theory approaches a second order phase transition. Hence, for field theory purposes, one is interested in the *critical* properties of lattice theories. In the case of QCD, the limit of interest is $g^2 \to 0$.

An outstanding problem is to prove that the two limits $g^2 = \infty$, where confinement of color sources can be rigorously demonstrated, and $g^2 \to 0$, where the theory is continuum QCD, are analytically connected. Such a demonstration would show that the physical string tension is finite and that color is confined in QCD. Whereas this has never been shown to be true analytically, there is very good numerical evidence that it is indeed true. This is from the beautiful work of the early pioneers of QCD numerical simulations [15]. However, many other features of QCD are not yet worked out such as a reliable calculation of the glueball spectrum and hadron masses. The most popular method to numerically study these issues is Monte Carlo simulation.

In the next Section however, I would like to describe a new method to study phase transitions that also dramatically verifies that there is no phase transition in lattice gauge theories separating strong and weak coupling. It turns out that this method is a very powerful alternative to standard numerical simulation methods and can be used very effectively to determine the order of phase transitions and accurately calculate critical exponents.

2. THE SPECTRAL DENSITY METHOD (SDM)

It is easiest to describe the SDM for a discrete theory although it can be applied (as we will see) to continuum theories as well. For concreteness, consider the $Z(2)$ gauge theory in four dimensions. The link variables U_μ now take only two values ± 1 so that the model may be called the Ising model of gauge theories. Define the action as,

$$S_2(U) = \frac{1}{2}\sum_{\mu\nu i}[1 - U_\mu(i)U_\nu(i+\hat{\mu})U_\mu(i+\hat{\nu})U_\nu(i)] \tag{16}$$

The partition function is

$$Z = \sum_{\{U\}} \exp[-\beta S_2(U)] \tag{17}$$

This model has been extensively studied both numerically and analytically [16,17]. It is known to be self dual and to have a first order phase transition at the self dual point $\beta_c = \ln(1+\sqrt{2})$.

Now, Eq. 17 can be rewritten as a sum over the action itself. Thus,

$$Z = \sum_{S_2} P(S_2)e^{-\beta S_2} \tag{18}$$

where $P(S_2)$ is the number of states in the ensemble of $U_\mu(i)$ with action S_2. The idea of the spectral density method [5] is to directly estimate the function $P(S_2)$ by generating configurations with S_2 values in a restricted set. Let us now consider how this is done in detail.

The contribution of any plaquette (one term in Eq. 16) to S_2 is either zero or unity. Hence, the total lattice action is restricted to the interval $[0, S_2^{\max} = N_P]$ where $N_P = 6L^4$ is the number of plaquettes in a hypercubic lattice of linear size L. It is also easy to see that the action can only take on even values. To compute $P(S_2)$, divide the total interval $[0, N_P]$ into $O(N_P/2)$ sets, each containing a fixed number of possible values of the total action. The sets are chosen so as to have the values of the action at the edges of successive sets equal. For example, one possible choice is $\{0,2,4,6\}$, $\{6,8,10,12\}$, $\{12,14,16,18\}\ldots$, etc. In principle, the number of action values in a set is irrelevant. In practice however, one makes a compromise between the need to have as few energies in a set as possible to get a good accuracy in measuring $P(S_2)$ and as many energies as possible to evolve rapidly in phase space during the updating and to ensure ergodicity. Since the action is integer valued, the number of allowed values of the action scales linearly with the system volume. This means that the number of sets also grows linearly with the volume.

Consider one of the sets of action values. To initialize the lattice, starting from an arbitrary configuration of $U_\mu(i)$ values, links are chosen at random and flipped so that the action is driven up (or down) to one of the values in the chosen set. The lattice is now ready to be updated. This is done by choosing links at random and again flipping the U's. If the change in the action from the link flip keeps the total lattice action within the set of chosen action values, the flip is accepted, otherwise not. As an aside, note that this algorithm seems intrinsically non vectorizable in a numerical calculation. However, it is possible to do a brute force vectorization by working simultaneously with a number of independent systems. After every trial update (or after a fixed number of trial updates), the lattice action is recorded. At the end of the simulation, one has a histogram of the number of times the system was found to be in the various action states in the set. This experiment is then repeated over all sets.

The occupation numbers within a set give an unbiased estimate of the relative number of states of the system at the corresponding values of the action. In other words, if $N(S_2)$ and $N(S_2')$ are the number of observations at action values S_2 and S_2' in a set, then $N(S_2)/N(S_2')$ is an unbiased estimator of the ratio $P(S_2)/P(S_2')$. Because the sets overlap in the values of the actions at the edges, the relative normalization of the P's between the sets can be determined. Since the overall normalization of the P's is known ($\sum_S P(S) = 2^{4L^4}$) and is in fact irrelevant, this procedure generates the full partition function (Eq. 18). We found that to obtain the smallest errors, the matching between the sets should be done from action values in the middle of the allowed range outwards towards the ends of the range. Note that if the lattice size is even and the gauge group has -1 as an element (as is true in this example), the density of states is symmetric ($P(S_2) = P(S_2^{\max} - S_2)$). One then needs to study only half the range of action values.

The partition function in Eq. 17 is a polynomial of degree $N_P + 1$ in $u = e^{-2\beta}$ with integer coefficients (the $P(S_2)$'s). Such a polynomial is uniquely specified by its zeros which come in complex pairs. For any finite volume, there cannot be a zero for real u. One possible way that a phase transition appears in Z in the thermodynamic limit is through a line of zeros pinching the real u axis. Indeed, this can rigorously be shown to happen in the two dimensional Ising model. At this point, let us digress for a moment to discuss the finite size scaling of these zeros.

3. FINITE SIZE SCALING OF Z AND ITS ZEROS [6,18]

The finite size scaling theory for the partition function discussed below is valid both for spin as well as gauge theories. Consider the behavior of Z as a function of u and the lattice size L. One can distinguish a region around the infinite volume critical point where the finite lattice is indistinguishable from the infinite volume theory. Here, $L >> \xi \sim 1$. Then, there is a scaling region, $L >> \xi >> 1$ where the finite lattice has the same scaling behavior as the infinite volume system. This is where we would like to be but seldom are. Finally, there is the region where $L \sim \xi >> 1$ where significant finite size effects are observed. The renormalization group can be used to study the latter two limits which may be said to define the scaling region. Most numerical work is done in the boundary between the last two regions.

Let $K_1, K_2, ...$ be the coupling constants of the theory. By the standard theory of finite size scaling [18], in the scaling region, under a change of scale by a factor of b, the free energy transforms as [6],

$$F_L(K_1, K_2, ...) = F_{L/b}(K_1 b^{y_1}, K_2 b^{y_2}, ...) + \epsilon_0(K_1, K_2, ...) \tag{19}$$

where ϵ_0 is a regular, smooth function of the couplings $K_1, K_2 ....$ For the partition function, we have then,

$$Z_L(K_1, K_2, ...) = Z_{L/b}(K_1 b^{y_1}, K_2 b^{y_2}, ...) e^{\epsilon_0} \tag{20}$$

In these expressions, the K_i's are the relevant and marginal couplings and the y_i's are the scaling dimensions of the associated operators. If b is a fixed fraction λ of L then the zeros of Z_L are the same as the zeros of,

$$Z_{1/\lambda}(K_1 L^{y_1} \lambda^{y_1}, ...) = Q(x_1, x_2, ...) \tag{21}$$

where

$$x_i = K_i L^{y_i} \tag{22}$$

are the scaling variables. The temperature and the magnetic field are always relevant couplings below the upper critical dimension. The analogues of these couplings in field theories like QCD are the coupling constant and the mass respectively. The first two leading scaling dimensions then are for the thermal and magnetic operators. Thus,

$$y_t = 1/\nu \tag{23}$$

and

$$y_h = \beta\delta/\nu \tag{24}$$

For large enough L, one may neglect the effect of the other operators. In this approximation (which is valid close to the infinite volume critical point), the zeros of Z are the zeros of the simple function,

$$Q(x_t, x_h) = Q([u - u_c]L^{1/\nu}, hL^{y_h}) \tag{25}$$

A formal solution of this equation is

$$h^2 L^{2y_h} = f_i([u - u_c]L^{1/\nu}) \tag{26}$$

and the reason that the solution is given in terms of h^2 is because the partition function is always invariant under $h \to -h$. Inverting this relationship, we find that the position of the zero closest to the infinite volume critical point scales with L as,

$$|u - u_c| \sim L^{-1/\nu} \tag{27}$$

Thus, a finite size scaling of the zeros of the partition function should yield an estimate of the correlation exponent ν. This is indeed numerically possible. An

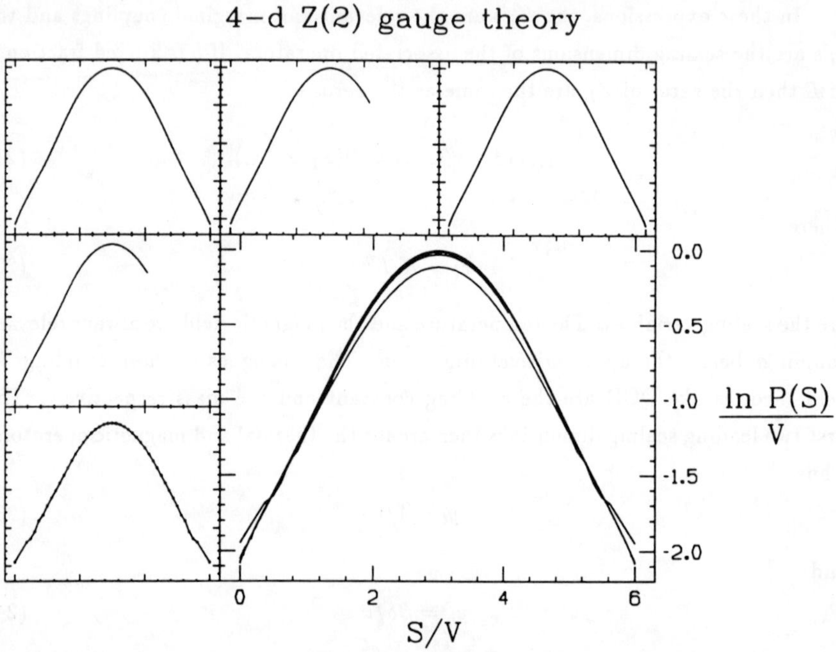

Figure 1.

Entropy density $\frac{\ln P(S)}{V}$ versus action per unit volume $\frac{S}{V}$ for the $Z(2)$ gauge theory on various lattice sizes. Reading clockwise from left bottom, the lattice sizes are $2^4, 3^4, 4^4, 5^4$ and 6^4. The larger plot is a superposition of data from all lattice sizes. Note the approach to a universal curve in the thermodynamic limit. Note also that on odd lattices, the periodic boundary conditions prevent values of S larger than a certain maximum value.

analysis of the three dimensional Ising model using an adaptation of the method described above to estimate the partition function [19] gave the result $\nu = 0.6295(10)$.

Let us now apply these ideas to the gauge theory we were discussing, namely $Z(2)$ gauge theory. Figure 1 shows the entropy density $\frac{\ln P(S_2)}{V}$ versus the action per site $\frac{S_2}{V}$ for the $Z(2)$ theory. From the data of Figure 1, one can easily compute any derivative of the free energy. For example, the average action per plaquette $< s >$, (normalized to range from unity to zero as β ranges from zero to infinity) is given

by:

$$1-<s>=-\frac{1}{N_P}\frac{\partial \ln Z}{\partial \beta}=\frac{1}{N_P}\frac{\sum_{S_2} S_2 P(S_2)e^{-\beta S_2}}{\sum_{S_2} P(S_2)e^{-\beta S_2}} \tag{28}$$

In Figure 2 I plot $< s >$ versus β for the various lattice sizes obtained from the data of Figure 1. Unlike the case of the usual Metropolis or heat bath simulation, where a strong hysteresis effect in this model makes it difficult to pinpoint the transition point, the spectral density method gives a very precise determination of the phase transition point (highlighted by a vertical line).

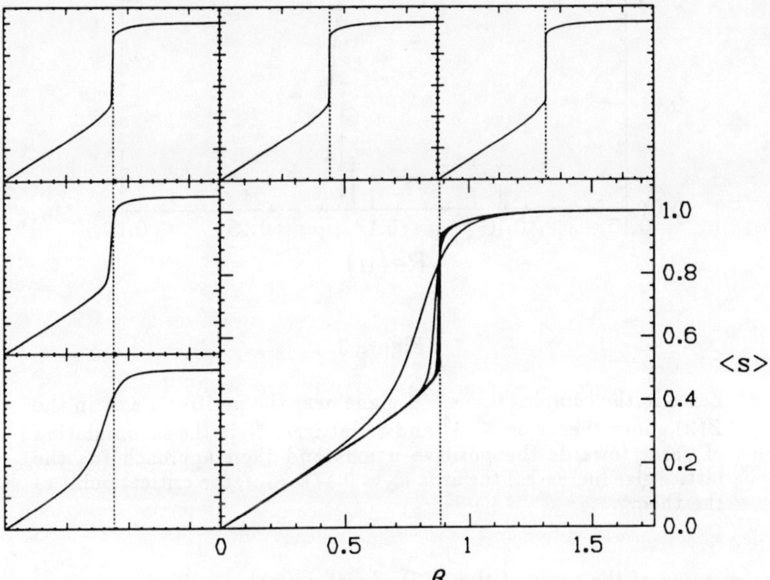

Figure 2.

The order parameter $< s >$ plotted as a function of β on various lattice sizes for the $Z(2)$ model. The vertical line is the location of the infinite volume phase transition point. Once again, the linear lattice size increases from 2 to 6 clockwise from bottom left.

As we have mentioned before, there is more information in the density of states $P(S_2)$ than merely the derivatives of the free energy. For example, one can study the analytic structure of Z via its zeros. Lee and Yang [20] and later, Fisher [21] have emphasized the study of these zeros of the partition function. In Figure 3

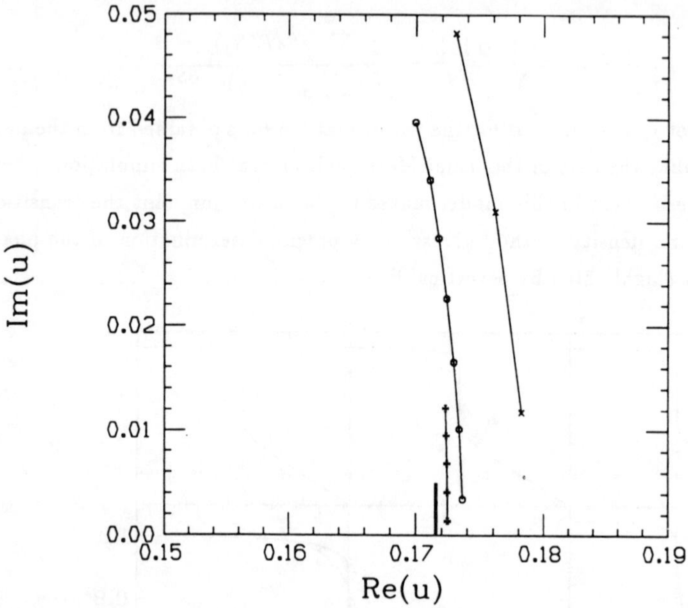

Figure 3.

Zeros in the complex $u = e^{-2\beta}$ plane near the positive u axis in the $Z(2)$ gauge theory on $3^4, 4^4$ and 5^4 lattices. Note the accumulation of zeros towards the positive u axis and their approach (as the lattice size increases) towards $u_c = 0.171573...$, the critical point of the theory.

we show some of the zeros of the $Z(2)$ model near the positive $u = e^{-2\beta}$ axis for lattice sizes $3^4, 4^4$ and 5^4. Since ν is effectively $1/d$ for a first order transition [22], one expects $|u - u_c| \sim L^{-1/4}$. Where $u_c = \frac{1}{3+2\sqrt{2}}$ is the critical value of u in the thermodynamic limit. In Figure 4 we show the fit of the data to this relation. One finds $\frac{1}{\nu} = 4.06(10)$ in agreement with expectations.

This direct measurement of $\nu = 1/d$ from the zeros is a confirmation of the first order nature of the phase transition in the $Z(2)$ gauge theory. Whereas this identification is not of great value in the present case, it is worth noting that one succeeded in finding ν from relatively small lattices. The spectral density method might therefore be very useful for systems with more subtle transitions. Recently, a study of the $U(1)$ gauge theory in four dimensions using the spectral density method

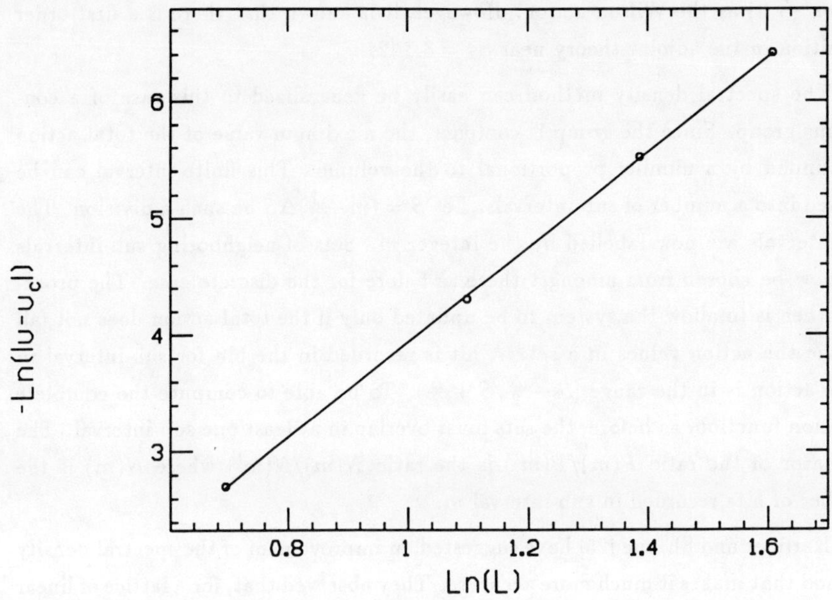

Figure 4.

Finite size scaling of the distance $|u - u_c|$ for the zero closest to the real positive u axis. The slope of the graph gives $1/\nu = 4.06(10)$.

has given unambiguous evidence of the first order nature of this transition [23] even though previous direct simulation methods could not completely resolve this issue.

4. GENERALIZATION OF THE METHOD TO QCD

Let us now consider the case of continuous groups [7]. In particular, consider the $SU(2)$ theory with two choices for the action, one identical with Eq. 7 with $n = 2$, which we will call the Wilson action, and another (the adjoint action) given by,

$$S_a(U) = \frac{1}{4} \sum_{\mu\nu i} \mathrm{Tr}|U_\mu(i) U_\nu(i + \hat{\mu}) U_\mu^\dagger(x + \nu) U_\nu^\dagger(x)|^2 \tag{29}$$

To distinguish it from the coupling β, we will call the adjoint coupling β_a. It is easy to show that the adjoint action has the same naive continuum limit (the

limit $a \to 0$) as the Wilson action. However, it is known that there is a first order transition in the adjoint theory near $\beta_a = 3.3$ [24].

The spectral density method can easily be generalized to this case of a continuous group. Since the group is compact, the maximum value of the total action is bounded by a number proportional to the volume. This finite interval can be divided into a number of sub-intervals. Let $S = (m + \frac{1}{2})\Delta S$ be such a division. The sub-intervals are now labelled by the integer m. Sets of neighboring sub-intervals can now be chosen from amongst these as before for the discrete case. The procedure then is to allow the system to be updated only if the total action does not fall outside the action values in a set. A hit is recorded in the bin for sub-interval m if the action is in the range $[S - \frac{\Delta}{2}, S + \frac{\Delta}{2})$. To be able to compute the complete partition function, as before, the sets must overlap in at least one sub-interval. The estimator of the ratio $P(m)/P(m')$ is the ratio $N(m)/N(m')$ where $N(m)$ is the number of hits recorded in sub-interval m.

Karliner and Sharpe [25] have suggested an improvement of the spectral density method that makes it much more accurate. They observed that, for a lattice of linear size L,

$$P_L(S) \sim \exp[V s(\frac{S}{V}) + O(1)] \tag{30}$$

where $s(x)$ is the entropy density which is a smooth function of its argument. Hence, given $P_L(m)$ for one volume V, one can estimate the thermodynamically relevant part of it for any other volume V'. If one has such a guess (say $P_L^g(m)$ for $P_L(m)$) then the updating procedure is revised as follows: When a variable is changed and the action changes from S to S' in a set, the change is not automatically accepted. It is only accepted conditionally with probability $P_L^g(m)/P_L^g(m')$. This has the effect of making the $N(m)$'s in a given set more or less equal and this increases the accuracy of the final answer. This is particularly true in the case of a transition at $g^2 = 0$ where the tail of the distribution $P_L(S)$ is very important. With the new acceptance method, the estimator of $P_L(S)/P_L(S')$ is $\frac{P_L^g(m)}{P_L^g(m')}\frac{N(m)}{N(m')}$. Even a crude initial estimate of P_L^g makes a dramatic difference in the accuracy [25].

An important point to discuss is what the size of the sub-intervals should be. In practice, we have found it sufficient to determine the size of the sub-intervals from a trial simulation on a small volume. Keeping that sub-interval size constant as the volume increases increases the number of sub-intervals linearly with the volume. This we found to be sufficiently accurate. Of course, one should also do a check of the extent to which the discretization of the action interval affects observed quantities and the zeros. We find that it is easily possible to do an accurate simulation such that the systematic error from the finite size of the interval is smaller than the statistical error even when the latter is one part in 10^4 or less. These issues have been studied in detail by Carter for the two dimensional XY model and will be reported in his Ph.D. thesis [26].

An issue to address is how the computer time required to calculate the $P(S)$'s to a fixed relative accuracy $\frac{dP}{P}$ scales with the volume. One possible concern might be that since the number of states is exponential in the volume, to get a fixed relative accuracy in $P(S)$ it might be necessary to increase the computer cost exponentially in volume. This expectation is incorrect. The computation of all the P's for a fixed accuracy scales at most like V^2 [4,5,7,19]. Another point to make is that the method is good for more than just computing the free energy and zeros. If O is any operator and we define the average value of O for fixed action S to be $O(S)$, then the ensemble average of O is clearly given by, $< O >= \frac{\sum_S O(S)P(S)e^{-\beta S}}{\sum_S P(S)e^{-\beta S}}$. Obviously, $O(S)$ can be computed during the numerical experiment that estimates $P(S)$.

The results from studying the Wilson and adjoint actions on rather small volumes [7] are shown in Figures 5 and 6. It is clear from the figure that the specific heat for the Wilson theory has a gentle bump suggesting crossover behavior while for the adjoint action there is a distinct discontinuity developing. The results are even more dramatic if one plots the zeros. Figure 7 shows these zeros. The adjoint action has a band of zeros pinching the real $u = e^{-\beta/8}$ axis near $\beta \sim 3.3$ while the Wilson theory has no such band. In particular, note that in both cases, there is a line of zeros pointing to the origin in the u plane, indicating the expected singularity (see Eq. 12) at $\beta = \infty$. As is clear from Eq. 12, this is an essential singularity and hence there is no real definition of ν as in the usual case. However, the distribution of zeros indicates clearly that the strong coupling limit $u = 1$ is analytically connected to the weak coupling limit $u = 0$ which is a visual illustration of quark confinement.

Finite size scaling of the zeros close to the phase transition can also yield useful information. Indeed, if g_L^2 and $g_{L'}^2$ are the locations of the zeros nearest to $g^2 = 0$ on lattices of linear size L and L', then

$$\frac{L}{L'} = \exp[\int_{g_L^2}^{g_{L'}^2} \frac{dg^2}{B(g^2)}] \tag{31}$$

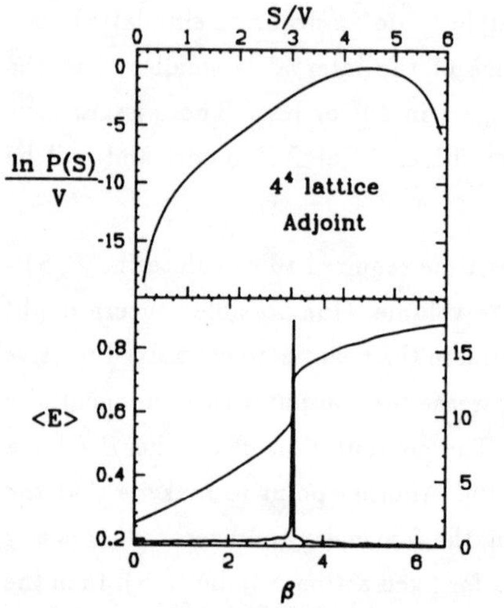

Figure 5. The partition function, $< E >$ and C_v for the adjoint theory on a 4^4 lattice. Notice the abrupt change in the order parameters corresponding to the expected first order transition near $\beta = 3.3$.

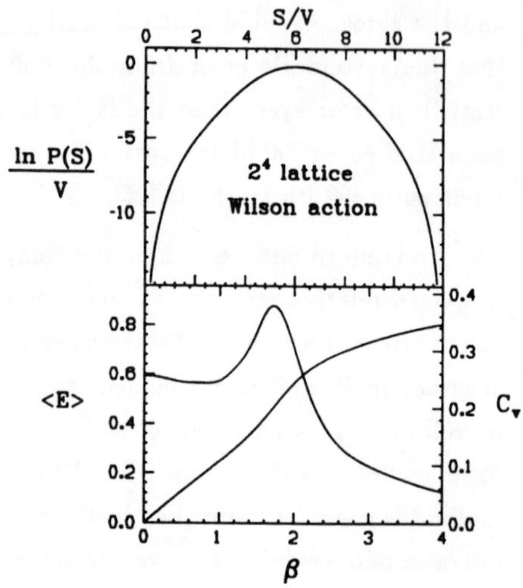

Figure 6. The same as Figure 5 but now for the theory with the Wilson action (Eq. 7).

where $B(g^2)$ is the full Beta function of the theory, defined as $B(g^2) = \frac{dg^2}{d\ln(a)}$. Thus one can use the spectral density method to compute the Beta function of QCD. This application of the spectral density method has not yet been attempted to my knowledge.

Crudely speaking, the spectral density method replaces the Monte Carlo importance sampling method of evaluating integrals by a version of the Riemann sum approximation. This is particularly useful when the action is complex. In QCD, the action is indeed complex for non-zero vacuum angle θ in instanton physics and for non-zero chemical potential when $n > 2$. In such situations, one expects the spectral density method to be (probably) the only method that might be useful.

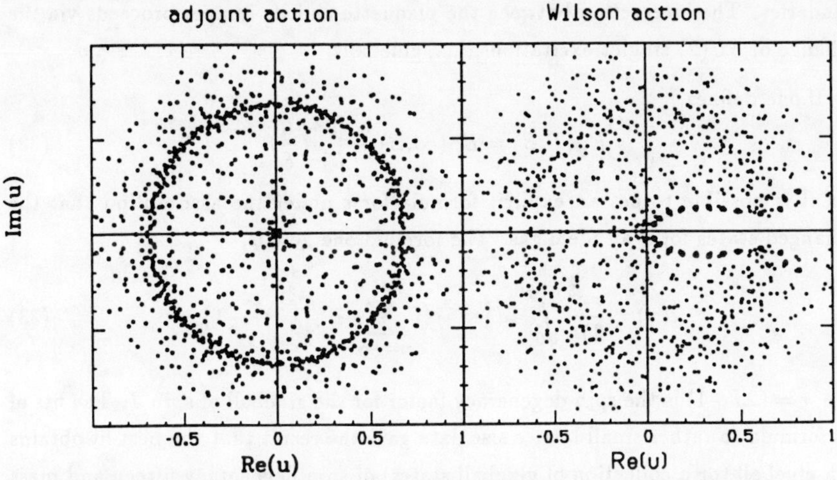

Figure 7.

The zeros for a 2^4 lattice for the adjoint action $SU(2)$ and the Wilson action. The variable plotted is $e^{-\beta/8}$ for the adjoint action theory and $e^{-\beta/4}$ for the Wilson theory. Notice that for the adjoint action, where one expects a first order transition [24], there is a line of zeros almost pinching the positive u axis. For the Wilson theory, there is no such line except near $u = 0$ ($\beta = \infty$) corresponding to the asymptotically free fixed point of the theory. Note also the presence of a shoulder of zeros near $\beta = 2.0$ corresponding to the crossover from weak to strong coupling.

5. VOLUME DEPENDENCE OF THE MASS GAP IN QCD

The thermodynamic limit is always the limit of interest. However, reality forces us to work on finite systems. This introduces finite size corrections to asymptotic formulae. Quite often, these finite size corrections may even give useful information about the theory. In such an attempt, Brower, Creutz and Nauenberg tried to estimate the mass gap in $SU(2)$ gauge theory by looking at the size dependence of the average plaquette expectation value,
$E_P(L,\beta) = \frac{1}{2}\mathrm{Tr} < U_\mu(i)U_\nu(i+\hat{\mu})U_\mu^\dagger(i+\hat{\nu})U_\nu^\dagger(i) >$, which is the average of the trace of the directed product of four $U_\mu(i)$ fields around a unit lattice plaquette. The idea was that on a finite system with periodic boundary conditions, the finite size corrections come from self correlations of the plaquette with its images over the

boundaries. The interaction between the plaquette and its images proceeds via the exchange of $SU(2)$ singlet excitations, i.e., glueballs.

If one defines,

$$E_P(L,\beta) = E_P(\infty,\beta) + \frac{1}{L^4}\epsilon(\frac{L}{\xi}) \tag{32}$$

then it is possible to get an explicit formula for ϵ under the assumption that the exchanged states form an ideal gas. The formula one gets is,

$$\epsilon(x) = \frac{rx^3}{176}\int_0^\infty d\lambda e^{-\frac{x}{2\lambda}}\{(\sum_{n=-\infty}^{\infty} e^{-\frac{n^2 x\lambda}{2}})^4 - 1\} \tag{33}$$

Here, $r = (2J+1)$ is the spin degeneracy factor for the glueball of spin J. The fits of this formula to rather small lattice size data gave the result that the best fit obtains for a glueball (or a collection of glueball states) of spin degeneracy fifteen and mass around $m_0 = (1.9 \pm 0.2)\sqrt{\sigma_{phys}}$. It is now clear from the analytic work of Lüscher and collaborators (see below) and the subsequent numerical studies that indeed with periodic boundary conditions and small system sizes, the 2^{++} and 0^{++} states (the superscripts correspond to parity and charge conjugation) are almost degenerate in mass and this validates the results of Ref.[27]. The formulae of Eqs. 32 and 33 should be applied by the Monte Carlo enthusiasts to their new data on larger systems.

I now turn to the results of Lüscher on the corrections to masses in periodic boxes [8]. Let $\phi(x) = \mathrm{Tr}F_{\mu\nu}F_{\mu\nu}$ or some lattice analogue of this such as the average plaquette. One could imagine an effective field theory of the ϕ field which for concreteness, one can think of as a ϕ^4 theory with quartic coupling λ_4. One usually computes the mass gap m_0 in the gauge theory by looking at the correlation function,

$$C(t) = < \phi(t)\phi(0) > \tag{34}$$

along the time (T) direction of an L^3T system with $T >> L$. On a finite lattice, the exponential fall off of $C(t)$ gives a lattice estimate M_0 of m_0. Thus, for t large,

$$C(t) \sim e^{-M_0 t} \tag{35}$$

The deviation $\delta_0 = (M_0 - m_0)/m_0$ is a function only of the Fisher scaling variable $z = m_0 L$ and can be related to the forward scattering amplitude for glueball-glueball scattering. It is possible to express δ_0 in terms of physically measurable quantities. The method is essentially to compute the finite volume contributions to the propagator of the ϕ field in the effective field theory. If one writes,

$$< \phi(x)\phi(0) >= \int_{-\infty}^{+\infty} \frac{dp_0}{2\pi} L^{-d} \sum_{\vec{p}} e^{ipx} \frac{1}{[m_0^2 + p^2 - \Sigma_L(p)]} \tag{36}$$

then the L dependence of $\Sigma_L - \Sigma_\infty$ can be computed in a loop expansion for the inverse propagator by replacing systematically one propagator in each diagram of the loop expansion by its finite lattice equivalent expression. In some sense, these are the contributions from particles that are exchanged around the lattice because of the finite size and the periodicity of the lattice. The one loop result of Lüscher is [8],

$$\delta_0 = \frac{3\lambda_4}{16\pi^2 m_0^2 L} \int_0^\infty dy e^{-\sqrt{m_0^2+y^2}L} \tag{37}$$

and generalized to all orders in perturbabtion theory,

$$\delta_0 = -\frac{3}{16\pi^2 m_0^2 L} \int_0^\infty dy e^{-\sqrt{m_0^2+y^2}L} F(iy) + O(e^{-\bar{m}L}) \tag{38}$$

Here $\bar{m} = \sqrt{3/2}m_0$ and F is the analytic continuation of the forward scattering amplitude. If there is an effective three point coupling in the theory for ϕ, there is an additional term in δ_0 which falls of as $\frac{1}{L}e^{-\frac{\sqrt{3}}{2}m_0 L}$ and which dominates as $L \to \infty$ over Eq. 38. Lüscher argues that these formulae are of a kinematical nature and therefore should apply to all theories, including QCD. They have indeed been verified in the non-linear σ-model.

Unfortunately, it is too early to test these results on QCD. The problem is that the lattice sizes that are accessible on present computers are too small. To explain why this might be, consider again Eq. 12 which gives the relationship between the correlation length and the bare coupling. In this equation Λ is a parameter that is specific to a particular choice of action and lattice. By numerical experiment, one finds the onset of behavior similar to Eq. 12 near $g^2 n \sim 4$ in $SU(n)$ gauge theories on hyper-cubic lattices. Near this coupling, the lattice correlation length is close to unity. One would instead like to work in a regime where the correlation length is bigger than a few lattice spacings *and* where the lattice size is bigger than a few correlation lengths. This means that one wants to go to smaller g^2 and increase the lattice size as much as possible. Unfortunately, Eq. 12 says that the signal for m_0 decreases exponentially as g^2 decreases. Moreover, the correlation length increases exponentially as g^2 decreases. Hence, to keep $z = L/\xi$ constant or increasing, one is forced to increase the lattice size exponentially. These limitations have prevented going to large z values on lattices with equal extent in all directions. Recently however, in an attempt to increase the effective z, most simulations have been forced to use asymmetric lattices. The numerical simulation situation has been summarized in recent articles [28] and I will not discuss it here. I would like to discuss however some other very interesting work by Lüscher that allows one to compute the glueball masses in a perturbation theory in $g^{2/3}$ for small z. It should be emphasized (and this is something that the enthusiasts sometimes forget) that the method described below is essentially perturbative and so *cannot* apply in the thermodynamic limit. However, that does not take away from its usefulness in understanding numerical simulation results and its potential of providing hints about the large volume limit.

6. THE SMALL Z EXPANSION [9,29]

For the moment I will assume that we have periodic boundary conditions. Consider the case when $z = m_0 L$ is small. This means that the scale over which physics happens ($\xi = \frac{1}{m_0}$) is much bigger than the size of the system. In QCD, this means that the whole system is 'small' compared to the size of hadrons or glueballs and because of asymptotic freedom, one should be able to compute everything in perturbation theory in the coupling constant g. The Hamiltonian of pure gauge theory on an L^3 box can be easily obtained using the transfer matrix formalism from the Lagrangian. Thus,

$$H = \int_0^L d^3x \{ \frac{1}{2} g^2 E_k^a(x) E_k^a(x) + \frac{1}{2g^2} B_k^a(x) B_k^a(x) \} \tag{39}$$

with $\vec{E}$ and $\vec{B}$ the analogue of the Electric and Magnetic fields in Maxwellian Electrodynamics, $E_i = F_{0i}$, $B_i = \frac{1}{2}\epsilon_{ijk} F_{jk}$. For small g^2, the magnetic energy B^2 is bounded only for configurations with $B_k = 0$. Such configurations are called Torons. Because of the non-trivial topology of the torus (lattice with periodic boundary conditions), the condition $B_k = 0$ is not the same as the statement that A_k is gauge equivalent to $A_k = 0$. The general solution that satisfies $B^2 = 0$ is parametrized by a set of angles, one for each group index in $SU(n)$. The toron solution is the diagonal matrix for the gluon field A_μ with these angles along the diagonals. The only constrain on the angles is that they add up to zero for each gauge field. It turns out however that the Hilbert space breaks up into non interacting sectors (at least in perturbation theory) with the same spectrum. Hence, it is sufficient to expand around $A_k = 0$ [9].

As we have alluded to before, perturbation theory requires that one eliminate the unphysical degrees of freedom associated with gauge invariance. This can be done in many ways. One popular choice is to work in the so called Coulomb gauge. One eliminates the gauge degrees of freedom by imposing the gauge constraint,

$$\partial_k A_k(x) = 0 \tag{40}$$

The Hamiltonian incorporating the Coulomb gauge constraint may now be expanded in momentum modes of the field $A_k^a(x)$. Thus,

$$A_k^a(x) = g^{2/3} L^{-d/3} c_k^a + g q_k^a(x) \tag{41}$$

where the mode c_k^a is a zero momentum mode (and therefore independent of x) and the modes q_k^a are normal modes with the constraint $\int d^d x q_k^a(x) = 0$. The canonical momenta of the A_k field are also expanded similarly. Define,

$$e_k^a = -i \frac{\partial}{\partial c_k^a} \tag{42}$$

and the canonically conjugate momenta p_k^a of q_k^a. p_k^a also satisfies a constraint that its space integral vanishes.

After substituting this into the Coulomb gauge Hamiltonian and doing degenerate perturbation theory, Lüscher [9] finds that the Hamiltonian can be expanded in powers of $g^{1/3}$. Thus,

$$H = H_0 + g^{2/3} \Delta H \tag{43}$$

with

$$\Delta H = \sum_{\nu=0}^{\infty} g^{\nu/3} H_{(1+\nu)} \tag{44}$$

It is instructive to write down H_0 and H_1. One gets,

$$H_0 = \frac{1}{2} \int_0^L d^d x \{ p_j^a p_j^a + \partial_l q_j^a \partial_l q_j^a \} \tag{45}$$

and

$$H_1 = L^{-d/3} [\frac{1}{2} e_k^a e_k^a + \frac{1}{4} (f^{abc} c_k^b c_l^c)^2] \tag{46}$$

Notice that only normal modes contribute to H_0 and only zero modes contribute to H_1. The reason for the normalization (funny powers of g) in Eq 41 is now clear. It was done to make sure that the terms in Eq. 46 have equal weight. The reason this must be so is that the presence of the interaction term $\frac{1}{4}(f^{abc}c_k^b c_l^c)^2$ causes the spectrum to be discrete whereas without it, it would be continuous. This means that both terms are equally important for any non-zero g and the field rescaling makes this explicit.

Now, states on the lattice transform as irreducible representations of the symmetry group of the lattice. This determines their spin quantum numbers on the lattice. From the perturbation expansion above, one can compute the ratio of masses of states of definite symmetry under the cubic group. These are now predictions for the way these ratios must behave in numerical simulations. One of the most striking such prediction of the perturbative analysis was that the 0^{++} state was almost degenerate with the 2^{++} state. This was indeed observed in the numerical calculations for 'small' z values. Since then, the numerical situation has improved somewhat with data available on larger z systems [28]. The analytic situation has also improved with significant contributions from Köller and van Baal [30]. Figure 8 is a compilation of all data to date for the glueball mass and the string tension in SU(2) and SU(3) lattice gauge theory. The solid curves in Figure 8 give the analytic predictions from the work of Köller and van Baal [30,31], Lüscher [9] and Vohwinkel [32]. Whereas the agreement between theory and simulation is impressive, it is not clear that we are at an asymptotic z value yet to make a statement about the infinite volume theory with any confidence. For SU(3) there exists a great deal of data on masses in the scaling region. However, the analogue of the calculations of Refs. 9, 30, 31 and 32 are not yet completed for SU(3) so a comparison between theory and simulation is not yet possible..

It has been pointed out that it might be more advantageous to study pure gauge theory on a lattice with twisted boundary conditions [33]. This is a generalization of the periodic boundary conditions under which, instead of mere periodicity, we have,

$$U_\mu(i + L_\mu \hat{\mu}) = \Omega_k U_\mu(i) \Omega_k^\dagger \tag{47}$$

where the Ω_k are constant $SU(n)$ matrices satisfying the ('t Hooft) consistency condition,

$$\Omega_k \Omega_l \Omega_k^\dagger \Omega_l^\dagger = e^{2\pi i \epsilon_{klj} m_j / n} \tag{48}$$

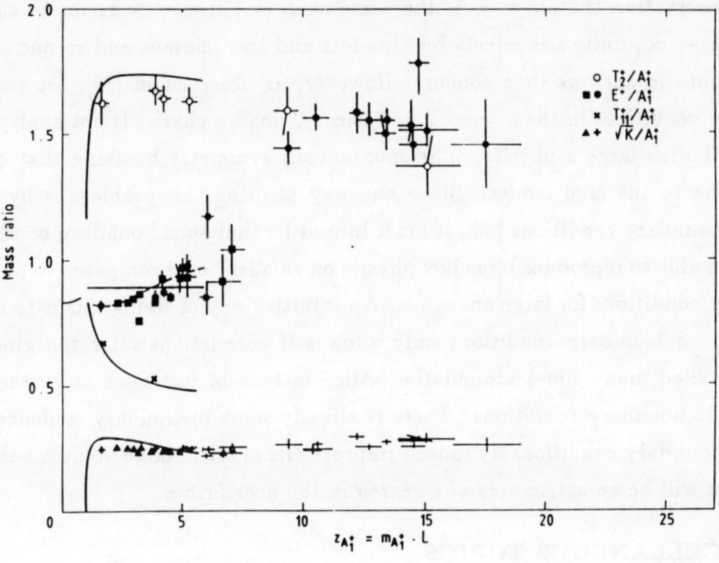

Figure 8.

The SU(2) lattice Monte Carlo data for the mass ratios and string tension $\sqrt{K}$ as a function of the Fisher scaling variable $z_{A_1^+}$.

with $\vec{m}$ the magnetic flux vector. The advantage of twisted boundary conditions is that the constant field modes (the zero momentum modes) are absent. It is indeed these modes that cause the small z spectrum to be quite different from the large z asymptotic results that one is looking for. The reason this must happen can also be understood from a large n argument.

Eguchi and Kawai [34] observed that the Schwinger Dyson equations of the $n = \infty$ theory in an infinite box are formally identical with those of the same theory on a one site lattice with periodic boundary conditions. It was pointed out that although this was true on a formal level, the solutions of these equations (which of course give the physics) did not respect this correspondence. The reason was that the zero modes of the fields caused a spontaneous breaking of the gauge symmetry. Two different cures [35,36] were suggested and one of these [36] was to use twisted boundary conditions in the one site model. With twist, the analogy of Eguchi and Kawai did indeed survive even in the solutions. Now on a naive level, the Eguchi-

Kawai observation that $L = \infty$ is the same as $L = 1$ if $n = \infty$,seems to suggest that as $n \to \infty$, finite size effects become less and less relevant and so one should see asymptotic answers in z sooner. However, as observed in [35], for periodic boundary conditions in the $n = \infty, L = 1$ theory, small z physics is not analytically connected with large z physics. The spontaneous symmetry breaking that causes this is due to the zero modes. Since one way of using this problem is by using twisted boundary conditions [36], it must indeed be that such boundary conditions are better able to reproduce large box physics on smaller boxes compared to periodic boundary conditions for large enough n. An intuitive way of seeing this is to notice that twisted boundary conditions only allow self correlations after the glueballs have travelled many times around the lattice instead of just once as in the case of periodic boundary conditions. There is already some preliminary evidence that twisted boundary conditions do indeed improve the onset of large volume behavior [37]. This will be an active area of research in the near future.

7. MISCELLANEOUS TOPICS

Finally, I would like to mention a few areas in which finite size scaling notions are beginning to be used in novel ways. In my opinion, these studies show great promise but are currently in a development stage and so cannot be reviewed. However, they are extremely interesting from the point of view of future research and therefore deserve mention.

The first such application is in QCD. Although QCD confines color charges so that only singlet states of color are seen in asymptotic states at zero temperature, when the temperature is high enough, there is a phase in which the color charges are free. This is the so called plasma phase of QCD. It is currently of interest to study this phase because of the possibility that the ion-ion experiments planned at Brookhaven Lab and CERN which might see such a phase [38]. An active issue is the one of the finite size scaling of the potential energy of two static color charges in the QCD vacuum with a Debye screening length (in the plasma phase). Unless this scaling form is known, it is not possible to extract useful information from numerical simulations. There has been a preliminary numerical study to find a scaling form [39]. It is important to understand this question further.

The next useful application is to compute the deconfinement temperature from finite size scaling analysis. Once again, some preliminary ideas exist [40] but they need much more work.

Finally, I would like to mention the problem of resonances and how to extract particle masses from simulations on finite lattices if they are unstable under decay into lower mass objects. This is discussed in a recent study by Wiese [41] but the kind of finite size scaling analysis suggested in this paper has never yet been performed numerically to my knowledge.

ACKNOWLEDGEMENTS

This research is supported in part by the Florida State University Supercomputer Computations Research Institute which is partially funded by the Department of Energy through contract number DE-FC05-85ER250000. I thank Martin Lüscher, Claus Vohwinkel, Michael Creutz, Pierre van Baal, Bernd Berg, Helmut Satz and Chris Korthals-Altes for pointing me to relevant literature for the review. I also thank Chris Korthals-Altes for patiently explaining the small z expansion to me. Figure 8 of this paper was obtained from P. van Baal and I thank him for that.

REFERENCES

1. K. Wilson, Phys. Rev. D24, 2445 (1974).
2. See C. Rebbi, "Lattice Gauge Theories and Monte Carlo Simulations," World Scientific, Singapore (1983) for a review of the subject and the literature.
3. For a discussion of the relevant background material, see M. Creutz, "Quarks, Gluons and Lattices," Cambridge University Press, 1983. Also see, M. Creutz, L. Jacobs and C. Rebbi, Phys. Rep. 95, 201 (1983).
4. G. Bhanot, S. Black, P. Carter and R. Salvador, Phys. Lett. B183, 331 (1987).
5. G. Bhanot, K. Bitar, S. Black, P. Carter, R. Salvador, Phys. Lett. B187, 381 (1987).
6. C. Itzykson, R. B. Pearson and J. B. Zuber, Nucl. Phys. B220, 415 (1983).
7. G. Bhanot, K. Bitar, R. Salvador, Phys. Lett. B188, 246 (1987).
8. M. Lüscher, in "Fields, Strings and Critical Phenomena," Lectures given at Summer School Les Houches, France, 1988, DESY preprint, DESY 88/156. Also see M. Lüscher in "Progress in Gauge Theory," Cargese, 1983), eds. G. 't Hooft, et al., Plenum (New York, 1984); M. Lüscher, Comm. Math. Phys. 104, 153 (1986); ibid, 177 (1986).
9. M. Lüscher, Nucl. Phys. B219, 233 (1983).
10. For a thorough introduction to quantum field theory, see, "Quantum Field Theory" by C. Itzykson and J.-B. Zuber, McGraw Hill (USA, 1980). For an introduction to QCD see W. Marciano and H. Pagels, Phys. Rep. 36, 137 (1978).
11. G. 't Hooft, Nucl. Phys. B33, 173 (1971); ibid., B35, 167 (1971).

12. K.G. Wilson and J. Kogut, Phys. Rep., C12, 77 (1974).

13. The current and evolving numerical situation in lattice gauge theory is discussed in: "Lattice Gauge Theory '86," eds. Satz, Harrity and Potvin, Plenum (New York, 1987); ' Field Theory on the Lattice', eds. Billoire, Lacaze, Morel, Napoly and Zinn-Justin, Nuclear Physics B (Proceedings Supplement 4), 1988. Also see Proceedings of 1988 Lattice Conference at Fermilab, to appear in Nucl. Phys. B.

14. D.J. Gross and F. Wilczek, Phys. Rev. Letts. 30, 1343 (1973); H. D. Politzer, Phys. Rev. Letts. 30, 1346 (1973).

15. M. Creutz, Phys. Rev. Letts. 45, 313 (1980); ibid, Phys. Rev. D21, 2308 (1980).

16. M. Creutz, L. Jacobs and C. Rebbi, Phys. Rev. Lett. 42, 1390 (1979).

17. R. Balian, J. M. Drouffe and C. Itzykson, Phys. Rev. D10, 3376 (1974); ibid, D11, 2098 (1975); ibid, D11, 2104 (1975); ibid, *Errata*, D19, 2514 (1979).

18. M.L. Glasser, V. Privman and L.S. Schulman, Phys Rev. B35, 1841, (1987).

19. G. Bhanot, S. Black, P. Carter, R. Salvador and R. Toral, Phys. Rev. Letts. 59, 803 (1987).

20. T. D. Lee and C. N. Yang, Phys. Rev. 87, 404, 410 (1952).

21. M. E. Fisher in "Lectures in Theoretical Physics," ed W. E. Brittin, (University of Colorado Press, Boulder, 1965) Vol 7c, p1.

22. B. Nienhuis and M. Nauenberg, Phys. Rev. Lett. 35, 477 (1975); M. E. Fisher and A. Nihat Berker, Phys. Rev. B 26, 2507 (1982).

23. See S. Sharpe, U. of Washington Preprint, Dec. 1988. To appear in the proceedings of the 1988 Fermilab Conf., Sept., 1988 and references therein.

24. G. Bhanot and M. Creutz, Phys. Rev. D24, 3212 (1981); I. G. Halliday and A. Schwimmer, Phys. Letts. 101B, 327 (1981); J. Greensite and B. Lautrup, Phys. Rev. Letts. 47, 9 (1981).

25. M. Karliner, S. Sharpe and Y. Chang, Nucl. Phys. B302, 204 (1988).

26. P. Carter, Ph.D. Thesis, Florida State University, 1989.

27. R. Brower, M. Creutz and M. Nauenberg, Nucl. Phys. B210, 133 (1982).

28. B. Berg and A. Billoire, Florida State University preprint, FSU-SCRI-88-136; B. Berg and C. Vohwinkel, Florida State University preprint, FSU-SCRI-88-137; P. van Baal, "Gauge Theory in a Finite Volume," CERN preprint, CERN-TH.5141/88; P. van Baal and A. Kronfeld, CERN preprint, CERN-TH.5241/88.

29. M. Lüscher and G. Münster, Nucl. Phys., B232, 445 91984); P. Weisz and V. Ziemann, Nucl. Phys. B284, 157 (1987).

30. J. Köller and P. van Baal, Nucl. Phys. B273, 387 (1986); ibid, Annals of Physics 174, 299 (1987)

31. J. Köller and P. van Baal, Phys. Rev. Lett. **58**, 2511 (1987); Nucl. Phys. **B302**, 1 (1988); Nucl. Phys. **B4** (Proc. Suppl.), 47 (1988). For references to the simulations where the data in Figure 8 comes from, see P. van Baal and

A. Krönfeld, CERN-TH.5241/88, to be published in the proceedings of 'Lattice 88', Fermilab, 1988.

32. C. Vohwinkel, Phys. Lett. **213B**, 54 (1988).

33. J. Groenveld, J. Jurkiewicz and C. P. Korthals Altes, Physics Scripta 23, 1022 (1981); C. P. Korthals Altes, CNRS Marseilles preprint, CPT-88/P.2190, Dec. 1988.

34. T. Eguchi and H. Kawai, Phys. Rev. Lett. 48, 1063.

35. G. Bhanot, U. Heller and H. Neuberger, Phys. Letts. 113B, 47 (1982).

36. A. Gonzales-Arroyo and M. Okawa, Phys. Letts. 120B, 174 (1983).

37. see Fig. 2 and related references in the conference proceedings paper of Ref. 31.

38. For a recent review, see the Proceedings of the Quark Matter '87 Conference, Z. Phys. C38; also see, J. Cleymans, R. V. Gavai and E. Suhonen, Phys. Rep. 130, 217 (1986).

39. J. Engels, F. Karsch and H. Satz, CERN preprint, CERN-TH.5024/88.

40. B. Berg, C.P. Korthals Altes, C. Vohwinkel, Phys. Letts. B209, 319 (1988).

41. U.-J. Wiese, Jülich Preprint, HLRZ Jülich 88-08.

X. SIMULATIONS OF SPIN GLASS SYSTEMS

A. Peter Young

TABLE of CONTENTS

Simulations of Spin Glass Systems

A. Peter Young

Physics Department
University of California Santa Cruz
Santa Cruz, CA 95064

1. INTRODUCTION

Spin glasses are magnetic systems which are disordered and "frustrated", by which one means that the interactions between the magnetic moments are in conflict with each other. Since this article will be rather short, the reader is referred to the reviews [1] for detailed background on spin glasses and for examples of experimental systems. For frustrated systems it is non-trivial even to determine the ground state configuration. In fact there is plenty of evidence that there are many states with energy close to the ground state but which differ in the orientation of a large number of spins. This degeneracy is characteristic of frustrated systems and is one of the reasons that analytical work has been very hard so most of the theoretical effort has used one sort of numerical approach or another. Experimentally, one observes a rapid change in behavior below a fairly well defined freezing temperature. The susceptibility suddenly drops and relaxation times increase astronomically, becoming longer than the time of the experiment so a remanent magnetization is observed when the applied field is reduced to zero. It seems clear that the large relaxation times come from the system having to make a fluctuation from one local minimum of the free energy to another by passing over a barrier whose height is much larger than the temperature.

No conventional type of long-range order, such as ferromagnetism or antiferromagnetism, can arise in spin glass systems, but the question of whether a sharp phase transition to a new type of "ordered" state can occur has been hotly debated. One of the techniques that has been used successfully to investigate this problem is the combination of Monte Carlo simulations and finite-size scaling. This article will discuss the finite size scaling approach and describe the results that have been obtained with it for spin glass models. Gratifyingly, the results agree fairly well with experiments.I will not discuss the basics of Monte Carlo simulations, since these are well described elsewhere [2].

2. THE SPIN GLASS PHASE TRANSITION

To see how a spin glass transition may arise it is helpful to make an analogy with ferromagnetism. At a second order phase transition there is a correlation length which diverges. For a ferromagnet this is simply the distance over which the spins have a net parallel alignment. Since correlations decay exponentially, to within a good approximation, we can write

$$<S_i\ S_j> \sim \exp(-R_{ij}/\xi_F) \tag{1}$$

which defines the ferromagnetic correlation length ξ_F. Here, R_{ij} is the distance between lattice points i and j. For convenience we are assuming Ising spins, where $S_i = \pm 1$, but this is not essential. As the transition temperature, T_c, is approached from above the ferromagnetic correlation length diverges with a power of $t = (T - T_c)/T_c$, i.e.

$$\xi_F \sim t^{-\nu} \quad . \tag{2}$$

While ξ_F can be measured by neutron scattering it is easier to determine the ferromagnetic susceptibility χ_F, which is proportional to the sum of the spin-spin correlation functions over all distances, i.e.

$$\chi_F = \sum_j <S_i\ S_j> \quad . \tag{3}$$

(Strictly this is T times the susceptibility but the factor of T is unimportant for critical phenomena unless the transition occurs at $T = 0$.) Notice that χ_F is a measure of the number of spins that are ferromagnetically correlated with a given spin and so also diverges at T_c, with a critical exponent called γ, i.e.

$$\chi_F \sim t^{-\gamma} \quad . \tag{4}$$

In a spin glass there is no net tendency towards ferromagnetism or antiferromagnetism so ξ_F stays short range. Nonetheless, if one looks at all pairs of spins a given relative distance apart, they will have substantial correlation, but some pairs will be ferromagnetically correlated and others will have an antiferromagnetic correlation, so the *average* is close to zero. This cancellation coming from random signs will not occur if we square the correlation function, so we define the spin-glass correlation length, ξ_{SG} by

$$[<S_i\ S_j>^2]_{av} \sim \exp(-R_{ij}/\xi_{SG}) \quad , \tag{5}$$

where the symbol $[\ \cdots\ ]_{av}$ denotes an average over different realizations of the random interactions. This is more convenient theoretically than the equivalent procedure of considering a single set of interactions and averaging over all pairs of spins a given distance apart. Just as with ferromagnets, there is a corresponding susceptibility, known here as the spin-glass or Edwards-Anderson susceptibility, obtained by summing the squared correlation function over all distances, i.e.

$$\chi_{SG} = \sum_j [<S_i\ S_j>^2]_{av} \quad . \tag{6}$$

A spin glass transition is characterized by the divergence of ξ_{SG} and χ_{SG} as the temperature is lowered, i.e.

$$\xi_{SG} \sim t^{-\nu} \tag{7}$$

$$\chi_{SG} \sim t^{-\gamma}$$

without a divergence of ξ_F or χ_F. Unfortunately, unlike ξ_F, the spin-glass correlation length cannot be measured by neutron scattering but it turns out that χ_{SG} can be determined by direct magnetic measurements. When a magnetic field, h, is applied one usually determines the *linear* response of the magnetization, controlled by χ_F. This is, however, only the first term of a series, i.e.

$$m = \chi_F h - \chi_{nl} h^3 \quad \cdots \quad . \tag{8}$$

The non-linear susceptibility, χ_{nl}, is the fourth derivative of the free energy with respect to h and so is a four spin cumulant average. Now clearly χ_{SG} is just one piece of the four spin cumulant. It is not difficult to show, for the models we will be discussing below with a symmetric distribution of exchange interactions, that the other terms in the four spin cumulant vanish so

$$\chi_{SG} \propto \chi_{nl} \quad . \tag{9}$$

Even for more realistic systems, where these two susceptibilities are not simply proportional to each other, one argues that the difference comes from the residual ferromagnetic, or antiferromagnetic, correlations in the system. Since these are of short range χ_{SG} and χ_{nl} should still both diverge with the same exponent at T_c. It is, of course, much more difficult to measure the non-linear response than the linear susceptibility because it is generally a small correction to the linear term. Close to T_c it becomes larger but then the higher order terms in eq. (8) become large too because they diverge with exponents larger than γ. Despite these difficulties one can determine χ_{nl} if the experiments are done with sufficient care and most of our knowledge of critical behavior at the spin glass transition has come from such experiments [3,4]. In fact the difficulty of subtracting the linear response can be circumvented by doing an a.c. measurement and looking for the response at the third harmonic [5].

Below the transition one expects an order parameter to develop. For ferromagnetism this is, of course, the magnetization

$$m = \langle S_i \rangle \quad . \tag{10}$$

For spin glasses the analogous quantity would vanish because the spins will freeze in directions which appear random (though they are of course determined by the random interactions through the requirement that the free energy be minimized). In fact the expectation value of the spin at any wavevector will be zero (more precisely of order $N^{-1/2}$, where N is the number of spins) but, once again, one circumvents the cancellations occurring from random signs by squaring the thermal average. This leads to the Edwards-Anderson [6] order parameter

$$q_{EA} = [\langle S_i \rangle^2]_{av} \quad . \tag{11}$$

Historically, the first measurement which indicated the possibility of a phase transition was the a-c susceptibility experiment of Cannella and Mydosh [7]. This showed a simple Curie, or Curie-Weiss, law at high temperatures, but, at a fairly well defined freezing temperature, the curve showed a cusp and the susceptibility dropped on lowering the temperature further. By itself, though, this did not show whether a sharp transition occurs at the cusp temperature, in which case the susceptibility is expected to drop at lower temperatures as in an antiferromagnet, or whether the cusp simply shows that relaxation times become very long and the system drops out of equilibrium on time scales of the measurement. The difference is that there is a divergent length, ξ_{SG} in the first scenario while, in the second picture, the correlation length stays fairly small but relaxation times become very large, though strictly not infinite. From the above discussion, it is clear that this issue can be resolved by looking at the non-linear susceptibility. Careful measurements [3-5] show that this becomes enormous and an analysis of the temperature dependence indicates fairly conclusively that it really diverges at a temperature close to the freezing temperature. While the earlier measurements [3] gave a value for γ near 3, more recent experiments [4,5], which were able to probe temperatures closer to T_c, give a value near 2.

3. THE MODEL

In order to study this problem theoretically it is useful to have a relatively simple model which incorporates the essential physical ingredients of randomness and frustration. The most commonly studied model is that of Edwards and Anderson [6] in which the Hamiltonian, for Ising spins, is

$$H = -\sum_{<i,j>} J_{ij} S_i S_j \quad , \tag{12}$$

where the spins lie on a regular lattice and the interactions, assumed nearest neighbor here, are independent random variables with a probability distribution $P(J_{ij})$. We will consider the "pure spin glass case" where the mean of the distribution is zero, and will also set the standard deviation to unity to define the temperature scale. Two distributions are especially popular, the $\pm J$ distribution in which

$$P(J_{ij}) = \frac{1}{2}[\delta(J_{ij} - 1) + \delta(J_{ij} + 1)] \quad , \tag{13a}$$

and the Gaussian distribution where

$$P(J_{ij}) = \frac{1}{\sqrt{2\pi}} e^{-J_{ij}^2/2} \quad . \tag{13b}$$

One interesting question is whether systems with different bond distributions lie in the same universality class. In addition, one can also study models with vector spins rather than Ising spins. For example the isotropic Heisenberg spin glass is obtained from eq. (12) by replacing $S_i S_j$ by the scalar product $\mathbf{S}_i . \mathbf{S}_j$. It is also of interest to study models with interactions of longer range than nearest neighbor. The RKKY interaction, for example, which is appropriate for most of the commonly studied spin glasses, can be mimicked within this type of model by letting the standard deviation of the interaction between spins separated by a distance r fall of like r^{-3}. Further generalizations to vector models with anisotropy can obviously be made, but the analysis is often then tricky because of the different crossover phenomena which occur simultaneously.

4. EQUILIBRATION

Several theoretical techniques have been used to study the spin glass problem including, high-temperature series expansions, domain wall renormalization group methods and Monte Carlo simulations . Here I will discuss the application of the Monte Carlo technique, combined with finite size scaling, which I have been working on in collaboration with Ravin Bhatt [8-10]. Unfortunately, one immediately runs into several difficulties when applying Monte Carlo techniques to spin glasses.

(a) Although the usual ideas of "self-averaging" indicate that, in the thermodynamic limit, one sample should give the same answer as an average over all samples with interactions chosen from the given distribution, this is not the case for the finite systems studied in simulations and one typically finds that an average of several hundred or several thousand samples must be simulated to do the configuration average, $[\cdots]_{av}$, accurately. In fact, self-averaging may not even apply below the spin glass transition temperature (so many samples would need to be averaged over even in the thermodynamic limit) since it is known not to hold [11] for the infinite range model [12]. Since many samples must be averaged over the simulations require substantial computer time.

(b) Another reason that the simulations require a lot of CPU time is that, just as in the actual experiments, relaxation times become very long as the transition temperature is approached. It is therefore imperative to have a sound criterion for deciding if the simulation ran for enough sweeps for the system to be in equilibrium. The necessary number of sweeps will, of course, vary with temperature and system size.

(c) Since the spin glass correlation length diverges at T_c, the results close to the transition will be in the regime where $\xi_{SG} > L$, where L is the linear dimension of the system. These will inevitably be affected by finite size effects. While the standard approach to dealing with this problem, finite size scaling, is well known and is indeed the subject of this book, it is not completely obvious how to apply finite size scaling in the optimum way for the spin glass problem to determine if and where the transition occurs and to obtain the values of critical exponents.

The solution to point (a) is of course just the brute force one, simply to run the simulation over and over again for many choices of interactions, but points (b) and (c) require a certain amount of discussion.

Let us first discuss how the criterion for equilibration is established [8-9]. It turns out to be convenient to simulate two identical copies of the system, with no coupling between them. After allowing t_0 sweeps for equilibration one then computes the instantaneous mutual overlap

$$Q'(t) = \frac{1}{N}\sum_i S_i^{(1)}(t_0+t)S_i^{(2)}(t_0+t) \tag{14}$$

for "times" t between 0 and t_0 (so the total number of sweeps is $2t_0$), Here the superscripts "1" and "2" refer to the two different systems. For each system one also computes the overlap between the state at times t_0 and $2t_0$

$$Q(t_0) = \frac{1}{N}\sum_i S_i(t_0)S_i(2t_0) \tag{15}$$

In the limit of $t_0 \to \infty$ both $Q(t)$ and $Q'(t)$ have the same distribution (since the second system can be regarded as a realization of the first after a long time), given by

$$P(q) = \frac{1}{t_0}\sum_{t=1}^{t_0} [\, \delta(q - Q'(t))]_{av} \quad . \tag{16a}$$

This time average is, of course, equivalent to the statistical mechanics average

$$P(q) = [<\delta(q - q_{12})>]_{av} \quad , \tag{16b}$$

where $q_{12} = N^{-1}\sum_i S_i^{(1)}S_i^{(2)}$. Clearly q_{EA} and χ_{SG} are related to the first two moments of $P(q)$, i.e.

$$q_{EA} = <q> \tag{17}$$

$$\chi_{SG} = N<q^2>$$

where the brackets here mean an average over $P(q)$. Suppose, however, that t_0 is less than the time needed to equilibrate the system. In this case the distributions of Q and Q' will be different. In particular, χ_{SG}, computed from the distribution of Q will be *larger* than the equilibrium value, since states of the system at neighboring times are very similar. Furthermore, the value of χ_{SG} obtained from the distribution of Q' will be *too small* at short times since we start the two sets of spins in independent random configurations. Hence we have two estimates of χ_{SG} which approach the equilibrium value *from opposite directions*. The simulation is run for values of t_0 which increase logarithmically, e.g. 10, 30, 100, ..., and one checks if the two results for χ_{SG} agree for the largest value of t_0. If they do not, the results are discarded and, where possible, a new run is performed for a larger value of t_0, to see if convergence can be obtained. For each size there is a minimum temperature below which one can not achieve convergence in available computer time. Note that this approach is much better than looking at a single quantity and checking for equilibration by requiring that its value seems to be independent of time. This is unsatisfactory because we know that correlations decay roughly logarithmically with time in spin glasses and a quantity which is still varying on a logarithmic scale can look flat on a linear scale. In fig. 1 we show some results, actually for a 4-dimensional model, which illustrate that the two estimates for χ_{SG} really do vary monotonically, one decreasing and the other increasing, and that the two only agree when they have reached the equilibrium (i.e. time independent) value.

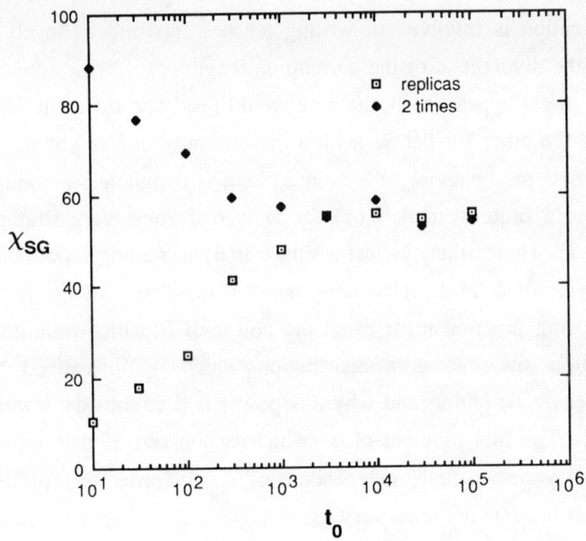

FIG. 1. Plot of χ_{SG} for the 4-dimensional model with Gaussian interactions at $T = 1.4$ for $L = 4$, averaged over 100 sets of bonds. We compute χ_{SG} both from the overlap between two replicas and from spins of one replica at two different times, as described in the text, as a function of equilibration time t_0. One sees that the replica value increases to the equilibrium value monotonically whereas the "two-time" value decreases monotonically. Once the two estimates agree, they do not change on further increasing t_0 showing that this is the equilibrium value. From ref. 8.

5. FINITE SIZE SCALING

Fortunately, the answer to point (c) above, the influence of finite size affects, can also be conveniently answered by looking at the distribution $P(q)$. Since q is the order parameter of the problem, its distribution should satisfy the finite-size scaling ansatz

$$P(q) = L^{\beta/\nu}\bar{P}(L^{\beta/\nu}q, L^{1/\nu}t) \quad , \tag{18}$$

where β is the order parameter exponent and $\bar{P}$ is a scaling function independent of L and t. The basic assumption in deriving eq. (18) is that finite size affects are controlled entirely by the ratio of the size of the system to the correlation length of the infinite system. Since from eq. (7) we have $\xi \sim t^{-\nu}$, then $L/\xi \sim Lt^{\nu}$. Taking this to the power $1/\nu$ we see that the reduced temperature always appears in the combination $L^{1/\nu}t$, which is just the second argument of the scaling function in eq. (18). Actually, another assumption is involved in writing the scaling form as in eq. (18) since the amplitude of the divergence of the correlation length is different above and below the transition. It might appear, then, that one would need two different functions, one for above T_c and the other for below, with a discontinuity in $P(q)$ at T_c. However, this is not so because the behavior of a finite system is completely smooth and, since eq. (18) applies to a finite system, $P(q)$ for a fixed q must vary smoothly even as T goes through T_c. Hence there is just a single analytic scaling function. To understand the other powers of L in eq. (18) note first that the power of the factor of L multiplying the scaling function must equal the power of L which multiplies q inside the scaling function since P is a distribution which is normalized to unity when integrated over q. To understand why this power is β/ν, consider a very large system just below T_c. The first moment of q (which is non-zero if one applies a symmetry breaking field) varies as $|t|^{\beta}$, independent of L. In order to reproduce the correct dependence on $|t|$, $\bar{P}(x,y)$ must vary as

$$\bar{P}(x, y) = |y|^{\beta}\,\tilde{P}(x)$$

and, to cancel the L dependence, the power of L multiplying q in eq. (18) must be β/ν. Here $\tilde{P}$ is another scaling function, whose precise form is unimportant here.

Eq. (18) has a rather large number of parameters and the analysis can be simplified following Binder [13] by looking at ratios of moments such as

$$g = \frac{1}{2}\left[3 - \frac{\langle q^4\rangle}{\langle q^2\rangle^2}\right] \tag{19}$$

It follows from eq. (18) that g satisfies the finite size scaling form

$$g = \overline{g}(L^{1/\nu}(T - T_c)) \quad , \tag{20}$$

where $\overline{g}$ is a scaling function. Note that g is dimensionless (since the powers of q in the numerator and denominator of eq. (19) are equal) so there is no power of L multiplying the scaling function in eq. (20). This makes g a particularly convenient quantity to study, as we shall see. In the paramagnetic phase for $T > T_c$ and for sizes $L \gg \xi$, $P(q)$ tends to a Gaussian centered around $q = 0$ of width $N^{-1/2} \sim L^{-d/2}$, so $g \to 0$. On the other hand, for $T < T_c$, where the infinite system develops a nonzero order parameter, then $g \to 1$ as $L \to \infty$. From eq. (20) one sees that the range of temperature in which the value of g changes from 0 to 1 varies with L as $L^{-1/\nu}$. Eq. (20) also shows that g is independent of size at T_c so curves of $g(T)$ for different sizes must *all cross at* T_c, with g being a decreasing function of L above T_c and an increasing function of L below T_c. This should provide a very convenient criterion for locating T_c. In addition, the exponent ν may be obtained from determining for which value the data are best fitted by the scaling form, eq. (20). The second independent exponent, η, can then be obtained from the finite size scaling form for the spin glass susceptibility

$$\chi_{SG} = L^{2-\eta}\, \overline{\chi}(L^{1/\nu}t)) \tag{21}$$

which can also be obtained from eq. (18) and the result that χ_{SG} is just L^d times the second moment of q. The exponent γ is related to η and ν by the scaling law

$$\gamma = (2 - \eta)\nu \tag{22}$$

and β is given by

$$\beta = \frac{1}{2}(d - 2 + \eta)\nu$$

One can in fact determine all the other exponents from ν and η using scaling laws. Our procedure for obtaining T_c and the critical exponents from the data is therefore as follows: (a) use g to locate T_c from the intersection of the curves for different sizes, (b) determine ν by fitting the data for *all* sizes to the scaling form of eq. (20), and (c) determine η from the fit of the data for χ_{SG} to the scaling form, eq. (21).

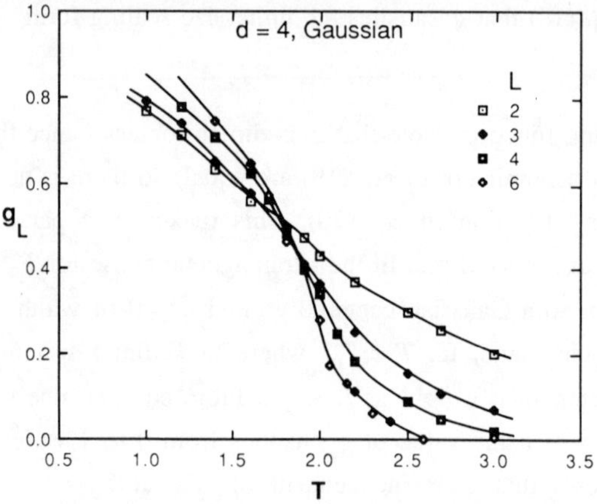

FIG. 2. Results for g against T in 4-dimensions for the Gaussian distribution. T_c is estimated to be 1.75 ± 0.05 from the intersection of the curves. The curves for different sizes splay out below T_c showing that the spin glass order parameter is non zero. The lines are just guides to the eye. From ref. 8.

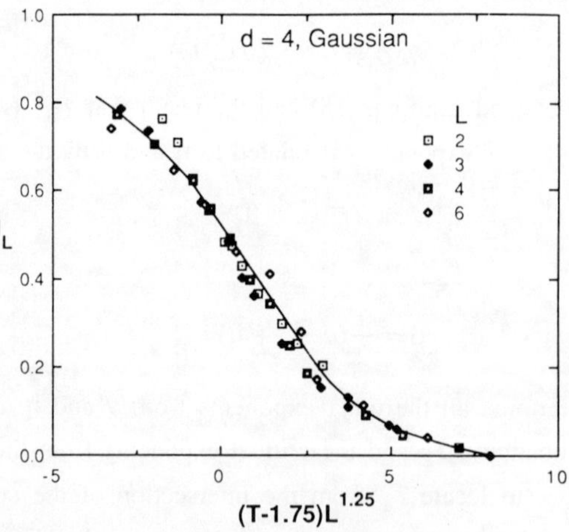

FIG. 3. Scaling plot of the data in fig. 2, fitted to eq. (20), with $T_c = 1.75$, and $1/\nu = 1.25$.

6. RESULTS FOR ISING SYSTEMS

These finite size scaling ideas all work out very cleanly in 4-dimensions [8]. We show, in fig. 2, data for g obtained on a hypercubic lattice with a Gaussian distribution of bonds for $2 \le L \le 6$, where each point is an average of 200 - 800 samples. The curves do intersect at (or at least close to) a single temperature, from which we estimate that $T_c = 1.75 \pm 0.05$, below the mean field estimate of $T_c^{MF} = \sqrt{8} \cong 2.83$. A scaling plot of the data for g to the expected finite size form discussed above, is shown in fig. 3 and the scaling plot for χ_{SG} is presented in fig. 4. This leads to the following estimates of the exponents: $\nu = 0.8 \pm 0.15$ and $\eta = -0.3 \pm 0.15$, which, combined with the scaling law, eq, (22), gives

$$\gamma = 1.8 \pm 0.4 \tag{23}$$

This value for γ agrees well with the estimate of $\gamma = 2.0 \pm 0.4$ obtained for a $\pm J$ distribution by Singh and Chakravarty [14]. This provides some evidence that the Gaussian and $\pm J$ models lie in the same universality class, though the error bars are too large to make a firm statement on this.

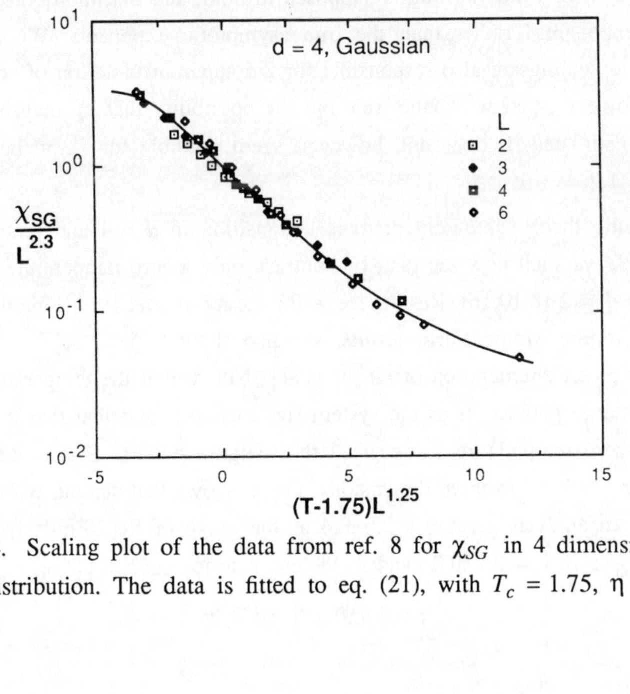

FIG. 4. Scaling plot of the data from ref. 8 for χ_{SG} in 4 dimensions with a Gaussian distribution. The data is fitted to eq. (21), with $T_c = 1.75$, $\eta = -0.3$ and $1/\nu = 1.25$.

Unfortunately, the results in three dimensions [8-10,15] did not work out quite so cleanly. Data for g from ref. 8 for the $\pm J$ distribution is shown in fig. 5. While the data at higher temperatures show the expected behavior and the curves come together at about $T = 1.2$, which is therefore the estimate of T_c, the curves for different sizes do not splay out again below T_c. The data above T_c scale quite nicely, as shown in fig. 6, from which one gets an estimate of $\nu = 1.3\pm0.3$. In addition, results for χ_{SG} [8-10] give $\eta = -0.3\pm0.2$ and hence

$$\gamma = 3.0\pm0.9 \quad (d = 3) \quad . \tag{24}$$

These values agree well with the high temperature series results [7] of $T_c = 1.2\pm0.1$ and $\gamma = 2.9\pm0.5$ and Ogielski's results [15] that $T_c = 1.175\pm0.025$, $\nu = 1.3\pm0.1$ and $\gamma = 2.9\pm0.3$.

The gamma values also agree rather well with the older experimental results [3] which generally obtained a value of about 3. However more recent experiments, done with data in smaller fields and closer to T_c, get smaller numbers. For example Bouchiat [4] finds $\gamma = 2.2\pm0.2$ and Levy [5] gets $\gamma = 2.3$. It is therefore quite possible that the numerical results are still rather far from the asymptotic limit, $L \to \infty$, $t \to 0$, in which scaling is supposed to hold, and that the quoted results are "effective exponents" rather than the true asymptotic exponents. We suspect that corrections to scaling are also responsible for the apparent violation of scaling below T_c. Note, however, that we cannot rule out the possibility that T_c may be somewhat lower than estimated. It does not, however, seem possible for T_c to be any higher than around 1.2.

While the theory definitely predicts a transition in $d = 4$ and almost certainly one in $d = 3$, we shall now see that, by contrast, only a zero temperature transition is predicted in $d = 2$ [8,10,16]. Results for g for the $\pm J$ model [8] are plotted in fig. 7. In addition to the Monte Carlo results, we also show a point at $T = 0$ for $L = 4$ obtained by direct enumeration of all the states. Notice that the temperature at which the curves merge gets lower as the system size increases and that that g has already reached its zero temperature value when the merging occurs. This is different from the behavior observed in three dimensions. Fig. 8 shows that scaling works well with $T_c = 0$ and an analysis gives $\nu = 2.6\pm0.4$ as the range of fits. Similar results [8] in fig. 9 for χ_{SG} give $\eta = 0.2\pm0.05$ and so the estimate for gamma is

$$\gamma = 4.6\pm0.5 \quad (d = 2) \tag{25}$$

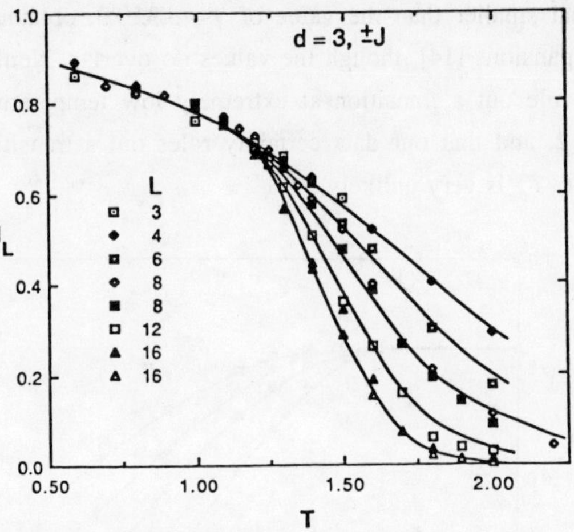

FIG. 5. Plot of g against T in 3-dimensions for the $\pm J$ distribution. The lines are just guides to the eye. The data for the open diamonds $L = 8$ and the filled triangles $L = 16$ are from A. T. Ogielski (private communication). The remaining data is from ref. 8.

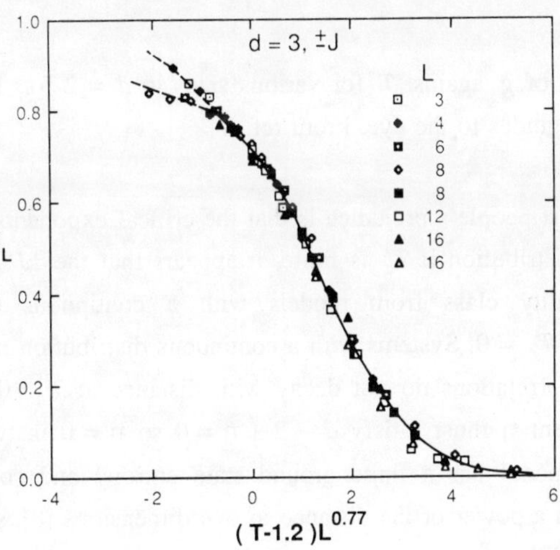

FIG. 6. Scaling plot of the data in Figure 5, fitted to Eq. (20), with $T_c = 1.2$, and $1/\nu = 0.77$.

which is somewhat smaller than the value of $\gamma = 5.3 \pm 0.3$, obtained from high temperature series expansions [14], though the values do overlap. Neither of these calculations definitely rule out a transition at extremely low temperature but, bearing in mind that $T_c^{MF} = 2$, and that our data certainly rules out a transition with $T_c > 0.3$, we feel that a finite T_c is very unlikely.

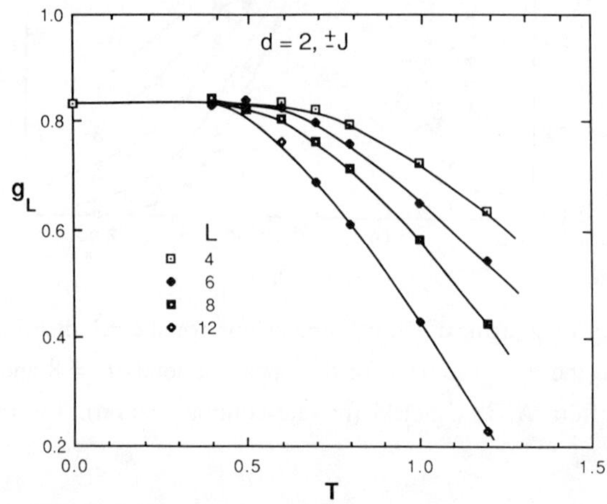

FIG. 7. Plot of g against T for various sizes in $d = 2$ for the $\pm J$ distribution. The lines are just guides to the eye. From ref. 8

Although most people's prejudice is that the critical exponents do not depend on the form of the distribution if T_c is finite, it appears that the $\pm J$ distribution is in a different universality class from models with a continuous distribution in 2-dimensions where $T_c = 0$. Systems with a continuous distribution must have a unique ground state so correlations do not decay with distance at $T = 0$. Consequently, if $T_c = 0$ the exponent η must satisfy $d - 2 + \eta = 0$ so $\eta = 0$ in two-dimensions. By contrast the $\pm J$ model has a finite ground state entropy and correlations seem to decay to zero with a power of the distance in two dimensions [8], so η is non-zero as noted above. Additionally, the value of ν also seems to be larger for the Gaussian model [17] that for the $\pm J$ model.

Since the calculations give very different predictions for 2 and 3-dimensional systems, it is obviously desirable to find an experimental system which is a 2-

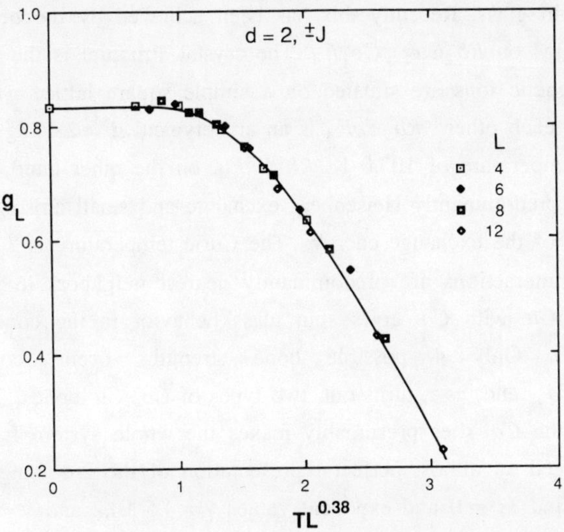

FIG. 8. A fit of the data in fig. 7 to the finite size scaling formula, eq. (20), with $T_c = 0$ and $1/\nu = 0.38$ for the $\pm J$ distribution in $d = 2$.

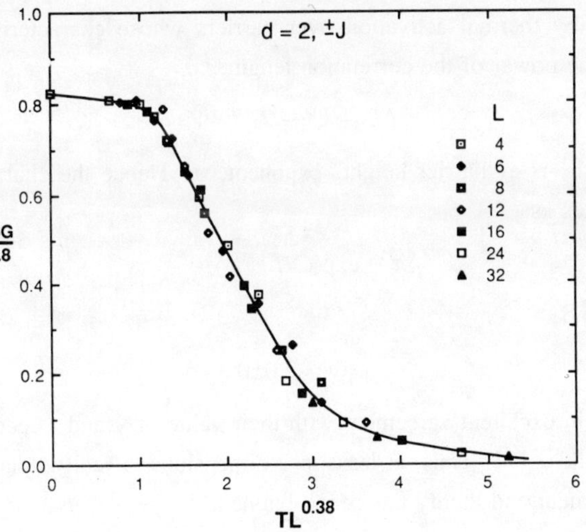

FIG. 9. A scaling plot of the data from ref. 8 for χ_{SG} in 2 dimensions with a $\pm J$ distribution. The data is fitted to eq. (21), with $T_c = 0$, $\eta = 0.2$ and $1/\nu = 0.38$.

dimensional spin glass. Recently this has been achieved by the beautiful work of Dekker et al [18] on $Rb_2Cu_{1-x}Co_xF_4$. The crystal structure is the K_2NiF_4 type in which the magnetic ions are situated on a simple square lattice which are widely separated from each other. Rb_2CoF_4 is an archetypical $d = 2$ Ising antiferromagnet with a Néel temperature of 103.0 K. Rb_2CuF_4, on the other hand, is a $d = 2$ ferromagnet with predominantly Heisenberg exchange and small anisotropy whose size is of order 1% of the exchange energy. The Curie temperature is $T_c = 6.05 \pm 0.09$K. The magnetic interactions are predominantly nearest neighbors in the plane. Randomly mixing Cu with Co gives spin glass behavior in the concentration range $0.18 \le x \le 0.40$. Only 4 possible bond strengths occur, corresponding to $Cu-Cu$, $Co-Co$, and, as it turns out, two types of $Co-Cu$ bonds. The strong Ising anisotropy on the Co sites presumably makes the whole system Ising-like, so this material is indeed an almost perfect representation of the 2-d Ising $\pm J$ spin glass. Dekker et al. find $T_c = 0$ and exponent values $\gamma = 4.5 \pm 0.2$ and $\nu = 2.4 \pm 0.3$, which are in excellent agreement with the Monte Carlo predictions. They also look at the dynamics and show that a characteristic relaxation time, τ, varies exponentially as $T \to 0$, also as expected from theory [19]. Essentially the theory predicts that relaxation proceeds by thermal activation over barriers whose characteristic height, ΔE, grows with some power of the correlation length, i.e.

$$\Delta E \sim \xi^{\psi} \sim T^{-\psi\nu} \tag{26}$$

which defines the is a "barrier height" exponent, ψ. Hence the characteristic relaxation time τ varies as

$$\tau \sim \exp(C/T^{1+\psi\nu}) \tag{27}$$

Dekker at al. find

$$\psi\nu = 2.0 \pm 0.3 \tag{28}$$

which is again in excellent agreement with their value of ν and a recent prediction of Cieplak [20] that $\psi = 1$. Thus, at least in 2-dimensions, a really detailed comparison between experiment and theory has been obtained.

7. RESULTS FOR VECTOR SPIN GLASSES

While $Rb_2Cu_{1-x}Co_xF_4$ is very anisotropic, and so is well described by an Ising model, most of the three-dimensional systems studied, such as $CuMn$, have relatively little anisotropy, so one might expect a Heisenberg model to be more appropriate. Surprisingly, several calculations [21] on 3-d isotropic Heisenberg spin glass models indicate that $T_c = 0$ and that the lower critical dimensional, d_l, (the dimension below which $T_c = 0$) is close to 4.

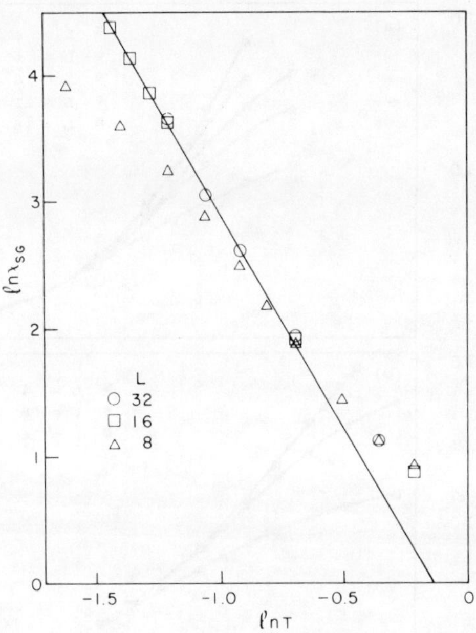

FIG. 10. A plot of $\ln\chi_{SG}$ against $\log T$ for the 3-dimensional nearest neighbor Heisenberg spin glass for sizes $L = 8$, 16, and 32. The straight line has slope $-\gamma$ where $\gamma = 3.42$. From Olive et al., ref. 21.

Evidence of this from Monte Carlo simulations was obtained by Olive et al. [21] and their results for the spin glass susceptibility are shown in fig. 10. From this plot one sees that even a "raw" log-log plot of the data against temperature shows a sizable region of straight line behavior for the size $L = 16$. Finite size effects are apparently negligible for this size down to the lowest temperatures which could be equilibrated so the data shows clear evidence of a phase transition at zero temperature.

It is curious, then, that the Ising model gives better agreement with experiment than the isotropic Heisenberg model even though the commonly studied spin glass systems have relatively small anisotropy. Presumably, the anisotropy, though small, cannot be ignored and must be included to explain the experiments in detail.

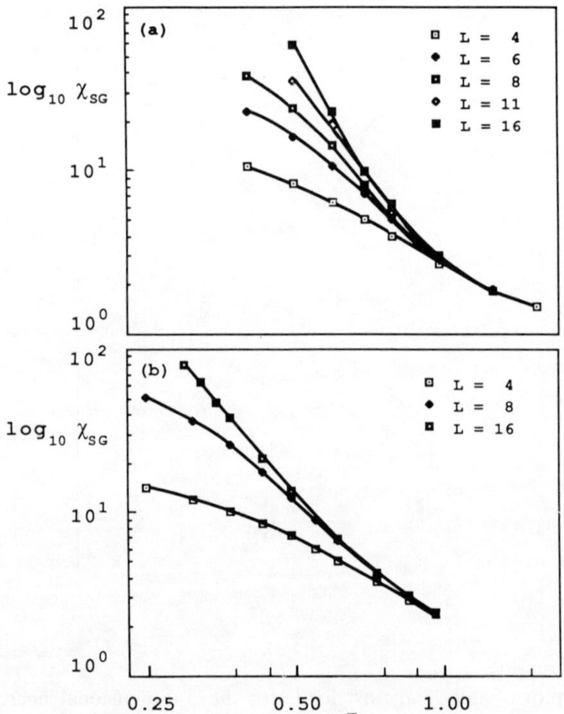

FIG. 11. (a) A log-log plot of the spin-glass susceptibility as a function of temperature for different system sizes for a Heisenberg spin glass model in which the standard deviation of the bond distribution falls of as r^{-3} to mimic the RKKY interaction. (b) Same as (a) but for the short range model (Olive et al, ref. 21). Note the slower increase of χ_{SG} with decreasing T for the short range model. T is in units of the mean field transition temperature.

Why a small anisotropy should have such a big effect is not entirely clear though. One possible explanation is that the commonly studied spin glasses have the RKKY interaction, which falls of as r^{-3}, whereas the models studied by simulation have short range couplings.

It has been argued [22] that RKKY Heisenberg systems are *at* their lower critical dimension, as opposed to short range systems which are below d_l. This means that any small anisotropy *will* have a large effect. For example, the transition temperature will increase from zero with the log of the anisotropy, D, whereas, for systems below d_l, T_c increases with a power of D.

To investigate this possibility a numerical study of an isotropic Heisenberg spin glass in which the distribution of couplings has zero mean and standard deviation which falls off as r^{-3} was carried out by Reger and the author [23]. The spin glass susceptibility increases much faster at low temperatures than for the nearest neighbor model, as shown in fig. 11. Reger and the author argued that the data is consistent with the system being at its lower critical dimension, though it did not scale extremely well with any of the standard forms, presumably indicating large corrections to scaling. It should be pointed out, though, that large corrections to scaling *are* expected at the lower critical dimension.

Presumably any detailed comparison with experiment would have to include both anisotropy and RKKY interactions, but it appears a prodigious task to disentangle all the different crossover effects that occur in this case and the project has not, to our knowledge, been attempted. Incidentally, a further complication is that the interaction in real materials does not have precisely the r^{-3} behavior at intermediate distances because of band structure effects [24]. This provides an additional discouragement to those who would like to simulate a truly realistic model of a particular material.

8. CONCLUSIONS

The question of a spin glass transition in three dimensions, which had remained controversial for so long, seems now to be settled in favor of a finite temperature transition [9,14,15,25] Several techniques have played a valuable role in clarifying the situation. The approach described here, Monte Carlo simulations combined with finite size scaling has been one of them. In fact, in the area of dynamics, which we have not discussed here, virtually the *only* results have come from Monte Carlo simulations, particularly the work of Andrew Ogielski [15] using the Bell labs special processor. It is especially gratifying to see how well the 2-d simulations agree with recent experiments on quasi two-dimensional systems, since these simulations were historically the first to be done, but there were no experimental results to compare with for a long time. Agreement between theory and experiment in 3-dimensions is reasonable, but not very precise. Presumably the discrepancy arises from the simulation (and possibly the system) not being in the asymptotic scaling region and from the various crossover effects mentioned above. The numerical methods also work very well in 4-dimensions, where we will presumably not see any real experiments. It is frustrating but perhaps typical that the 3-dimensional case has been the hardest to solve.

Acknowledgements

This work was supported in part by NSF grant DMR 87-21673. My own work in this area was largely done in collaboration with R. N. Bhatt. I would like to think him for contributing many of the ideas presented here and for many stimulating discussions. In addition I have benefited from a collaboration with J. D. Reger on the the work on RKKY interactions and would also like to thank him for many helpful interactions. J. A. Olive and D. Sherrington collaborated on the work on short range Heisenberg systems and I am grateful to them and to S. Jain for many useful discussions.

References

1. K. Binder and A. P. Young, Rev. Mod. Phys. **58**, 801 (1986); See also the articles in *Heidelberg Colloquium on Glassy Dynamics*, edited by J. L. van Hemmen and I. Morgenstern Springer, Berlin, 1987).
2. K. Binder ed. *Monte Carlo Methods in Statistical Physics*, 2nd ed. (Springer, Berlin, Heidelberg, New York, 1986); K. Binder ed. *Applications of the Monte Carlo Methods in Statistical Physics*, 2nd ed. (Springer, Berlin, Heidelberg, New York, 1987).
3. A. P. Malozemoff, Y. Imry and B. Barbara, J. Appl. Phys. **52**, 7672 (1982);R. Omari, J. J. Prejean and J. Souletie, J. Phys. (Paris) **44**, 1079 (1983); P. Beauvillain, C. Chappert and J. P. Renard, J. Phys. (Paris) Lett. **45** L665 (1984).
4. H. Bouchiat, J. Phys. (Paris) **47**, 71 (1986).
5. L. P. Levy and A. T. Ogielski, Phys. Rev. Lett. **57**, 3288 (1986); L. P. Levy, Phys. Rev. B**38**, 4963 (1988).
6. S. F. Edwards and P. W. Anderson, J. Phys. F, **5**, 965 (1975).
7. V. Cannella and J. A. Mydosh, Phys. Rev. **B6**, 4220 (1972).
8. R. N. Bhatt and A. P. Young, Phys. Rev. **37**, 5606 (1988).
9. R. N. Bhatt and A. P. Young, Phys. Rev. Lett. **58**, 924 (1985).
10. R. N. Bhatt and A. P. Young, in *Heidelberg Colloquium on Glassy Dynamics*, edited by J. L. Van Hemmen and I. Morgenstern, (Springer, Berlin, 1987), p. 215.
11. A. P. Young, A. J. Bray and M. A. Moore, J. Phys. C **17**, L149 (1984); M. Mézard, G. Parisi, N. Sourlas, G. Toulouse and M. Virasoro, Phys. Rev. Lett. **52**, 1156 (1984); J. Phys. (Paris) **45**, 843 (1984).
12. D. Sherrington and S. Kirkpatrick, Phys. Rev. Lett. **35**, 1972 (1975).
13. K. Binder, Z. Phys. **B48**, 319 (1982).
14. R. R. P. Singh and S. Chakravarty, Phys. Rev. Lett. **57**, 245 (1986).
15. A. T. Ogielski, Phys. Rev. **B32**, 7384, (1985); A. T. Ogielski and I. Morgenstern, Phys. Rev. Lett. **54**, 928 (1985).
16. I. Morgenstern and K. Binder, Phys. Rev. Lett. **43**, 1615 (1979); A. P. Young, J. Phys. C **17**, L517 (1984); W. L. McMillan, Phys. Rev. B.**28**, 5216, (1983); H.

F. Cheung and W. L. McMillan, J. Phys. C **16**, 7027 (1983); R. R. P. Singh and S. Chakravarty, Phys. Rev. Lett. **57**, 245 (1986).

17. D. Huse and I. Morgenstern, Phys. Rev. **B32**, 3032 (1985).

18. C. Dekker, A. F. M. Arts and H. W. de Wijn, J. Appl. Phys.]**63**, 4334 (1988); Phys. Rev. **b38**, 8985 (1988); C. Dekker, A. F. M. Arts, H. W. de Wijn, A. J. van Duyneveldt and J. A. Mydosh, Phys. Rev. B. (to be published).

19. D. S. Fisher and D. A. Huse, Phys. Rev. Lett. **56**, 1601 (1986); A. J. Bray and M. A. Moore,, in *Heidelberg Colloquium on Glassy Dynamics*, edited by J. L. Van Hemmen and I. Morgenstern, Lecture Notes in Physics, Vol. 275 (Springer, Heidelberg, 1987), p. 121; A. J. Bray, Comments on Cond. Mat. Phys. **14**, 21 (1988); D. S. Fisher and D. A. Huse, Phys. Rev. **B36**, 8937, (1987).

20. M. Cieplak (unpublished).

21. J. R. Banavar and M. Cieplak, Phys. Rev. Lett. **48**, 832 (1982); W. L. McMillan, Phys. Rev. **31**, 342 (1985); B. W. Morris, S. G. Colborne, M. A. Moore, A. J. Bray and J. Canisius, J. Phys. C **19**, 1157 (1986); J. A. Olive, A. P. Young, and D. Sherrington, Phys. Rev. **B34**, 6344 (1986).

22. A. J. Bray, M. A. Moore and A. P. Young, Phys. Rev. Lett. **56**, 2641 (1986).

23. J. D. Reger and A. P. Young, Phys. Rev. **B37**, 5493 (1988).

24. P. M. Levy and Q. Zhang, Phys. Rev. **B33**, 665 (1986).

25. A. J. Bray and M. A. Moore, Phys. Rev. **B31**, 631 (1985); W. L. McMillan, Phys. Rev. **31**, 340 (1985).

XI. SYSTEM-SIZE EFFECTS IN METASTABILITY

L. S. Schulman

TABLE of CONTENTS

SYSTEM-SIZE EFFECTS IN METASTABILITY

L.S. Schulman
Physics Department
Clarkson University
Potsdam, NY 13676 U.S.A.

The thermodynamic limit appears to be a central concept in modern statistical mechanics, but for understanding metastability it may be inappropriate. This article examines both static and dynamic finite-size features of systems that exhibit metastability. After an Introduction, Sections 2 and 3 deal with calculating and understanding $P_V(m)$, the finite-size probability for finding magnetization m in a two dimensional Ising ferromagnet, especially for m far from its equilibrium value. Section 4 treats metastability as a dynamical phenomenon. In it we define a probability distribution for metastable states and propose it as a replacement for the Boltzmann distribution. The last section is a discussion.

1. INTRODUCTION

The existence of metastable states in statistical mechanics is a common but nevertheless profound reflection of a finite size effect. The concept of phase transition nowadays depends on the idealization of a passage to the thermodynamic limit. In that limit, unfortunately, metastable phases of systems with short range forces* simply do not exist, as we will see below. In a sense, the supercooling of steam can occur because the phenomenon is characterized by *two* large numbers, both the number of constituents (V) and the lifetime for homogeneous nucleation (τ). Letting $V \to \infty$ would eliminate metastability, but it happens that (in the appropriate units) for situations of physical interest $\tau \gtrsim V$ or $\tau \gg V$. For metastability, the now axiomatic "thermodynamic limit" ($V \to \infty$) is thus unsuitable. In this article we will work with finite V in two different contexts: static and dynamic. In the static treatment we will deal with an idea that has traditionally

* I am not considering spin glasses or the like where for all anyone knows metastable features may persist in the thermodynamic limit.

been associated with metastability, namely that the probability distribution (on the space of states) has some sort of local maximum. This idea is fine for mean field theories, but as we will shortly see, making any sense of it in systems with short range forces requires restriction to finite, perhaps even small, volumes. Our dynamic discussion proposes a replacement for the Boltzmann distribution and this new distribution is *defined* in terms of the dynamics. Again, there is reason to believe that for realistic systems the definitions break down but in this case that breakdown is for extremely large volumes where indeed our physical ideas about metastable states may no longer be valid. The view that we will finally advocate in this article relies on the fact that the breakdown volume for the definition goes to infinity rapidly as $h \downarrow 0$, far more rapidly than the size of the critical droplet. This leads to the suggestion that a mathematical idealization for metastability—a replacement for the thermodynamic limit—should involve joint $h \downarrow 0$, $V \to \infty$ asymptotics in which the path in the h-V plane along which $V \to \infty$ can be selected from a broad range of functions larger than the critical droplet size and smaller than the breakdown volume.

For the statics we will study details of the V-dependence in a particular model [1]. The model describes a well studied phase transition in which the finite size aspects of the metastable state are made manifest. The system is the two dimensional Ising model and the quantity on which we focus is $P_V(m)$, the probability that a V-spin system is found to have magnetization m for a system in equilibrium (at some given temperature T and external field h). Clearly the finiteness of V plays a role since for $V \to \infty$ this function must become a (sum of) δ-function(s).

Our interest will center on $T < T_c$, where T_c is the transition temperature for spontaneous magnetization. $P_V(m)$ will have two maxima but the drop between them will depend on V. This allows us to take a position between the opposite camps on this subject: Those who naively expect the two-peak behavior to survive the thermodynamic limit, as it does for long range forces, and those who dismiss the entire venture by observing that the convexity of the free energy (which is related to $P_V(m)$) forbids the very phenomenon that we are studying. In Fig. 1 these ideas are illustrated. Subsequent to [1] there were studies published [2] of

Figure 1. Conflicting ideas about metastability motivated by differing views of the function, $P_V(m)$ = Probability that a ferromagnetic system of V spins has total magnetization m. Where the curves differ, the upper curve reflects the convexity of the free energy in the thermodynamic limit; in this perception, there is no such thing as metastability. The lower curve corresponds to a Curie-Weiss model with long range forces and the less prominent peak is the metastable state. The short range, finite V curve, to be given below, differs significantly from both of these.

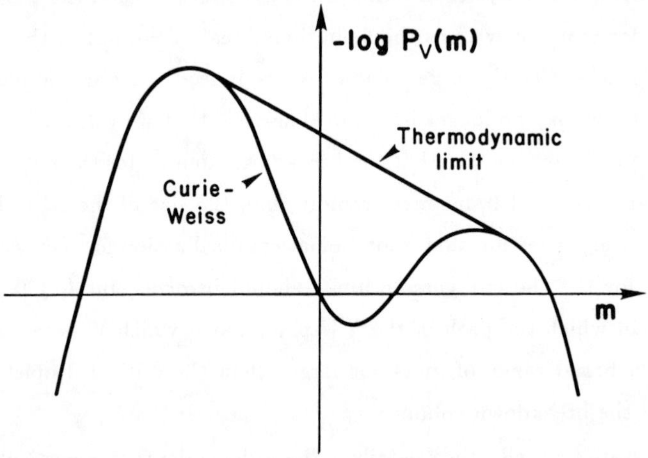

functions quite similar to our $P_V(m)$ for various dimensions and with particular emphasis on finite size scaling issues associated with these functions.

Calculating $P_V(m)$ involved an extension of standard Monte Carlo stochastic dynamics because of our particular interest in far from equilibrium values of m. Even for small systems occupancy of these states will be prohibitively rare. The method that was developed for this calculation [1] bears a strong resemblance to what is now known [3] as "umbrella sampling," and is explained in detail below.

Before introducing our dynamic methods, we mention other approaches to metastability that work mainly from the statics. Since the phenomenon presents us with a pseudo-equilibrium it is possible to get information from techniques that imitate those used for true equilibrium. For example, from the numerical standpoint, truncated sums [4] for the partition function would surely provide good estimates for "metastable magnetization." Such truncations exclude droplets larger than critical and are generally based on an educated intuition. Another approach to metastability has involved analytic continuation and a concrete implementation of this idea was based on the droplet model [6]. The finite-V aspect of metastability made the transfer matrix an attractive vehicle for carrying out this continuation [7, 8] because, although the partition function is entire for any V, the transfer matrix eigenvalues have branch points and multivaluedness. The numerical evidence for this hypothesis, based, say, on comparison with stochastic dynamics, is good, at least far from criticality. There is, however, some suspicion [9] that this may be due to the free energy's providing an asymptotic series near the first order transition. Another feature of the analytic continuation idea is that the imaginary part of the (continued) free energy should provide decay rates. Here too my current thinking [10, 11] is that the best one could hope for is that the "Im f" hypothesis be true in an asymptotic sense as one approaches the phase transition from the metastable side. It is known that the free energy for the Ising ferromagnet is singular as a function of external field exactly at the first order transition [12–15] but this does not preclude analytic continuation around that point. For example, Privman [16, 17] has found ways to use the transfer matrix to deduce the coefficients of the power series expansion of the free energy about points of nonzero magnetic field, bypassing the weak singularity at zero.

Although study of the function $P_V(m)$ is not yet a solution of the metastability problem, it will be seen to provide or confirm important insights into the situation. It will debunk both the extreme views expressed by the upper and lower graphs of Fig. 1, and in particular show that metastability is a finite size effect. In the next section we will show how the function $P_V(m)$ is calculated and following that (Sec. 3) the results of that calculation will be interpreted physically.

There are various reasons for believing that a dynamical approach is to be preferred for understanding metastability. The physical problem after all has non-trivial time dependence. As mentioned earlier, it is because *both* τ and V are neither small nor infinite that difficulties arise. Furthermore, static methods are necessarily inadequate for dynamical predictions. Even if you believe Im f gives lifetimes, there is arbitrariness in the overall dimensional constant, [Time/Energy].

Our context [5] for the dynamical treatment of metastability is the Master equation governing evolution under stochastic dynamics, although the idea can be applied to Hamiltonian or other dynamics. We allow the dynamics to define the metastable state for us. If the system, for whatever reason, tends to get stuck in some region of its space of configurations, then all one needs is a vague characterization of the outer limits of that region in order to define the associated metastable state. The idea is to use future conditioning and to look at time T positions of evolutionary histories that do not depart from the aforementioned outer limits before time $2T$. A rationale for this characterization and an indication of where it could succeed are given in detail in Sec. 4.

In Sec. 5 we discuss the way in which the dynamically defined probability distribution could be used in the context of an asymptotic theory to provide a mathematically well defined concept of metastability.

2. CALCULATION OF $P_V(m)$, THE FINITE SIZE PROBABILITY FOR FINDING MAGNETIZATION m

We consider the 2-dimensional Ising model on an $N \times N$ lattice ($N^2 = V$) with periodic boundary conditions. Configurations are labeled $\mu = (\sigma_1, \ldots, \sigma_V)$, $\sigma_j = \pm 1$ and the energy is

$$E_h(\mu) = -\sum_{\langle j,k \rangle} \sigma_j \sigma_k - h \sum_{k=1}^{V} \sigma_k \tag{2.1}$$

where $\langle j, k\rangle$ indicates a sum over nearest neighbor pairs, h is an external magnetic field and the ferromagnetic coupling constant (the coefficient of the nearest neighbor sum) is taken to be one. The magnetization of a configuration μ will be characterized in one of two equivalent ways. The quantity $m(\mu)$ will denote the per unit spin magnetization, $m(\mu) = \sum_{i=1}^{V} \sigma_i / V$. At times it will be convenient to characterize μ by the number of its "down" spins $k(\mu) = \sum_{j=1}^{V} (1 - \sigma_j)/2$ and it follows that $m(\mu) = (V - 2k(\mu))/V$. The usual free energy F_h (suppressing the temperature and volume dependence) is defined through

$$e^{-\beta F_h} = Z = \sum_\mu \exp(-\beta E_h(\mu)) \tag{2.2}$$

The important quantity in the present work is the constrained free energy $F_h(m)$. This quantity is defined by restricting the sum in (2.2) to those configurations having magnetization m

$$\begin{aligned} \exp(-\beta F_h(m)) &= \sum_{\substack{\mu \\ m(\mu)=m}} \exp(-\beta E_h(\mu)) \\ &= \exp(\beta h V m) \sum_{\substack{\mu \\ m(\mu)=m}} \exp(-\beta E_0(\mu)) \end{aligned} \tag{2.3}$$

By the usual assumptions of statistical mechanics, the probability that a system in equilibrium has magnetization m is given by

$$P_V(m; h, \beta) = \mathcal{N} \exp(-\beta F_h(m)) \tag{2.4}$$

where the normalization factor is $\exp(\beta F_h)$. Because $F_h(m)$ is trivially related to $F_0(m)$, by Eq. (2.3), and since our interest is in the m dependence of P_V, we will restrict ourselves to calculating $P_V(m; 0, \beta)$. This quantity will be called $P_V(m)$ and equals $\exp(-\beta F_0(m) + \beta F_0)$.

By an abuse of notation, we will also write $P_V(k)$ or $F_h(k)$ where k is the number of down spins corresponding to a particular m.

The real work of calculating $P_V(m)$ will be done by modified stochastic dynamics, but it is also worthwhile to see how far the actual sum in Eq. (2.3) can be carried. For $k = 0$ there is but one summand and we have

$$\exp(-\beta F_0(0)) = \exp(-\beta E_0(\text{all up})) = Q^{-V/2} \tag{2.5}$$

with

$$Q \equiv e^{-4\beta} \tag{2.6}$$

Having one spin flipped does not much complicate matters so that

Table 1. Values of $\exp(-\beta F_h(k))$ for $k < N = \sqrt{V}$ on the $N \times N$ Ising lattice. $Q = \exp(-4\beta)$. Let $\exp(-\beta F(k)) = VQ^{-V/2}H(k)$ and let $H(k) = \sum_{i=0}^{2k} Q^{2k-i}G_{ki}$. Listed below are non-zero G_{ki}, $k \leq 6$.

k	G_{kl}
0	$G_{00} = 1/V$
1	$G_{10} = 1$
2	$G_{20} = (V-5)/2$, $G_{21} = 2$
3	$G_{30} = (V^2 - 15V + 62)/6$, $G_{31} = 2V - 16$, $G_{32} = 6$
4	$G_{40} = [V[V[V-30] + 323] - 1254]/24$, $G_{41} = V^2 - 21V + 118$,
	$G_{42} = 8V - 85$, $G_{43} = 18$, $G_{44} = 1$
5	$G_{50} = [V^4 - 50V^3 + 995V^2 - 9370V + 35424]/120$,
	$G_{51} = V^3/3 - 13V^2 + 536V/3 - 872$, $G_{52} = 5V^2 - 132V + 926$,
	$G_{53} = 30V - 400$, $G_{54} = V + 43$, $G_{55} = 8$
6	$G_{61} = [V^4 - 62V^3 + 1511V^2 - 17242V]/12 + 6520$,
	$G_{62} = 2V^3 - 189V^2/2 + 3129V/2 - 9144$,
	$G_{63} = 67V^2/3 - 711V + 17978$, $G_{64} = V^2/2 + 189V/2 - 1651$,
	$G_{65} = 10V + 30$, $G_{66} = 40$, $G_{67} = 2$,
	$G_{60} = \frac{1}{V}\binom{V}{6} - \sum_{j=1}^{7} G_{6j}$

$$\exp(-\beta F_0(1)) = Q^{-V/2}VQ^2 \tag{2.7}$$

For two down spins there are two kinds of configurations and the complexity rises rapidly with k. Until 1979 no one had bothered to go beyond $k = 3$ (see [18]). In Table I we give the results for $k \leq 6$, derived by means of computer assisted counting. For the purposes of the present paper these numbers are useful as a check on the stochastic work, to which we now turn.

The sum in Eq. (2.2) calculates the *constrained* free energy and if one uses a stochastic dynamics that respects this constraint one will find the class of configurations that dominates the sum. For the constraint $m(\mu) = m$ one uses double spin flip stochastic dynamics [19], that is, the only allowed transitions are those in which a pair of oppositely oriented spins both change. Various implementations of this process are possible. In the one on which we here report the two spins were randomly selected (not necessarily adjacent to one another) and ΔE for the proposed double flip was calculated. For $\Delta E < 0$ the flip was implemented. For $\Delta E > 0$ it was implemented with probability $\exp(-\beta \Delta E)$. Since we worked at T well below T_c the system reached equilibrium rapidly.

This process finds the class of chief contributors to the sum—it does not yet give $F_0(k)$. This is where a trick was needed, a trick that is closely related to "umbrella sampling." Let $W(\mu \to \mu')$ be the transition probabilities for stochastic dynamics *without* the $m =$ const. constraint. Under these dynamics the system relaxes to the Boltzmann distribution, the $P_V(k)$'s that we seek. For each pair of microstates (μ, μ') the process satisfies detailed balance

$$\text{Prob}(\mu) W(\mu \to \mu') = \text{Prob}(\mu') W(\mu' \to \mu) \tag{2.8}$$

Now sum Eq. (2.8) over all μ with $k(\mu) = k$ and over all μ' with $k(\mu') = k'$. This yields

$$P_V(k) \sum_{\mu'} W(\mu \to \mu') = P_V(k') \sum_{\mu'} \text{Prob}(\mu'|k') \sum_{\mu} W(\mu' \to \mu) \tag{2.9}$$

where the notation $\text{Prob}(\mu|k)$ means the conditional probability that the configuration is μ, given that the system has k spins down. The double sum

$$\sum \text{Prob}(\mu|k) \sum W(\mu \to \mu')$$

is a transition probability for going from k to k' spins down, given that the configurations μ are selected with weights appropriate for the constrained equilibrium state. Call this rate $W(k \to k')$. By adding a fictitious step to the constrained stochastic dynamics, $W(k \to k \pm 1)$ can be calculated. First allow the constrained system to come to equilibrium. Then randomly select *one* spin and test it for a spin flip under the *unconstrained* stochastic dynamics. For the implementation described above this would mean: If $\Delta E < 0$ definitely flip, if $\Delta E > 0$, then flip with probability $\exp(-\beta \Delta E)$. *But do not flip it*, just keep a record of whether

Figure 2. $-\log P_V(k)$ against k (number of spins down) for $T = 2.0$, $N = 19$.

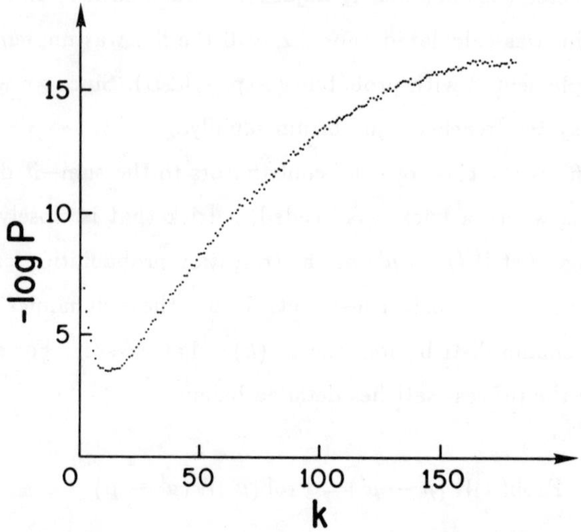

it would or would not have flipped. Now allow the system to continue to evolve under the constrained dynamics and then after a suitable interval again test the system for whether it would or would not flip, given the opportunity. By doing this for each k, $0 \le k \le V$, we collect statistics on the transition rates $W(k \to k \pm 1)$. From Eq. (2.9) we have $P_V(k)W(k \to k') = P_V(k')W(k' \to k)$, so that

$$\frac{P_V(k+1)}{P_V(k)} = \frac{W(k \to k+1)}{W(k+1 \to k)} \tag{2.10}$$

This, together with the normalization condition

$$\sum_k P_V(k) = 1 \tag{2.11}$$

allows us to compute $P_V(k)$.

The results of a typical calculation are shown in Fig. 2. For the lowest few k values the exact information was used. We compared the exact and the simulation results and found good agreement. Another check on the simulation was the value deduced for the spontaneous magnetization. For $T = 2$ ($T_c = 2.27$) this was good

Figure 3. $[-\log P_V(m) - (-\log P_V(m))_{\min}]/N^2$ against m for $T = 2.0$, various N ($V = N^2$).

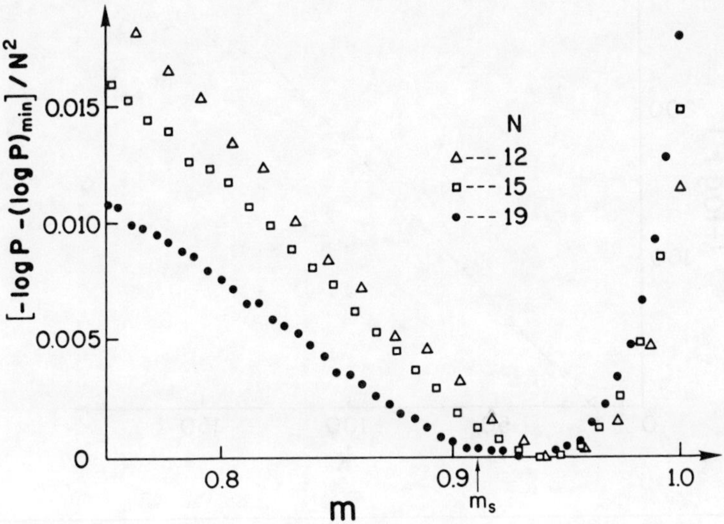

---

to better than 1% while for $T = 1.8$ the errors were in the fourth decimal place.

The matters of interest with respect to metastability concern the V dependence of the probabilities. In particular, for physical reasons (given below) we expected $\frac{1}{V}\log P_V(m)$ to be asymptotically constant for $|m| \geq m_s$ where m_s (> 0) is the spontaneous magnetization at the given temperature. On the other hand, for $|m| < m_s$ it should be $(1/\sqrt{V})\log P_V(m)$ that is asymptotically constant.

These expectations are realized and in Figure 3 we plot $\frac{1}{V}\log P_V(m)$ for various $N = \sqrt{V}$. The difference in V dependence above and below m_s is striking.

Another issue is the m dependence for fixed V, for $|m| < m_s$. We note from Fig. 4 that over a broad range of m, this quantity *squared* is linear in m. We will later show how this is related to surface tension.

3. HEURISTICS OF $P_V(m)$

As V grows the probability distribution for configurations peaks sharply on those with the "right" value(s) of m, in our case either $+m_s$ or $-m_s$. Fluctu-

Figure 4. $-\log P_V(k)^2$ against k for $T = 2.0$, $N = 19$ ($V = N^2$).

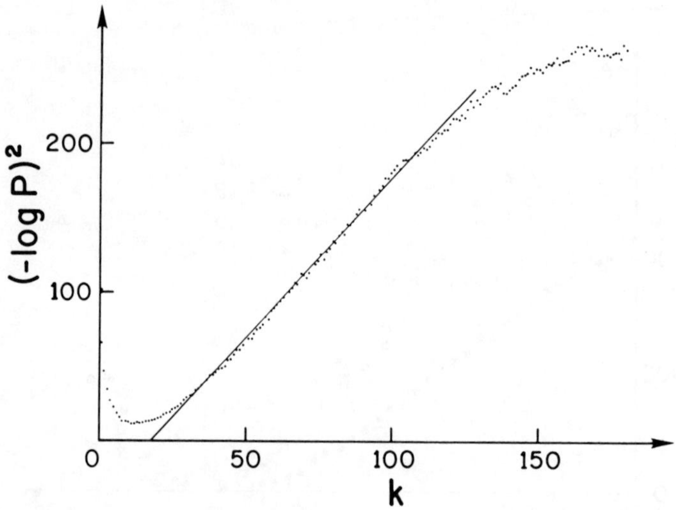

ations away from these values are rare. This article is concerned not only with the question, "How rare?" but with ascertaining the *least unlikely* of those rare fluctuations as well. For $|m| > m_s$, the way the system achieves this value of m is to be "wrong" everywhere, that is, throughout the volume. That such spatially uniform distortion is the cheapest (in free energy currency) way to accomplish this is borne out by the V independence of $\frac{1}{V}\log P_V(m)$ for $|m| > m_s$, as illustrated in Fig. 3.

On the other hand, for $|m| < m_s$ there is an inexpensive strategy, namely to go over to a two phase configuration: One region at $+m_s$, one at $-m_s$ with an interface between them. Only the interface penalizes the system for $|m| < m_s$. For sufficiently large V and for a range of m, $0 \le |m| < m_s$, the shape of the interface should be more or less independent of m and will be determined by a Wulff construction [20].

Let V_1 be the volume in the $-m_s$ phase and assume the actual interface thickness to be zero. We use the constraint $m_s(V - V_1) - m_s V_1 = mV$ to solve for V_1 and for $m > 0$ we write the length of the interface as $g\sqrt{V_1}$ with g a geometrical factor derivable in principle from the Wulff construction. Letting σ be the surface tension, the free energy cost of the interface is

$$F_0(m) - F_0(m_s) = g\sigma\sqrt{V_1} = g\sigma\sqrt{V}\sqrt{\frac{m_s - m}{2m_s}} \tag{3.1}$$

The linearity of $[F_0(m) - F_0(m_s)]^2$ with respect to m was checked in Fig. 4. The $\sqrt{V}$ in (3.1) is what allows metastability for finite V, but the fact that for $V \to \infty$ this term will be dominated by hVm (an external field term), no matter how small h, is what causes the trouble in the foundations of the subject. If one can disentangle the geometric factor, (3.1) can also be used to measure surface tension.

4. DYNAMIC METASTABILITY

As discussed above, metastability arises because there is a time scale (for homogeneous nucleation) that is comparable to the size scale (Avogadro's number). I believe therefore that progress in understanding this phenomenon requires the introduction of time into the problem. There are two levels at which this can be done: the Master equation, which is in principle exact, and a phenomenological Fokker-Planck equation which is useful despite its defects. Most of our remarks below relate to the exact formulation but first we will outline the results of the simpler approach.

One replaces the random walk on the space of all configurations by a random walk involving the variable m only. In an approximation that neglects certain correlations, the potential seen by this walker is found to be the (logarithm of the) constrained free energy $F_h(m)$, defined above. One can see from our method of computing $F_h(m)$ (in Sec.2) that it represents a collective transition probability for the variable m, so our approximation is not completely off the mark. With this approach one can show [1] that the transition rate (for leaving the metastable state) is substantially given by the Arrhenius formula, which for our problem*

* Note a slight change of notation compared to [1]. In [1], Γ was a time scale constant; here it is the dynamical decay rate. In [1] one can also see that the way in which one arrives at the Fokker-Planck equation from the exact equation involves significant approximation. In particular, dealing with the collective variable m is

takes the form

$$\Gamma \equiv \text{Decay rate}$$
$$\sim V \cdot (\text{Relatively slowly varying quantities}) \cdot \exp\left(\frac{-\beta g^2 \sigma^2}{4 m_0 h}\right) \tag{4.1}$$

To discuss the exact formulation we introduce some notation. Let μ_t be the system configuration at time t. We take an unconstrained stochastic dynamics to be the fundamental physical process so that μ_t is a random variable with transition rates given by the details of the stochastic dynamics. Earlier we mentioned particular examples (1 or 2 spin flips, transition with probability $\min[1, \exp(-\beta \Delta E)]$) but for now we will not be specific except to forbid simultaneous large scale rearrangements of the spins. The Master equation involves the function

$$\rho(\mu, t) = \text{Prob}(\mu_t = \mu)$$

and in terms of the transition probabilities is given by

$$\rho(\mu, t+1) = \rho(\mu, t) + \sum_{\mu'} \Big[\rho(\mu', t) W(\mu' \to \mu) - \rho(\mu, t) W(\mu \to \mu')\Big] \tag{4.2}$$

which we also write with obvious notation as

$$\rho_{t+1} = \rho_t U \tag{4.3}$$

Our transition rates are based on the existence of an energy function $E(\mu)$ and are defined in such a way that the Boltzmann distribution is an invariant or stable equilibrium state of the system. For this it was enough to have required

$$\exp(-\beta E(\mu)) W(\mu \to \mu') = \exp(-\beta E(\mu')) W(\mu' \to \mu) \tag{4.4}$$

---

inadequate exactly in the neighborhood of the transition. A small critical droplet in a large volume is invisible as far as m is concerned but is in fact the specific mechanism of decay.

Therefore when

$$\overline{\rho}(\mu) = \mathcal{N} \exp(-\beta E(\mu)), \qquad \mathcal{N} = \text{const} \tag{4.5}$$

is recognized as the equilibrium distribution, Eq. (4.4) becomes the statement of detailed balance. In terms of the operator U of Eq. (4.3), the function $\overline{\rho}(\mu)$ is an eigenvector and it is the eigenvector of largest eigenvalue, one.

Eigenstates with eigenvalue less than one will decay, so it is tempting to identify the eigenstate of next largest eigenvalue as the metastable state. Sometimes this works and sometimes it almost works, but in all interesting cases there are complications. The first complication is that if $\overline{\rho}$, which is positive everywhere, is normalizable, the next eigenstate is orthogonal to it and necessarily negative somewhere, an unsuitable property for a probability. One way to handle this difficulty is to send the main support for the stable state far away and to decrease the associated energies, rendering $\overline{\rho}$ nonnormalizable. This is natural in some models but in general calls for tinkering. A second complication is that although the metastable state must be slow to decay in order to have any meaning, there may be even slower processes elsewhere in the vast space of configurations that have nothing to do with the metastable state in which we are interested (we are thinking more generally than the 2-dimensional Ising ferromagnet). These will be embarrassing insertions in the spectrum of U.

A way that has been proposed [5] to define metastability and which ameliorates these problems is dynamic future conditioning. Some tinkering is still needed. You have to recognize configurations that are definitely *not* to be in the metastable state but we expect successful definition even with fairly vague characterization; certainly we will not have to give an a priori definition of the "critical droplet," as has sometimes been done [4]. Recall that the task of defining a metastable state is the giving of a (nonnegative) probability distribution on the space of configurations. We build this distribution in the following way: Let $\mathcal{A}$ be a set of allowed configurations. $\mathcal{A}$ includes configurations we think should belong to the metastable state, but much more as well. What it should not include are configurations belonging to the stable state. So all we need is a vague characterization of the class of configurations of interest. Let $\hat{\mu}$ be a fixed configuration that bears reasonable resemblance to what a metastable configuration is expected to look like. Define

$$\theta(\mu; \mathcal{A}, T) = \text{Prob}\Big(\mu_T = \mu \;\Big|\; \mu_0 = \hat{\mu},\ \mu_t \in \mathcal{A} \text{ for all } t, 0 \le t \le 2T\Big) \tag{4.6}$$

In words, we look at those sample paths (individual runs in a Monte Carlo simulation, say) that up to time *two* T have not stepped out of the set $\mathcal{A}$. Then we ask, where were they at time T. The idea here is that $\mathcal{A}$ includes not only candidates for the metastable state but much more as well. (This is after all the point: We don't know how to define precisely the metastable state, but we are assuming a vague knowledge.*) Now if there is any metastable state at all, there is some dynamical reason for its existence. That means that here *is* some collection of configurations that the sample path will tend to stay in for a long time, an attractor for this stochastic system. Now T is a large number; in fact, we will later let it go to infinity. The system must remain within $\mathcal{A}$ for the long time $2T$. For $T \to \infty$ this is extremely unlikely since we want $\mathcal{A}$ to exclude the stable state. The chief contributors to $\theta(\mu; \mathcal{A}, T)$ will then consist of those paths that have found the *least* unlikely way to achieve this unlikely goal. If there is an attractor in $\mathcal{A}$, and if its basin of attraction is in some sense distant from the boundary of $\mathcal{A}$, then the least unlikely way would be to stay within that basin for most of the time. Then, shortly before $2T$, the system would move away from the attractor and leave $\mathcal{A}$. Outside the basin of attraction progress toward the boundary of $\mathcal{A}$ would be unhindered. Thus at time T the system should be within the attractor, in a typical metastable configuration, a concept now defined in terms of a smearing of the basin of attraction. From (4.6) we see that this time T position is what

* This vagueness seems to me analogous to the vagueness that exists in discussions of *order parameters*. Specifically, there *is* a physically recognizable phenomenon but a general, abstract mathematical definition suitable for all systems has not, perhaps cannot, be given.

is used in θ. In this way the dynamics automatically will define the state. Now let T and $\mathcal{A}$ go to infinity. The meaning to be attached to $\mathcal{A} \to \infty$ will depend on the system. For the one dimensional model of Ref. [5], in which there was no stable state, there was a spatial cutoff that did go to infinity. Below we will give a possible meaning for the Ising model. Assuming then that meaning has been given to "$\mathcal{A} \to \infty$," we define the metastable state to be

$$\theta(\mu) = \lim_{\mathcal{A}\to\infty} \lim_{T\to\infty} \theta(\mu; \mathcal{A}, T) \tag{4.7}$$

We address four questions: Does this ever work? Does it work for the Ising model? How does $\theta(\mu)$ relate to solutions of the Master equation, eigenvectors, etc.? How—in principle at least—can $\theta(\mu)$ be used to calculate physical quantities?

Before taking up these questions we would like to argue in favor of the physical reasonableness of the definition. A properly constituted metastable state should look like an equilibrium state, that is, it should sit there unchanging. When it does change, it will generally go suddenly and drastically. Until a critical droplet forms, nothing happens. Once it forms, it grows, usually rapidly but even if not rapidly at least in some way that makes the overall state of the system a nontrivial function of time. This is no longer pseudo-equilibrium; there is change.* Now suppose one wanted to know the index of refraction (or whatever) of (supercooled) water at $-2°$C. If, immediately following the measurement the water crystallized, the experimenter would have good reason to invalidate the measurement. Perhaps the transcritical ice crystal was already present during the measurement and what was measured was not a property of "true" metastable supercooled water but that of a system in transition. Thus the future or later behavior of the system is used as an indicator of the validity of the earlier procedure. Analogous ideas were used implicitly in Monte Carlo studies of metastability in [8]. In the context of

* At this point, we emphasize that in our study of the metastable state we are not considering the yet more complex process whereby the system makes its way to the stable state. There is naturally a great deal of interest in this process and it is sometimes lumped together with the problem we here study, but for our purposes we restrict ourselves to the pseudo-equilibrium problem.

a simulation one can see the entire history of a system. For example, in [8] we considered the magnetization of a two dimensional Ising model as a function of time. Soon after the beginning of a run the system reached a value, call it m_{ms} ("metastable"), where it stayed for a long time. Then it would suddenly depart from m_{ms} (at a random time that we found to be exponentially distributed) and changed rapidly until it reached the value of the magnetization of the stable phase for the particular value of external field. The quantity m_{ms} was calculated from this numerical experiment by averaging the time dependent m after discarding the initial transient regime *and after discarding a* substantial stretch of *time before the decay.* There did not seem to be sensitivity to the amount of time discarded just as we do not expect our definition (4.6) to be sensitive to whether $2T$ or $1.1T$ is used.

Here is a model system in which the definition (4.6) works: The system is called an urn or a droplet [5, 21]. Our version is an outgrowth of droplet models for metastability developed by Fisher [12] and Fisher and Felderhof [22]. Because we study *dynamics* as well as statics we simplify their model yet further to consideration of a single droplet. The only state variable is the size n of the droplet, a nonnegative integer. An energy E_n is assigned to each droplet size (usually we take $E_n = \log n$ for $n \geq 1$, $E_0 = 0$, following [12, 22]) and a chemical potential μ is given so that the free energy for a droplet is $E_n - n\mu$. The partition function is

$$Z = 1 + \sum_{n=1}^{\infty} \frac{e^{\beta\mu n}}{n^{\beta}} \tag{4.8}$$

and is analytic for $\operatorname{Re}\mu < 0$. The phase transition takes place at $\mu = 0$. The chemical potential here plays the role that the external field h does in the Ising model. In other versions of the droplet model one may take $E_n \sim n^{(d-1)/d}$ with d the dimension. This associates E_n with the surface energy. The chemical potential term $\exp(n\beta\mu)$ dominates the energy factor and again the phase transition occurs at $\mu = 0$. For $\mu > 0$, Z diverges. In this simplified picture the droplet grows indefinitely and the stable phase (for $\mu > 0$) has been pushed to infinity. Nevertheless, we will see that even for $\mu > 0$ something quite interesting is happening at small

n—the system gets stuck in a way that is fully analogous to physical metastability, which is after all what we are trying to model. Of course "getting stuck" is meaningless unless one defines a dynamics and we do so in a way that is consistent with the statics. Consistency means that the equilibrium state of the dynamical system, when it exists, is the Boltzmann factor $\exp(-\beta E_n + \beta\mu n)/Z$.

As usual, we take the dynamics to be a stochastic dynamics associated with the Boltzmann factor and for $\mu > 0$ find that the motion can be described abstractly as a random walk on the nonnegative integers in a potential $E_n - \mu n$. For $\mu < 0$ the system is confined to a neighborhood of 0; this represents the distribution of other-phase droplets within a stable phase. For $\mu > 0$ (but small) the potential rises away from 0 but finally reaches a maximum (at $n = 1/\mu$ for a logarithmic energy) and then goes to $-\infty$. The temporary confinement—more or less—within the local minimum is the metastability phenomenon. The general formalism defined above can be applied to this model. It gives precise meaning to the term "more or less" with no tinkering whatever. The integer valued random variable representing the system configuration is called X and its time t value is denoted X_t. (This corresponds to "μ_t" above.) As in Eq. (4.2) or (4.3), a Fokker-Planck equation can be defined and the spectrum of eigenstates of its generator studied. For $\mu < 0$ the Boltzmann distribution is a normalizable (L^1 here) eigenfunction with eigenvalue one; it is the invariant equilibrium state. For $\mu > 0$ that distribution is still formally an eigenstate but it is not normalizable. The next largest eigenvalue is just below one (for small μ) and its eigenvector is a good candidate for having some relation to the metastable phase—especially since in this particular case it is strictly positive, orthogonality with respect to the inadmissible Boltzmann distribution no longer being a requirement. (But this is an artifact of our disregard of the stable phase so that this eigenstate itself has flawed credentials as a metastable state candidate. As we will see, it is only its square that finally appears in the distribution.) Our general future conditioning formalism can be applied to this model. Characterization of the set of admissible states, $\mathcal{A}$, is naturally accomplished by a cutoff on the droplet size. Let

$$\mathcal{I}_N \equiv \{0, 1, ..., N\}$$

correspond to $\mathcal{A}$ and let "$\mathcal{A} \to \infty$" mean $N \to \infty$. Our earlier definitions become

$$\theta(n; \mathcal{I}_N, T) = \text{Prob}\Big(X_T = n \,\Big|\, X_0 = \hat{X} \in \mathcal{I}_N, \\ X_t \in \mathcal{I}_N \text{ for all } t, 0 \le t \le 2T\Big) \tag{4.9}$$

and the final candidate for the probability distribution in the metastable state is

$$\theta(n) = \lim_{N\to\infty} \lim_{T\to\infty} \theta(n; \mathcal{I}_N, T) \tag{4.10}$$

In [5] we show the fundamental connection between $\theta(n)$ and the largest normalizable eigenstate of the Fokker-Planck equation (call it $g(n)$). That connection is equivalent to[†]

$$\theta(n) = \frac{g^2(n)}{\exp(-\beta E_n + \beta\mu n)} \tag{4.11}$$

This is an eminently plausible result. Inside the metastable well g looks like the Boltzmann factor so that $\theta(n)$ itself is basically the Boltzmann factor. Outside the well, not only does g vanish, but the formerly problematic growth of the Boltzmann factor aids in the convergence. In the transition range, $n \sim$ critical droplet size $\sim 1/\mu$, $g(n)$ provides a smooth transition that suppresses but does not sharply eliminate transcritical droplets. The function $\theta(n)$ can be used to produce a metastable partition function and the decay rate is given by the negative of the logarithm of the eigenvalue associated with g. See [5] for details.

We next discuss whether our proposal for defining the metastable state can be expected to work for realistic systems. "Realistic," for our purposes means a collection of particles coupled by short range forces[‡] and again the archetype is the

---

[†] g in the present article is associated with the forward Fokker-Planck equation. In [5] the eigenfunctions discussed were those of a transform of this equation.

[‡] ... which is not to say that these are the only interactions of physical relevance. Besides References [23–25], recent analysis of nucleation in a quenched ferroelectric suggests the importance of long range forces [26].

2-dimensional Ising model. We take $T < T_c$ and suppose there is a nonzero external magnetic field h so that the stable state has positive magnetization. Experience (e.g., [8]) indicates that for h not too large* there is a metastable state of negative magnetization. The set of allowed configurations can be defined as (using the notation of Sec. 2)

$$\mathcal{A} = \{\mu \mid m(\mu) < \hat{m}\} \tag{4.12}$$

where $\hat{m}$ is a fixed constant magnetization value between the stable and metastable values, for example, 0. Strictly speaking, for any fixed V we will not implement any $\mathcal{A} \to \infty$ limit. Rather we will consider V growing large, which in a sense makes the $m(\mu) = \hat{m}$ increasingly distant from the $m(\mu) \cong m_{ms}$ states. The claim, or perhaps conjecture, that I make is that our definition will work in an asymptotic sense. The asymptotics will involve $V \to \infty$ and $h \to 0$, and as for any asymptotic theory the usefulness of the ideas will depend on there being a broad range of values over which it well describes natural phenomena.

Consider fixed, small h. If V is smaller than the critical droplet, there should be no problem in defining metastability, but for our purposes this case is too easy.† Now let V be considerably larger than the critical droplet. In principle, during the long time $2T$ droplets could grow larger than critical but those that do will have the (nearly) overwhelming tendency to continue growing and therefore will be excluded as contributors to θ. Thus the metastable state will give little—but

* Disappearance of the metastable state for large h and the nature of any singularities associated with the disappearance ("spinodals") is a topic we will not go into, although speculations relating it the spectrum of the transition matrix have appeared [5].

† For systems with long range forces, this circumstance may indeed be important. Also the proofs may not be all that "easy." The constraints used in [23] to study the Kac-Uhlenbeck-Hemmer model prevented nucleation. This allowed a definition of metastability and showed the significance of the "metastable" branch of the van der Waals equation of state for that model. Unfortunately it seems that these techniques do not carry over to systems with short range forces [24]. Similarly, in [25] the range of the forces is sufficient to allow metastability (with some mean field like properties) even for fairly large h.

not zero—weight to configurations with droplets larger than critical. This is the dynamically induced cutoff that we sought.

From a mathematical point of view there is a defect in the fact that the tail of the distribution is not precisely zero near the edges of $\mathcal{A}$ so that the definition is not completely independent of $\hat{m}$. I believe that this tail is on the scale of the effects of the compulsory sign change in the second eigenstate, when the first eigenstate is normalizable. Both effects are extremely small. For the eigenstate approach it leads to a problem of principle, for us it just a slightly premature cutoff.

The real problem concerns the limit $V \to \infty$. The argument [27] runs as follows: We claimed that for very long times the best way for the system to avoid exit from $\mathcal{A}$ is *not* to form a critical droplet. If the per spin droplet formation rate γ ($\equiv \Gamma/V$) is independent of V for large enough V (which it should be for short range forces), then the probability of not forming a critical droplet (per time step) in volume V is

$$(1-\gamma)^V \sim e^{-\Gamma} \tag{4.13}$$

On the other hand, suppose that at some time well before $2T$, well before T in fact, the system forms a droplet larger than critical. According to the lore it should begin to grow almost deterministically: every time one of the interface spins is chosen it will favor a flip in the direction of the stable magnetization. The interface thus is expected to move inexorably so as allow the stable phase to take over the entire volume. However, the movement is not inexorable, only likely. Suppose each interface spin that is tested has probability p of flipping, $1-p$ of not flipping with p finitely separated from $\frac{1}{2}$ (one expects $p \sim \frac{1}{2} + h\cdot\text{const}$ for small h).

Let us consider the likelihood that despite its "inexorable" tendencies the interface motion comes to a halt. Time is measured in units in which every spin in V is sampled once per time step (this time unit was implicit in (4.13) and earlier). Thus in one time step every spin along the interface must make the "wrong" decision in order that the interface not advance. The probability of this event is

$$(1-p)^{\alpha\sqrt{V}} \tag{4.14}$$

where α is a geometrical factor and I have assumed the interface has reached macroscopic dimensions. This number will be extremely small compared to other possible interface motions, roughly by the ratio $[(1-p)/p]^{\alpha\sqrt{V}}$ (ignoring entropic factors). The peculiar fact of life however is that when one takes the mathematical limit $V \to \infty$ absolutely seriously, it eventually happens that

$$(1-p)^{\alpha\sqrt{V}} \gg (1-\gamma)^V \tag{4.15}$$

even if γ is extremely close to 1 and even if p is well over $\frac{1}{2}$. For such V our future conditioning will not pick out what we intuitively hope to define as the metastable state. The *least unlikely* way to avoid crossing $m = 0$ is not altogether to avoid letting a critical droplet form. Rather, let it form but stop it from growing. The best size to let it grow to will involve a tradeoff between the penalty for halting interface growth and the penalty for leaving large volumes empty of the stable phase.

On the assumption that the halted interface is macroscopic, we will estimate the volumes at which our future conditioning definition breaks down. V must satisfy

$$\epsilon^{-\lambda\sqrt{V}} = (1-\gamma)^V \tag{4.16}$$

where $\lambda = \alpha \log(1-p)$ is order unity, and (from (4.1))

$$\gamma \sim e^{-k/2h} \tag{4.17}$$

with k again order unity in appropriate units. (In dimension d, droplet model arguments give the power of h in (4.17) as $d-1$.) Inserting (4.17) in (4.16) yields (for $\gamma \ll 1$)

$$V \sim \lambda^2 e^{k/h} \tag{4.18}$$

Figure 5. The right curve is the breakdown volume for future conditioning, as given by Eq. (4.18); the left curve is the volume of the critical droplet.

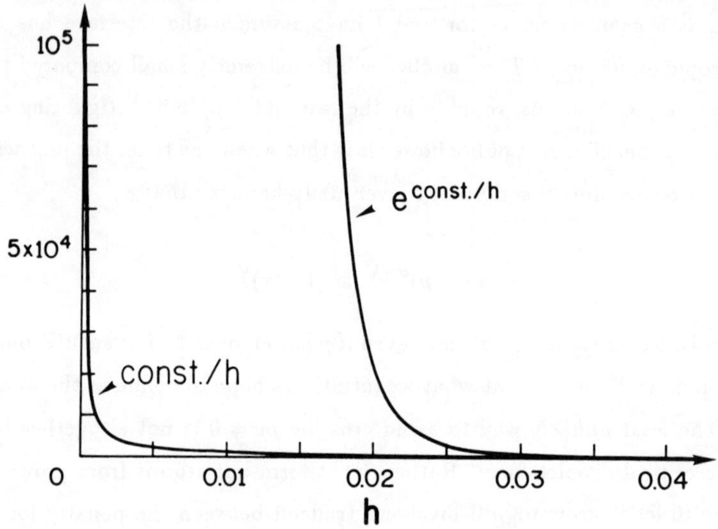

For small h, V will be enormous. Recall that the critical droplet size is algebraically large in h (again, const/h). In Fig. 5 we show a schematic plot of V versus h, indicating the critical droplet volume size and comparing it to the volume V of (4.18). The figure suggests that provided one suspends the ideal mathematical limit $V \to \infty$, the future conditioning definition will provide the physically meaningful state.

5. DISCUSSION

The physical phenomenon of metastability for macroscopic systems is one for which the standard idealization, the thermodynamic limit ($V \to \infty$), is, at least in its direct application, unsuitable. To students of finite-size scaling this news will be particularly underwhelming. When your interest is size dependence, you will surely not consider $V \to \infty$ the only way to make a theory well-defined. In this spirit, we studied one of the central functions associated with the two-dimensional Ising model and found that its size dependence gave insight into its metastable behavior. The function is $P_V(m)$, the probability to find a system of size V with magnetization m. (In the notation we suppress the temperature, which is below T_c, any external field, and other properties such as boundary conditions.) This function was computed by an elaboration of constrained stochastic dynamics that is related to the method of umbrella sampling. For simplicity we discuss $h = 0$. For $|m| > m_s^+$, with m_s = spontaneous magnetization, we found that $\log P_V(m)$ scales like V, the system size. For $|m| < m_s^-$, the scaling of $\log P_V(m)$ is with $\sqrt{V}$. The notation $m_s^\pm$ indicates that the scaling only holds outside a V dependent neighborhood (whose size goes to 0 as $V \to \infty$) of $|m| = m_s$.

It follows from this that for any finite V and for sufficiently small h, the curve $hVm + \log P_V(m)$ will have a minimum between the maxima near $\pm m_s$. However, for fixed $h \neq 0$, for $V \to \infty$ that minimum disappears and ultimately one obtains the standard result that in the thermodynamic limit the free energy (which is a multiple of $\log P_V(m)$) is a convex function of m.

The shallow minimum of $\log P_V(m)$ is a surface effect. These same considerations show why metastability can be successfully defined in models with long range interactions. The function $P_V(m)$ associated with those models has a minimum that scales more strongly with V; in the Curie-Weiss model $\log P_V(m)$ is proportional to V over the entire range of m values. There is no surface in the Curie-Weiss theory. For these systems dynamical studies show [7, 28] that the decay rates are not proportional to V but rather go like $\exp(-\text{const} \cdot V)$.

In studying dynamics, the function $P_V(m)$ may also be useful. It is possible to derive a Fokker-Planck equation in which the system behaves like a random walker in the potential $-\log P_V(m)$. However, although gross features may be describable in these terms, the approach is flawed in that the actual decay is via small critical droplets for which the collective variable m is a poor descriptor.

Our more ambitious attempt to understand metastability involves dynamical ideas. For stochastic dynamics one always has a Master equation. The spectrum and eigenfunctions of the generator of this equation are potentially useful in the description of metastability. However, to avoid certain problems in interpreting the eigenfunctions we took a slightly different tack.* The metastable probability

---

* Even when the candidate for metastable eigenstate is nonnegative, it is not

distribution was defined in terms of the time T configuration of those system evolutions that did not go too far from the metastable region up to some time $2T$; then we let $T \to \infty$. In Sec. 4 we argued for the physical suitability of this definition. We also stressed that for the purposes of our conditioning only a vague characterization of the metastable phase was necessary and that we could avoid the burden of having to provide a specific characterization of critical droplets or similar structures. The desirability of being free of such details is clear. For example, there are physical systems in which the usual picture of nucleation probably does not hold for small clusters [29].

The definition was shown to work on a simple urn-cluster model but we also anticipated problems with the $V \to \infty$ limit for systems with short range forces. We refer to Sec. 4 for the details of those problems, but we will comment here on the general picture and the overall prognosis that seems to develop from this.

The method should work for volumes $V \lesssim \exp(k/h)$ for some constant k. For small h these volumes will be enormous. Moreover, the arguments of Sec. 4 pinpointing the $V \to \infty$ problem encountered at $V \sim \exp(k/h)$ seem to me to apply to the physical system as well as to the mathematical definition. That is, if the mathematical definition breaks down for the reasons described above (interface growth halting being more likely than critical droplet nonformation), then indeed one would be loath to consider the corresponding physical system to be in a reasonable metastable state.

However, irrespective of physical consideration, does the procedure of Sec. 4 have mathematical meaning in any kind of $V \to \infty$ limit? We previously alluded to the suggestion [9] that the accuracy of transfer matrix and low temperature series predictions with respect to Monte Carlo simulation values of m in the metastable region was due to those analytic objects having *asymptotic* validity in the metastable regime. Thus the free energy $f(h)$ (in 2-d Ising) is known to be ana-

---

clear that by itself it is the appropriate probability distribution. For the urn model we found that the *square* of that eigenstate, divided by the Boltzmann factor, should be used. I do not know how one could (or whether one should) arrive at this conclusion if only the spectrum and the eigenstates of the evolution operator are considered.

lytic for $\mathrm{Re}\, h < 0$ (at fixed temperature $< T_c$), but is also known to be nonanalytic at $h = 0$ (albeit C^∞). The formal series around $h = 0$, although nonconvergent, is apparently still useful for magnetization predictions.† Does this mean that this small h, asymptotic theory ought to be the whole story of metastability? I believe not. Take a fixed positive h. If one lets $V \to \infty$ at this stage there is no known way to get a free energy that is the analytic continuation of the free energy function that is analytic for $\mathrm{Re}\, h < 0$. (And even if there were, the function may not be physically relevant.) So take some large fixed V. Now let $h \downarrow 0$. Eventually the critical droplet size exceeds V and the restriction of the system to V will give it properties different from what the expansion predicts. Not much different to be sure, but any discrepancy at all is enough to show this limiting process to be different from something that could in principle recover the low temperature expansion. The solution, I believe, is to consider the joint limit $h \to 0$, $V \to \infty$. The curve $(h, V(h))$ for $h \downarrow 0$ would use $V(h)$, some rapidly increasing function of h for $h \downarrow 0$. Its ratio to the critical droplet size should grow infinitely. However, this still leaves it plenty of room to be far below the right hand curve $V(h)$ of Fig. 5, representing the breakdown of our conditioning. In our view, a *mathematical* theory of metastability could consist of showing the validity of a low temperature or a perturbation expansion, where that expansion would hold on a broad class of curves $\{(h, V(h)) \mid h \in [0, \epsilon]\}$ for some $\epsilon > 0$. The *physical* theory would be based on noting the actual rapid growth of $V(h)$.

The use of the subjunctive in the foregoing paragraph relates to the anticipated difficulties in proving some of the conjectures. For example, the mechanism of future conditioning breakdown discussed in Sec. 4 may not be the worst, although it is the first that comes to mind. There may be some other, ordinarily speedy, process that becomes the limiting factor as $V \to \infty$. Problems of this sort may be expected to be severe for systems more complicated then the Ising model. In any case, whether or not the specifics of our conjecture succeed, we would like to leave the reader with the idea that for metastability the otherwise

---

† It is also closely related to the perturbation expansion for field theories with an unstable vacuum. See [30].

useful dogma of the thermodynamic limit is not appropriate. We have suggested that this dogma be replaced by the use of joint asymptotics, where the nature of the physical phenomenon ensures the quantitative usefulness of the asymptotic theory.

ACKNOWLEDGEMENTS

This work was supported in part by NSF grant PHY 88-11106.

REFERENCES

1. L. S. Schulman, Magnetization probabilities and metastability in the Ising model, J. Phys. A **13**, 237 (1980).
2 K. Binder, Z. Phys. B **43**, 119 (1981).
3. D. Chandler, *Introduction to Modern Statistical Mechanics* (Oxford, New York, 1987) and *Solution Manual*, by D. Wu and D. Chandler (Oxford, New York, 1988).
4. D. Capocaccia, M. Cassandro and E. Olivieri, Commun. Math. Phys. **39**, 185 (1974).
5. B. Gaveau and L. S. Schulman, Dynamical metastability, J. Phys. A **20**, 2865 (1987).
6. J. S. Langer, Ann. Phys. **41**, 108 (1967). See also J. D. Gunton and M. Droz, *Introduction to the Theory of Metastable and Unstable States* (Springer-Verlag, Lecture Notes in Physics 183, New York, 1983).
7. C. M. Newman and L. S. Schulman, J. Math. Phys. **18**, 23 (1977).
8. R. McCraw and L. S. Schulman, J. Stat. Phys. **18**, 293 (1978).
9. J. Lebowitz, private communication.
10. C. M. Newman and L. S. Schulman, J. Stat. Phys. **23**, 131 (1980).
11. B. Gaveau and L. S. Schulman, Lett. Math. Phys., to appear (1989).
12. A. F. Andreev, Sov. Phys. JETP **18**, 1415 (1964); M. E. Fisher, Physics **3**, 255 (1967).
13. G. A. Baker and D. Kim, J. Phys. A **13**, L 103 (1980).
14. W. Klein, D. J. Wallace and R. K. P. Zia, Phys. Rev. Lett. **37**, 639 (1976).
15. S. N. Isakov, Commun. Math. Phys. **95**, 427 (1984).
16. V. Privman and L. S. Schulman, J. Phys. A **15**, L 231 (1982).
17. V. Privman and L. S. Schulman, J. Stat. Phys. **29**, 205 (1982).
18. A. J. F. Siegert, Phys. Rev. **97**, 1456 (1955).
19. M. Kalos, J. L. Lebowitz, O. Penrose and A. Sur, J. Stat. Phys. **18**, 39 (1978).
20. See for example, D. P. Woodruff, *The Solid Liquid Interface* (Cambridge Univ. Press, London, 1973).

21. G. Roepstorff and L. S. Schulman, J. Stat. Phys. **34**, 35 (1984).

22. M. E. Fisher and B. U. Felderhof, Ann. Phys. **58**, 176 (1970); ——**58**, 217 (1970); B. U. Felderhof and M. E. Fisher, Ann. Phys. **58**, 268 (1970).

23. J. L. Lebowitz and O. Penrose, J. Math. Phys. **7**, 98 (1966); O. Penrose and J. L. Lebowitz, J. Stat. Phys. **3**, 211 (1971).

24. O. Penrose and J. L. Lebowitz, Molecular theory of metastability, appearing in *Fluctuation Phenomena*, E. W. Montroll and J. L. Lebowitz eds. (North Holland, Amsterdam, 1987).

25. W. Klein and C. Unger, Phys. Rev. B **28**, 445 (1983); N. Novotny, W. Klein and P. Rikvold, Phys. Rev. B **33**, 7729 (1986); T. S. Ray and W. Klein, J. Stat. Phys. **53**, 773 (1988).

26. P. Chandra, Phys. Rev. A **39**, 3672 (1989).

27. M. Aizenman and C. M. Newman, private communication.

28. R. B. Griffiths C.-Y. Weng and J. S. Langer, Phys. Rev. **149**, 301 (1966).

29. K. F. Kelton and A. L. Greer, Phys. Rev. B **38**, 10089 (1988).

30. J. Greensite and M. B. Halperin, Nucl. Phys. B **242**, 167 (1984).